Advances in Climatology

Advances in Climatology

Editor: Dale Sullivan

Callisto Reference
www.callistoreference.com

Callisto Reference,
118-35 Queens Blvd., Suite 400,
Forest Hills, NY 11375, USA

Visit us on the World Wide Web at:
www.callistoreference.com

ISBN: 978-1-64116-136-7 (Hardback)

Cataloging-in-Publication Data

Advances in climatology / edited by Dale Sullivan.
p. cm.
Includes bibliographical references and index.
ISBN 978-1-64116-136-7
1. Climatology. 2. Meteorology. 3. Climatic changes. I. Sullivan, Dale.
QC861.3 .A38 2019
551.6--dc23

Table of Contents

Preface

I am honored to present to you this unique book which encompasses the most up-to-date data in the field. I was extremely pleased to get this opportunity of editing the work of experts from across the globe. I have also written papers in this field and researched the various aspects revolving around the progress of the discipline. I have tried to unify my knowledge along with that of stalwarts from every corner of the world, to produce a text which not only benefits the readers but also facilitates the growth of the field.

Climatology is the scientific study of climate or weather conditions averaged over an extended period of time. It is instrumental in the forecasting of weather by using analog techniques like El Nino-Southern Oscillation, the Madden-Julian oscillation, Arctic oscillation, Pacific decadal oscillation, etc. Projections of future climates as well as study of the dynamics of weather are achieved by employing different statistical and mathematical climate models. The study of various climatic phenomena is approached from the multiple dimensions of paleoclimatology, paleotempestology and historical climatology. This book elucidates the concepts and innovative models around prospective developments with respect to climatology. It is a compilation of chapters that discuss the most vital concepts and emerging trends in this field. It aims to serve as a resource guide for students and experts alike and contribute to the growth of the discipline.

Finally, I would like to thank all the contributing authors for their valuable time and contributions. This book would not have been possible without their efforts. I would also like to thank my friends and family for their constant support.

Editor

1

Assessing freshwater life-stage vulnerability of an endangered Chinook salmon population to climate change influences on stream habitat

Jon M. Honea[1,4,*], Michelle M. McClure[2], Jeffrey C. Jorgensen[1,5], Mark D. Scheuerell[3]

[1]Conservation Biology Division, Northwest Fisheries Science Center, National Marine Fisheries Service, National Oceanic and Atmospheric Administration, 2725 Montlake Blvd E., Seattle, WA 98112, USA

[2]Fishery Resource Analysis and Monitoring Division, Northwest Fisheries Science Center, National Marine Fisheries Service, National Oceanic and Atmospheric Administration, 2725 Montlake Blvd E., Seattle, WA 98112, USA

[3]Fish Ecology Division, Northwest Fisheries Science Center, National Marine Fisheries Service, National Oceanic and Atmospheric Administration, 2725 Montlake Blvd E., Seattle, WA 98112, USA

[4]*Present address:* Emerson College, 120 Boylston Street, Boston, MA 02116, USA

[5]*Present address*: Ocean Associates, under contract to Northwest Fisheries Science Center, National Oceanic and Atmospheric Administration, 2725 Montlake Blvd E., Seattle, WA 98112, USA

ABSTRACT: We linked a set of climate, hydrology, landscape, and fish population models to estimate the relative influence of freshwater habitat variables on the abundance of a population of endangered stream-type Chinook salmon *Oncorhynchus tshawytscha* responding to a warming climate. The hydrology models estimated that increases in annual air temperature and winter precipitation would lead to increases in water temperature and changes in discharge, including higher flows during the egg-incubation period and lower flows during the summer rearing period. The spatially explicit population model estimated a resulting decline of 0 to 7% in the number of spawners, with 3 of 4 global climate models estimating a decline of 4 to 7%. Increased water temperature during the summer spawning period was the most limiting among habitat variables modeled, but our modeling suggested that aggressive habitat restoration (increasing forested area and reducing impervious area) could mitigate some spawner abundance reductions. Better knowledge of the links between climate changes and habitat response, including increased streambed scour due to the larger and more frequent winter high-discharge events predicted by our hydrology models, would improve our ability to estimate climate effects on populations. Future limitation by elevated summer water temperature, and potentially egg-pocket scour, would further stress an endangered population currently limited by the percentage of fine sediment around egg pockets. Identifying such changes demonstrates the utility of models that consider climate and integrate life-stage-specific habitat influences over a species' life cycle, thereby indicating restoration actions with the potential to benefit sensitive life stages.

KEY WORDS: Life-cycle model · Landscape model · Hydrology model · Downscale

1. INTRODUCTION

To effectively manage the recovery of imperiled species, there is an imperative need for understanding the cumulative effects of climate change on habitat and how species respond (Groves et al. 2012, Seney et al. 2013). Each stage of a species' life history may be affected differently by changes in its habitat. For example, Radchuk et al. (2013) found positive effects from warming temperatures in all stages of an

*Corresponding author: jon_honea@emerson.edu

endangered butterfly species except survival in its overwinter larval stage, which resulted in reduced population viability and increased extinction risk. Identifying such potential bottlenecks to recovery as a consequence of forecasted climate change, by itself and in combination with other factors, is becoming an increasingly important aspect of natural resource management (Finch et al. 2014, Runge et al. 2014).

Climate change presents a challenge to Pacific salmonids because they are philopatric and have a relatively low straying rate (Neville et al. 2006, Quinn et al. 2006, Keefer & Caudill 2014). This pattern promotes local adaptation (Quinn & Dittman 1990) and may limit salmonid resilience if changes in water temperature and discharge regimes outpace their capacity to adapt. Currently, maximum water temperatures during the summer in many western USA salmon-bearing streams approach and occasionally exceed thermal preferences or tolerances of salmonids (Torgersen et al. 1999, Richter & Kolmes 2005, Goniea et al. 2006). Future increases in air temperature and changes in mountain snowpack are projected to exacerbate high water temperatures and low flows during the summer (Mote & Salathé 2010) as well as increase the frequency of high discharge events in other seasons (Warner et al. 2014), putting additional environmental stresses on fishes (Mantua et al. 2010). An understanding of these effects can ensure that habitat restoration efforts are targeted to ameliorate the anticipated negative impacts (Beechie et al. 2013, McClure et al. 2013). However, only a few population studies have analyzed life-stage-specific responses to climate change in a life-cycle context (e.g. Battin et al. 2007, Crozier et al. 2008, Radchuk et al. 2013). Important life-stage bottlenecks could be overlooked if the population-level influences of individual life-stage responses to climate change are not assessed (Dupont et al. 2010, Radchuk et al. 2013).

To better understand the potential vulnerability of an endangered salmon population, we linked a set of global climate models (GCMs) to a set of hydrology models to estimate changes in water temperature and discharge at key periods in the freshwater life stages of stream-type Chinook salmon *Oncorhynchus tshawytscha* from the Wenatchee River. We estimated the population response with a spatially explicit life-cycle population model (Scheuerell et al. 2006, Honea et al. 2009), using the output (water temperature and flows) from the hydrology models as input, to ascertain which life stages are most influenced by the effects of climate change (Fig. 1). We focused on freshwater life stages because of our capacity to influence habitat changes there with restoration activities (Palmer et al. 2008, Bryant 2009, Yvon-Durocher et al. 2011). We examined climate effects on population dynamics for 2 habitat states as contrasting boundary conditions: a status quo (current) habitat and an approximation of historical habitat conditions (Jorgensen et al. 2009). We conducted a sensitivity analysis to determine which freshwater life stages were most vulnerable to climate change effects under each habitat condition and to assess the relative potential of different restoration actions for mitigating climate impacts.

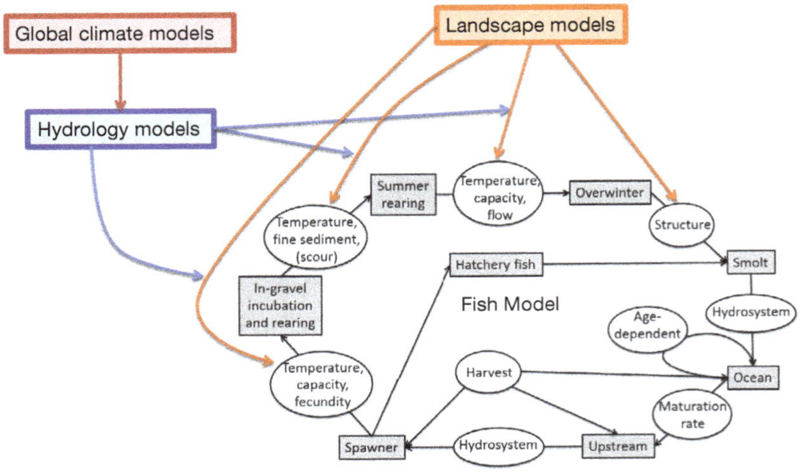

Fig. 1. Diagram of links between climate, hydrology, and landscape influences on a spring-run Chinook salmon population dynamics life-cycle model, showing where factors (ovals) influencing transitions between stages of the fish life cycle (shaded rectangles) are determined by outputs of the hydrology and landscape models. Each color represents a different model

2. MATERIALS AND METHODS

2.1. Population

Wenatchee River spring-run Chinook salmon *Oncorhynchus tshawytscha*, 1 of 3 populations that compose the endangered Upper Columbia spring-run Chinook salmon Evolutionarily Significant Unit (NMFS 1999), typically spend nearly 2 yr in freshwater after they hatch, migrate to the ocean in their second year, and return from the ocean as 4 and 5 yr old adult spawners to the river where they hatched.

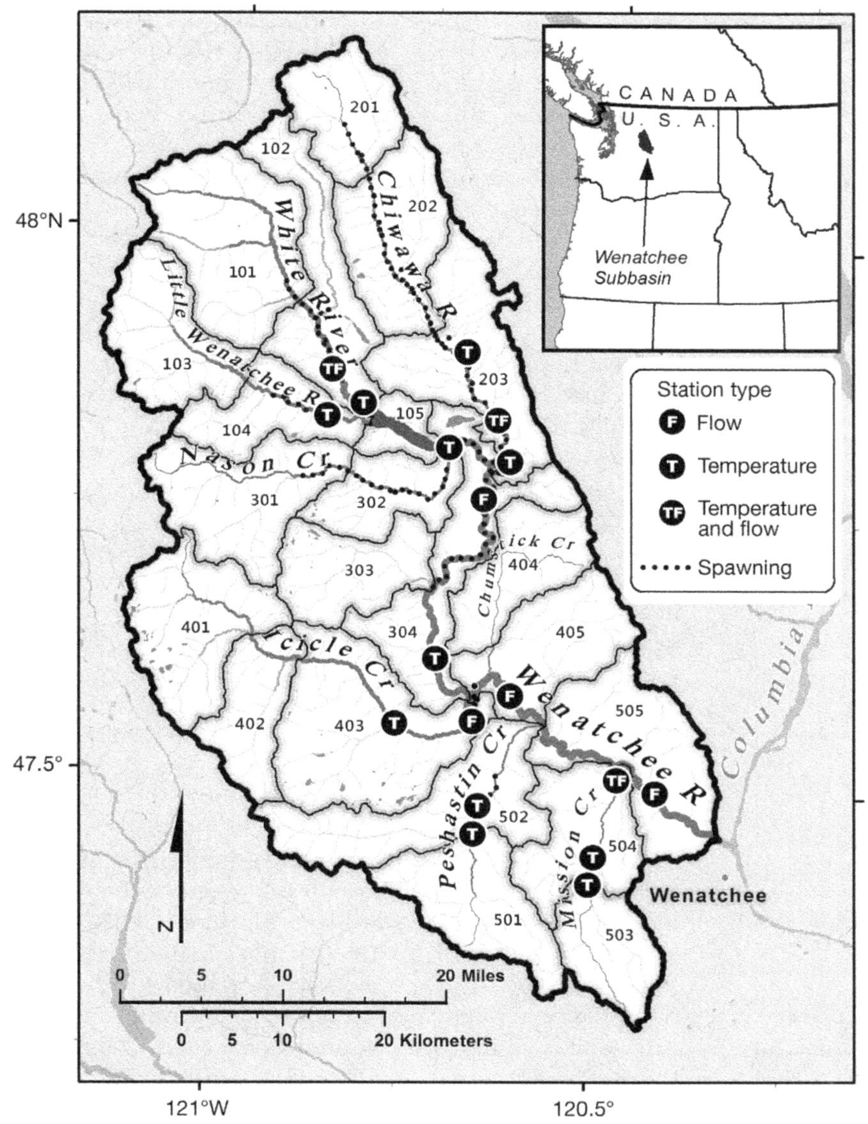

Fig. 2. The study area, showing the subbasins modeled (labeled by the final 3 digits of the sixth-field hydrologic unit code) and the temperature and stream discharge stations used for calibrating the hydrology models

The Wenatchee River Basin drains about 3400 km^2 of the eastern slopes of the Cascade Mountains and empties into the Columbia River, the largest North American river flowing into the Pacific Ocean (Fig. 2). Wenatchee River Basin hydrology is dominated by snowmelt with the highest flows occurring between May and July and the lowest in August and September. See Jorgensen et al. (2009) for a more detailed description of the basin. We modeled all of the major salmon-producing tributaries of the Wenatchee River, including the Chiwawa, White, and Little Wenatchee rivers, as well as Nason Creek and the upper reaches of the mainstem Wenatchee River. The Chiwawa River and Nason Creek both have fish supplementation programs producing salmon from local broodstock intended to augment the wild population. The Icicle River supports few wild Chinook salmon but is the location of a major Chinook salmon hatchery supporting commercial and recreational harvest using broodstock from fish that originated outside of the Wenatchee River Basin population.

2.2. Climate, hydrology, landscape, and population models linking climate to habitat condition and population response

We used outputs from 4 GCMs to simulate future climate scenarios. To estimate changes in stream temperature and discharge in response to potential future climate, we used air temperature and precipitation estimates from the GCMs — downscaled to 150 m, 3 h values — as inputs to a set of local hydrology models. The output of the hydrology models, water temperature and discharge, was used as input to a population model to project population response to predicted changes in climate. More details of the climate and hydrology models can be found in the Supplement at www.int-res.com/articles/suppl/c071p127_supp.pdf.

To assess the influence of climate change on salmon survival, we used a spatially explicit population model calibrated for the study area (Honea et al. 2009), using changes in mean adult spawner abundance as the population response measure. Survival through 3 freshwater stages (spawning, in-gravel egg incubation and rearing, and summer rearing) was driven by functional relationships that included water temperature (Fig. 3). Another freshwater life stage, the overwinter stage, was indirectly influenced by climate because lower discharge at the end of summer increased the number of fish moving downstream into new areas for their overwintering stage. The 3 remaining life

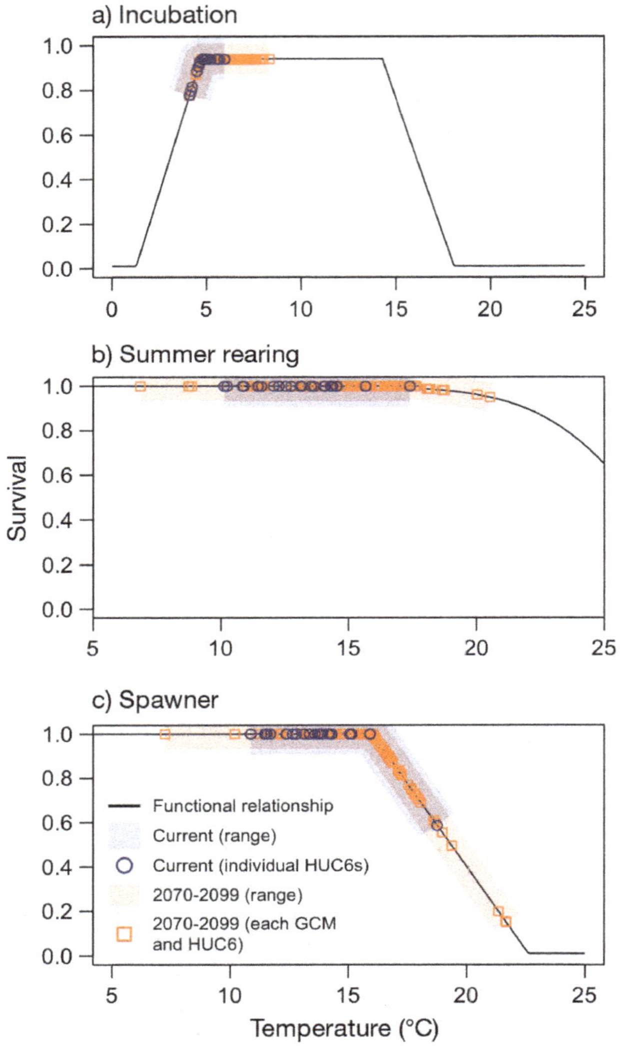

Fig. 3. Functional relationships incorporated in the population model that related water temperature to survival for several of the life stages, both at estimated current conditions and in the 2080s from outputs from hydrology modeling. These relationships included (a) in-gravel incubation and rearing survival (based on data from Velsen 1987 and Beacham & Murray 1989); (b) summer rearing (based on data from Brett 1952, McCormick et al. 1972, and Coutant 1973); and (c) spawner survival (based on data from Cramer 2001). Symbols represent estimates for each HUC6 subbasin that have spawning and rearing fish (Fig. 2)

stages included in the population model—juvenile outmigrants (smolts), adults in the ocean, and upstream migrating adults—occur in the mainstem Columbia or in the ocean. Although climate change will likely affect those areas as well, there is insufficient information about the projected changes in those environments and their effects on salmon. Thus, our focus was on freshwater spawning and rearing habitat where there is more information about climate change-induced changes (Fig. 1). While we use spawner abundance as the response variable, the purpose of the model is not to predict actual spawner numbers in future climate conditions. Rather, we are comparing the relative influence of different habitat variables for which data are available, to assess which of these variables known to affect survival through different life stages is likely to have the greatest influence on future spawner abundance. Additional details of the population model are described in the Supplement (www.int-res.com/articles/suppl/m071p127_supp.pdf).

We also considered the potential for more frequent high flows during the in-gravel egg incubation and rearing stage, due to increases in precipitation (Warner et al. 2014) and more transient hydrology dominated by both rain and snow (Mantua et al. 2010), to increase streambed scour and so also egg and larval fish mortality (Lapointe et al. 2000, Goode et al. 2013). To simulate the population response to greater scour due to higher flows during the in-gravel stage, we included a scenario with a 25% increase in mortality through this stage. We did not include a functional relationship directly linking streambed scour to survival through this stage in the population model because of limited data on the relationship between flow and scour for this basin.

We estimated terrestrial habitat variables that affect fish survival through changes to in-stream habitats at the sixth-field hydrologic unit code (HUC6 hereafter; Seaber et al. 1987) subbasin unit scale using a set of landscape models developed for this basin that quantitatively link landscape attributes to fish habitat characteristics (Jorgensen et al. 2009). Landscape attributes included total forest cover, riparian forest cover, road density, stream channel slope, elevation, precipitation, and drainage area. See Jorgensen et al. (2009) for additional details of the landscape models. For this analysis, we expanded on our previous work in Jorgensen et al. (2009) by re-examining the landscape models to determine the key landscape variables that could be targets for action to ameliorate climate impacts on stream habitat. We did this by assessing which landscape variables were most frequently present in the highest-ranked models predicting climate-linked stream habitat variables—i.e. water temperature during spawning, in-gravel egg incubation and rearing, and summer rearing life stages.

Therefore, in this study, we modeled a total of 17 different combinations of habitat conditions coupled with climate scenarios. These combinations consisted of 2 extreme landscape states, climate estimated by 4

Table 1. Scenarios modeled. 'Climate' refers to the drivers of the water temperature and flow data used in the population model (for details on the hydrology and global climate models, see the Supplement). 'Habitat' refers to the landscape state influencing stream habitat condition. 'Scour' refers to mortality during the in-gravel egg incubation and rearing stage

Scenario name	Climate	Habitat	Scour
CurrentClimate-CurrentHabitat-CurrentScour	Current	Current	Current
Warmest-CurrentHabitat-CurrentScour	ECHAM5	Current	Current
Warmer1-CurrentHabitat-CurrentScour	CCSM3	Current	Current
Warmer2-CurrentHabitat-CurrentScour	CGCM3.1	Current	Current
Warm-CurrentHabitat-CurrentScour	PCM1	Current	Current
Warmest-HistoricalHabitat-CurrentScour	ECHAM5	Historical	Current
Warm1-HistoricalHabitat-CurrentScour	CCSM3	Historical	Current
Warm2-HistoricalHabitat-CurrentScour	CGCM3.1	Historical	Current
Warm-HistoricalHabitat-CurrentScour	PCM1	Historical	Current
Warmest-CurrentHabitat-IncreasedScour	ECHAM5	Current	+25%
Warm1-CurrentHabitat-IncreasedScour	CCSM3	Current	+25%
Warm2-CurrentHabitat-IncreasedScour	CGCM3.1	Current	+25%
Warm-CurrentHabitat-IncreasedScour	PCM1	Current	+25%
Warmest-HistoricalHabitat-IncreasedScour	ECHAM5	Historical	+25%
Warm1-HistoricalHabitat-IncreasedScour	CCSM3	Historical	+25%
Warm2-HistoricalHabitat-IncreasedScour	CGCM3.1	Historical	+25%
Warm-HistoricalHabitat-IncreasedScour	PCM1	Historical	+25%

CGMs, and an additional factor to test the hypothesis that increased scour mortality may play a significant role in determining spawner abundance (Table 1). We modeled the 2 landscape states as contrasting boundary conditions: current and historical condition. The current condition included estimates from the landscape models of the current values of habitat variables in each HUC6 subbasin. Historical conditions for the study area were estimated with the landscape models by reverting the landscape variables influenced by human development to their expected historical values. For example, forest clearcuts were reverted to fully vegetated states and roads, and other anthropogenic impervious surfaces were converted to ecoregionally appropriate land-cover classes (Jorgensen et al. 2009). For each of the 2 landscape states, current and historical, we altered the resulting habitat variables associated with climate (i.e. water temperature and flow) by the change from current to future estimated by the hydrology models starting with air temperature and precipitation output from each of the 4 GCMs. Finally, to each of these 8 scenarios (2 landscape states × 4 GCMs), we added an additional variable: a 25% increase in mortality during the in-gravel egg incubation and rearing stage.

2.4. Life-stage vulnerability analysis

We conducted a sensitivity analysis to determine which predicted change in the habitat variables had the greatest influence on population dynamics due to their impacts at particular life-history stages, as well as to estimate which landscape variables were most strongly associated with the influential habitat variables. We did so by running the population model, altering only one of the climate-driven habitat variables at a time while holding all other variables constant at their estimated current values and comparing the change in spawner abundance (relative to current conditions) resulting from each model run. For water temperature during the in-gravel, summer rearing, and spawner periods and for low discharge at the end of the summer rearing period, we used the hydrology model output values for each of the GCMs. In this comparison of the relative influence of variables, we also included a model run using only 25% increased mortality during the in-gravel period to simulate the effects of increased streambed scour.

3. RESULTS

In response to habitat changes produced by the linked climate, hydrology, and landscape models (for climate and hydrology results, see the Supplement at www.int-res.com/articles/suppl/c071p127_supp.pdf), the population model forecasted either no change or a small decline in mean spawner abundance, depending on GCM. Three GCMs resulted in a 4 to 7% decline in spawner numbers (Warmest-CurrentHabitat-CurrentScour, Warmer1-CurrentHabitat-Current

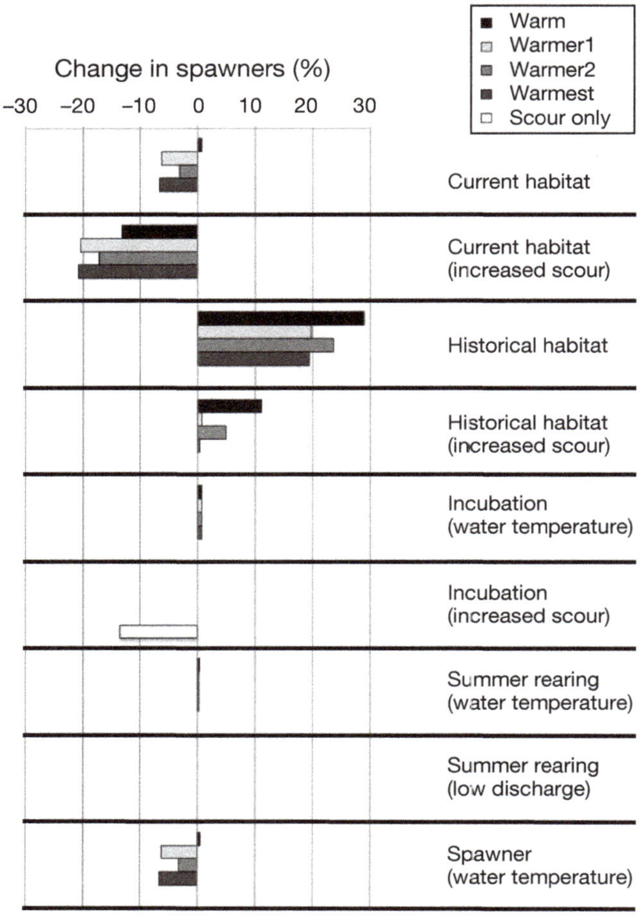

Fig. 4. Changes in wild spawner abundance (as percent change relative to estimated current numbers) resulting from scenarios of future climate under current habitat conditions, historical conditions, and to changes in individual habitat parameters in the sensitivity analysis. Increased stream-bed scour effects were simulated as a 25% increase in mortality during the in-gravel incubation and rearing period

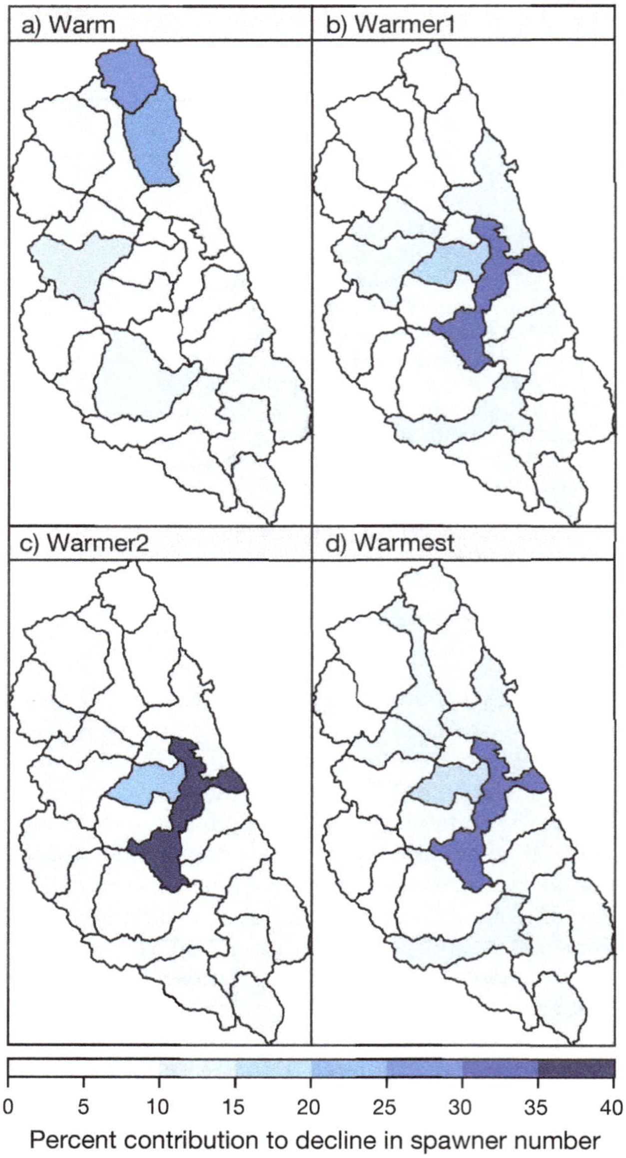

Fig. 5. Percent contribution of each HUC6 subbasin to the decline in spawner abundance, including effects of water temperature on incubation, summer rearing, and spawning stages as well as low discharge at the end of the summer rearing stage: (a) Warm (PCM1); (b) Warmer1 (CCSM3); (c) Warmer2 (CGCM3.1); (d) Warmest (ECHAM5)

Scour, and Warmer2-CurrentHabitat-CurrentScour in Table 1); however, Warm-CurrentHabitat-Current Scour, which included the GCM projecting the smallest changes in air temperature, resulted in little change in spawner numbers (Fig. 4). For the 3 GCM scenarios where a decline in wild spawner abundance was observed, the spatial pattern and magnitude of the change in spawner numbers was similar, with most of the decline occurring in lower Nason Creek and the upper Wenatchee River (Fig. 5), the lowest in elevation and warmest of the major spawning areas. Historical habitat conditions with end-of-century climate conditions estimated by all GCMs resulted in an increase in mean spawner abundance of nearly 30% versus future climate at current habitat conditions (Fig. 4). In the modeling of landscape links to stream habitat, upland and riparian forested area and anthropogenic impervious area appeared most frequently in the highest ranked models, indicating their greater influence on stream habitat characteristics than other landscape attributes modeled. This suggests that habitat restoration may have an important role in mitigating the effects of climate change for vulnerable populations.

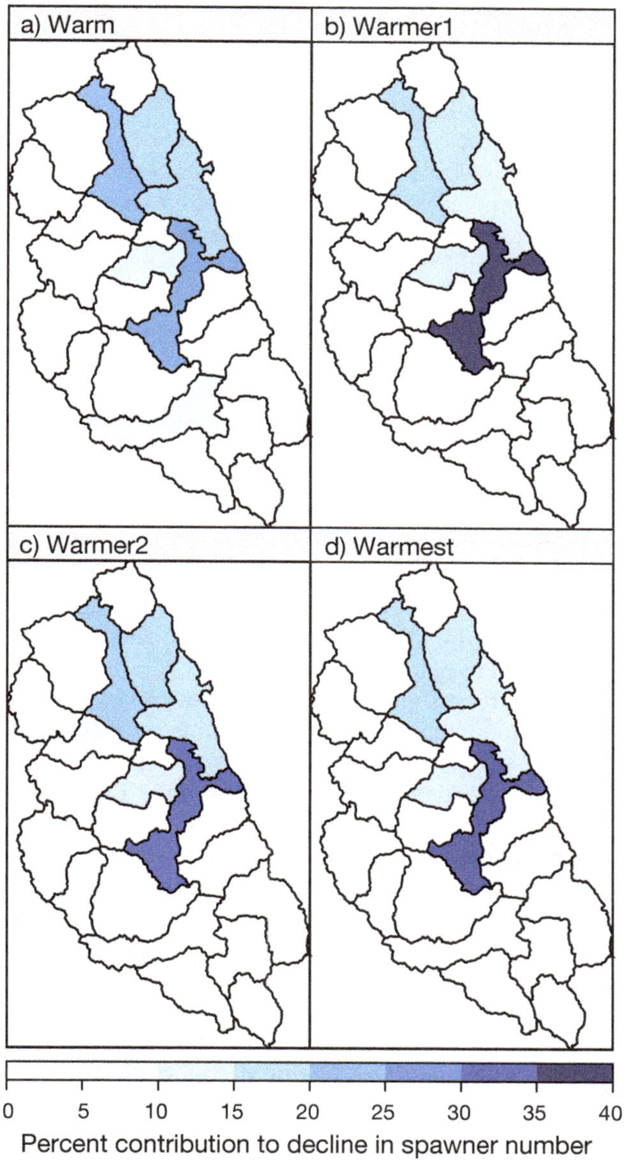

Fig. 6. Percent contribution of each HUC6 subbasin to the decline in spawner abundance, including Fig. 5 variables as well as 25% increase in mortality during the in-gravel egg incubation and rearing stage simulating response to increased streambed scour: (a) Warm (PCM1); (b) Warmer1 (CCSM3); (c) Warmer2 (CGCM3.1); (d) Warmest (ECHAM5)

Our sensitivity analysis, where we manipulated variables one at a time, revealed that the in-gravel egg and rearing life stage had the most influence on spawner abundance. When we reduced survival through the in-gravel stage by 25% to simulate the potential influence of increased streambed scour due to larger and more frequent high-discharge events during the in-gravel period, spawner abundance declined by an additional ~15% below numbers for the future climate and habitat scenarios (Figs. 4 & 6). Among the life stages that had explicit habitat–survival links in our model, the one with the greatest influence on population dynamics was the spawning life stage. For all GCMs, except again Warm, end of century higher water temperature had the most substantial and negative impact on fish abundance (Fig. 4). The in-gravel and summer rearing stages responded only weakly to increased temperatures and lower summer flows, having a negligible impact on spawner abundance (Fig. 4).

4. DISCUSSION

Predicted increases in air temperature and changes in precipitation are likely to alter aspects of stream habitat that are important to salmon at different stages of their life history. Our linked climate and hydrology models showed increased mean and maximum water temperatures in all months and increased average and maximum discharge during the incubation period. There was variability, however, among the climate models regarding low summer discharge. Three GCMs (Warmest, Warmer1, and Warmer2 in Table 1) resulted in a lower mean and minimum discharge, while a more moderate GCM (Warm) — in terms of air temperature change — predicted a higher mean and minimum discharge relative to current conditions (see the supplement at www.int-res.com/articles/suppl/c071p127_supp.pdf) due to a projected increase in precipitation during the summer months.

When we linked these conditions to a population model, future climate caused small population declines, with the spawning life stage most affected by climate-driven changes in freshwater habitat (Fig. 4). Survival during the spawning stage was sensitive to changes in water temperature in the range predicted for the century's end (Fig. 3; Richter & Kolmes 2005). Predicted changes in water temperature during the in-gravel and the summer rearing stages had little influence on spawner numbers (Fig. 4). The relationship between survival and water temperature during the in-gravel and summer rearing life stages was based on multiple observational experiments for each stage (see Honea et al. 2009) and is therefore likely robust. The link between survival and water temperature at the spawning life stage, in contrast, was based on only 1 data set and should be verified as more observations are made.

We found no change in spawner abundance due to the influence of reduced summer flows on juvenile movement downstream to overwintering habitat. The lower flows drove a greater number of juveniles to migrate downstream to reaches with more favorable flows and habitat for the overwinter period. Survival through the overwinter period was a function of the percent of cobble and boulders in downstream pool habitats, and observations indicate that availability of this type of habitat is not limiting the population (Honea et al. 2009). Therefore, when fish moved downstream as a consequence of lower flows, there was no resulting change in survival. However, more information is needed to better understand and to model survival through the overwintering stage.

With mean peak discharge during the autumn to early spring period increasing by 27 to 34% and the current 10 yr and 100 yr peak discharge events returning with frequencies of 4–5 yr and 9–15 yr, respectively (see the supplement), increased in-gravel egg and rearing mortality due to streambed scour is likely (Lapointe et al. 2000). Although we were unable to explicitly include in-gravel mortality from scour due to a lack of basin-specific scour estimates, spawner abundance declined by 14 to 21% below the estimates for the current climate and habitat scenario when we increased in-gravel mortality by 25% in the future climate scenarios as a sensitivity test. These estimates of scour effects are similar to estimates of Battin et al. (2007) for Chinook salmon *Oncorhynchus tshawytscha* in a basin where rain-on-snow events between November and February are more common than in the Wenatchee River Basin. Wenatchee River Basin hydrology is currently dominated by spring snowmelt; however, if climate estimates for the end of the century hold, it may potentially change to a more transient hydrology dominated by both rain and snow (Mantua et al. 2010), likely increasing the influence of streambed scour on in-gravel mortality.

We focused our modeling on the direct effects of climate on habitat variables with links to fish survivorship through successive life stages within freshwater habitats for salmon (Moussalli & Hilborn 1986, Scheuerell et al. 2006, Honea et al. 2009). While this is an important first step in beginning to consider the effects of climate on the vulnerability of at-risk species, there are other steps that could be included in the future. For example, numerous studies have associated a diverse set of ocean condition indices to salmon ocean survival (e.g. Mantua et al. 1997, Scheuerell & Williams 2005, Zabel et al. 2006); however, at this time, the net impact on conditions for salmon survival due to forecasted changes in ocean conditions and the marine food web is difficult to estimate (Bakun 1990, Snyder et al. 2003, Harley et al. 2006, Overland & Wang 2007, Branch et al. 2013). In addition, freshwater habitat is likely to be affected by climate change indirectly — for example, via indirect effects due to climate-induced changes in demands on water for irrigation (Donley et al. 2012) or to landscape changes, such as in vegetation patterns due to changes in precipitation, air temperature, fire regime (Flannigan et al. 2000), and insect pathogens (Dalton et al. 2013). Other considerations include climate change-induced discharge and water temperature changes that may affect interactions with other organisms, such as predators (Petersen & Kitchell 2001) or fish pathogens (Marcogliese 2001), and interactions with hatchery fish with the consequent genetic effects of interbreeding (e.g. McClure et al. 2008, Muhlfeld et al. 2014). The various potential influences not included in our study highlight the need for a better understanding of the functional forms of the relationships between habitat and survival at specific life stages and their consequences to population dynamics (Hilborn 2009).

Although we found a potential decline in Wenatchee spring Chinook salmon spawners under climate change, other regions and species may respond differently to climate change. For steelhead *Oncorhynchus mykiss*, some evidence suggests that decreasing summer discharge may be the most limiting factor in the Wenatchee River and elsewhere in the Upper Columbia River Basin (Wade et al. 2013). In contrast, in colder regions where cool water temperatures may be further below the high temperature tolerance threshold, warming may have the opposite effect and in some locations may lead to an increase in fish survival and production (Leppi et al. 2014).

Taking into account the influence of a changing climate on habitat will improve the likelihood that restoration actions will succeed in recovering at-risk species (Waples et al. 2009, Beechie et al. 2013). Our modeling found that the restoration actions that most addressed the impacts of increasing water temperature were restoring forests, both riparian and upland, toward estimated historical conditions and reducing anthropogenic impervious area. There is empirical and other modeling evidence that is consistent with these findings. For example, Pollock et al. (2009) found that the total amount of forest harvest within a basin was positively correlated with water temperature in 40 small basins in western Washington State. In the Wenatchee River Basin, Cristea & Burges (2010) used a riparian shade model linked to a stream channel heat-balance model to similarly conclude that riparian restoration would mostly offset the effects of future warming in Nason Creek, the Icicle River, and the Wenatchee River mainstem. In another tributary of the Columbia River near the Wenatchee River Basin, Ruesch et al. (2012) used a statistical model to predict stream temperature based on flow and stream unit connectivity. They concluded that riparian restoration would partially compensate for an estimated loss of 69 to 95% of Chinook salmon habitat as a result of climate change. Elsewhere, other restoration actions may be more appropriate. For example, Battin et al. (2007) concluded that increasing habitat capacity in the lower reaches of a transitional rain and snow-melt dominated river basin by reconnecting the main channel to side-channels and to floodplain habitat would be the most effective means of mitigating changes resulting from future climate because the lower reaches had been highly modified and disconnected from their floodplains while the higher elevation reaches were mostly already protected and pristine.

While we project that water temperature during the spawning period will be more limiting by this century's end than it currently is, and that streambed scour may become more important as well, our modeling predicts that the percentage of fine sediment in gravels during the incubation period will remain the most influential freshwater habitat variable limiting spawner abundance, as we found to be the case in our earlier study (Honea et al. 2009) for present conditions. Thus, restoration priorities for freshwater habitat could remain the same, particularly because some of the same restoration actions that impact fine sediments in incubation gravels have been shown to reduce high water temperature and high discharge extremes as well. For example, reforestation and reducing riparian grazing can promote riparian forests that shade stream channels as well as reduce runoff, leading to less erosion and reduced flashiness of discharge in response to precipitation (Beschta 1997, Medina et al. 2005). Restoring riparian forests will likely be crucial to preparing streams for the impacts of climate change (Seavy et al. 2009).

5. CONCLUSION

This work demonstrates the utility of examining the impacts of habitat changes on multiple life stages to identify potential bottlenecks that may drive long-term persistence. Such models may indicate a shift in the importance of different habitat variables and life stages — as with our work indicating the increasing importance of water temperature during the summer and perhaps high flows during winter and spring — or reveal a vulnerable life stage amid others responding positively to change — as with the overwinter larval stage in the butterfly life-cycle model investigated by Radchuk et al. (2013). Furthermore, the apparent importance of relationships about which we have little local information, such as scour and its influence on survival during the egg incubation and in-gravel rearing period, indicates worthwhile areas for further data collection and research to guide population recovery efforts.

Acknowledgements. The downscaled dataset of GCMN estimates was provided by Pablo Carrasco of the Climate Impacts Group at the University of Washington (in collaboration with the Washington State Department of Ecology, Bonneville Power Administration, Northwest Power and Conservation Council, Oregon Water Resources Department, and the British Columbia Ministry of the Environment). Erin Rogers, then at the University of Washington Department of Civil Engineering, calibrated the DHSVMs for the Wenatchee River Basin. Both Mr. Carrasco and Ms. Rogers were guided by Alan Hamlet, then at the University of Washington Department of Civil Engineering. We also thank Lisa Crozier, Corey Phillis, and 3 anonymous reviewers for providing comments on earlier drafts; Alan Hamlet and Ray Hilborn for helpful discussions; countless field crew members who collected habitat data; and Damon Holzer for creating the map in Fig. 2. J.M.H. was supported by a postdoctoral grant from UCAR (administered through NOAA) and by Emerson College.

LITERATURE CITED

Bakun A (1990) Global climate change and intensification of coastal ocean upwelling. Science 247:198–201

Battin J, Wiley MW, Ruckelshaus MH, Palmer RN, Korb E, Bartz KK, Imaki H (2007) Projected impacts of climate change on salmon habitat restoration. Proc Natl Acad Sci USA 104:6720–6725

Beacham TD, Murray CB (1989) Variation in developmental biology of sockeye salmon (*Oncorhynchus nerka*) and Chinook salmon (*O. tshawytscha*) in British Columbia. Can J Zool 67:2081–2089

Beechie T, Imaki H, Greene J, Wade A and others (2013) Restoring salmon habitat for a changing climate. River Res Appl 29:939–960

Beschta RL (1997) Riparian shade and stream temperature: an alternative perspective. Rangelands 19:25–28

Branch TA, DeJoseph BM, Ray LJ, Wagner CA (2013) Impacts of ocean acidification on marine seafood. Trends Ecol Evol 28:178–186

Brett JR (1952) Temperature tolerance in young Pacific salmon, genus *Oncorhynchus*. J Fish Res Board Can 9: 265–309

Bryant MD (2009) Global climate change and potential effects on Pacific salmonids in freshwater ecosystems of southeast Alaska. Clim Change 95:169–193

Coutant CC (1973) Effects of thermal shock on vulnerability of juvenile salmonids to predation. J Fish Res Board Can 30:965–973

Cramer SP (2001) The relationship of stream habitat features to potential for production of four salmonid species. Final report to the Oregon Building Industry Association. S. P. Cramer and Associates, Gresham, OR

Cristea NC, Burges SJ (2010) An assessment of the current and future thermal regimes of three streams located in the Wenatchee River basin, Washington State: some implications for regional river basin systems. Clim Change 102:493–520

Crozier LG, Zabel RW, Hamlet AF (2008) Predicting differential effects of climate change at the population level with life-cycle models of spring Chinook salmon. Glob Change Biol 14:236–249

Dalton MM, Mote PW, Snover AK (eds) (2013) Climate change in the Northwest: implications for our landscapes, waters, and communities. Island Press, Washington, DC

Donley EE, Naiman RJ, Marineau MD (2012) Strategic planning for instream flow restoration: a case study of potential climate change impacts in the central Columbia River basin. Glob Change Biol 18:3071–3086

Dupont S, Dorey N, Thorndyke M (2010) What meta-analysis can tell us about vulnerability of marine biodiversity to ocean acidification? Estuar Coast Shelf Sci 89:182–185

Finch T, Pearce-Higgins JW, Leech DI, Evans KL (2014) Carry-over effects from passage regions are more important than breeding climate in determining the breeding phenology and performance of three avian migrants of conservation concern. Biodivers Conserv 23:2427–2444

Flannigan MD, Stocks BJ, Wotton BM (2000) Climate change and forest fires. Sci Total Environ 262:221–229

Goniea TM, Keefer ML, Bjornn TC, Peery CA, Bennett DH, Stuehrenberg LC (2006) Behavioral thermoregulation and slowed migration by adult fall Chinook salmon in response to high Columbia River water temperatures. Trans Am Fish Soc 135:408–419

Goode JR, Buffington JM, Tonina D, Isaak DJ and others (2013) Potential effects of climate change on streambed scour and risks to salmonid survival in snow-dominated mountain basins. Hydrol Processes 27:750–765

Groves CR, Game ET, Anderson MG, Cross M and others (2012) Incorporating climate change into systematic conservation planning. Biodivers Conserv 21:1651–1671

Harley CDG, Hughes RA, Hultgren KM, Miner BG and others (2006) The impacts of climate change in coastal marine systems. Ecol Lett 9:228–241

Hilborn R (2009) Life history models for salmon management: the challenges. In: Knudsen EE, Michael JHJ (eds) Pacific salmon environmental and life history models: advancing science for sustainable salmon in the future. American Fisheries Society, Bethesda, MD, p 23–32

Honea JM, Jorgensen JC, McClure MM, Cooney TD, Engie K, Holzer DM, Hilborn R (2009) Evaluating habitat effects on population status: influence of habitat restoration on spring-run Chinook salmon. Freshw Biol 54: 1576–1592

Jorgensen JC, Honea JM, McClure MM, Cooney TD, Engie K, Holzer DM (2009) Linking landscape-level change to habitat quality: an evaluation of restoration actions on the freshwater habitat of spring-run Chinook salmon. Freshw Biol 54:1560–1575

Keefer ML, Caudill CC (2014) Homing and straying by anadromous salmonids: a review of mechanisms and rates. Rev Fish Biol Fish 24:333–368

Lapointe M, Eaton B, Driscoll S, Latulippe C (2000) Modelling the probability of salmonid egg pocket scour due to floods. Can J Fish Aquat Sci 57:1120–1130

Leppi JC, Rinella DJ, Wilson RR, Loya WM (2014) Linking climate change projections for an Alaskan watershed to future coho salmon production. Glob Change Biol 20: 1808–1820

Mantua NJ, Hare SR, Zhang Y, Wallace JM, Francis RC (1997) A Pacific interdecadal climate oscillation with impacts on salmon production. Bull Am Meteorol Soc 78: 1069–1079

Mantua N, Tohver I, Hamlet A (2010) Climate change impacts on streamflow extremes and summertime stream temperature and their possible consequences for freshwater salmon habitat in Washington State. Clim Change 102:187–223

Marcogliese DJ (2001) Implications of climate change for parasitism of animals in the aquatic environment. Can J Zool 79:1331–1352

McClure MM, Utter FM, Baldwin C, Carmichael RW and others (2008) Evolutionary effects of alternative artificial propagation programs: implications for viability of endangered anadromous salmonids. Evol Appl 1:356–375

McClure MM, Alexander M, Borggaard D, Boughton D and others (2013) Incorporating climate science in applications of the U.S. Endangered Species Act for aquatic species. Conserv Biol 27:1222–1233

McCormick JH, Hokanson KEF, Jones BR (1972) Effects of temperature on growth and survival of young brook trout *Salvelinus fontinalis*. J Fish Res Board Can 29:1107–1112

Medina AL, Rinne JN, Roni P (2005) Riparian restoration through grazing management: considerations for monitoring project effectiveness. In: Roni P (ed) Monitoring stream and watershed restoration. American Fisheries Society, Bethesda, MD, p 97–126

Mote PW, Salathé EP (2010) Future climate in the Pacific Northwest. Clim Change 102:29–50

Moussalli E, Hilborn R (1986) Optimal stock size and harvest rate in multistage life history models. Can J Fish Aquat Sci 43:135–141

Muhlfeld CC, Kovach RP, Jones LA, Al-Chokhachy R and others (2014) Invasive hybridization in a threatened species is accelerated by climate change. Nat Clim Change 4:620–624

National Marine Fisheries Service (NMFS) (1999) Endangered and threatened species: threatened status for three Chinook salmon Evolutionarily Significant Units (ESUs) in Washington and Oregon, and endangered status for one Chinook salmon ESU in Washington. National Marine Fisheries Service (NMFS), National Oceanic and Atmospheric Administration (NOAA), Department of Commerce, Washington, DC

Neville HM, Isaak DJ, Dunham JB, Thurow RF, Rieman BE (2006) Fine-scale natal homing and localized movement as shaped by sex and spawning habitat in Chinook salmon: insights from spatial autocorrelation analysis of individual genotypes. Mol Ecol 15:4589–4602

Overland JE, Wang M (2007) Future climate of the north Pacific Ocean. Eos Trans AGU 88:178–182

Palmer MA, Reidy Liermann CA, Nilsson C, Flörke M, Alcamo J, Lake PS, Bond N (2008) Climate change and the world's river basins: anticipating management options. Front Ecol Environ 6:81–89

Petersen JH, Kitchell JF (2001) Climate regimes and water temperature changes in the Columbia River: bioenergetic implications for predators of juvenile salmon. Can J Fish Aquat Sci 58:1831–1841

Pollock MM, Beechie TJ, Liermann M, Bigley RE (2009) Stream temperature relationships to forest harvest in western Washington. J Am Water Resour Assoc 45: 141–156

Quinn TP, Dittman AH (1990) Pacific salmon migrations and homing: mechanisms and adaptive significance. Trends Ecol Evol 5:174–177

Quinn TP, Stewart IJ, Boatright CP (2006) Experimental evidence of homing to site of incubation by mature sockeye salmon, *Oncorhynchus nerka*. Anim Behav 72:941–949

Radchuk V, Turlure C, Schtickzelle N (2013) Each life stage matters: the importance of assessing the response to climate change over the complete life cycle in butterflies. J Anim Ecol 82:275–285

Richter A, Kolmes SA (2005) Maximum temperature limits for Chinook, coho, and chum salmon, and steelhead trout in the Pacific Northwest. Rev Fish Sci 13:23–49

Ruesch AS, Torgersen CE, Lawler JJ, Olden JD, Peterson EE, Volk CJ, Lawrence DJ (2012) Projected climate-induced habitat loss for salmonids in the John Day River network, Oregon, USA. Conserv Biol 26:873–882

Runge CA, Martin TG, Possingham HP, Willis SG, Fuller RA (2014) Conserving mobile species. Front Ecol Environ 12: 395–402

Scheuerell MD, Williams JG (2005) Forecasting climate-induced changes in the survival of Snake River spring/summer Chinook salmon (*Oncorhynchus tshawytscha*). Fish Oceanogr 14:448–457

Scheuerell MD, Hilborn R, Ruckelshaus MH, Bartz KK, Lagueux KM, Haas AD, Rawson K (2006) The Shiraz model: a tool for incorporating anthropogenic effects and fish-habitat relationships in conservation planning. Can J Fish Aquat Sci 63:1596–1607

Seaber PR, Kapinos FP, Knapp GL (1987) Hydrologic unit maps. United States Geological Survey, Reston, VA

Seavy NE, Gardali T, Golet GH, Griggs FT and others (2009) Why climate change makes riparian restoration more important than ever: recommendations for practice and research. Ecol Res 27:330–338

Seney EE, Rowland MJ, Lowery RA, Griffis RB, McClure MM (2013) Climate change, marine environments, and the U.S. Endangered Species Act. Conserv Biol 27: 1138–1146

Snyder MA, Sloan LC, Diffenbaugh NS, Bell JL (2003) Future climate change and upwelling in the California Current. Geophys Res Lett 30:1823

Torgersen CE, Price DM, Li HW, McIntosh BA (1999) Multi-scale thermal refugia and stream habitat associations of Chinook salmon in northeastern Oregon. Ecol Appl 9: 301–319

Velsen FPJ (1987) Temperature and incubation in Pacific salmon and rainbow trout: compilation of data on median hatching time, mortality, and embryonic staging. Can Data Rep Fish Aquat Sci 626:1–58

Wade AA, Beechie TJ, Fleishman E, Mantua NJ and others (2013) Steelhead vulnerability to climate change in the Pacific Northwest. J Appl Ecol 50:1093–1104

Waples RS, Beechie TJ, Pess GR (2009) Evolutionary history, habitat disturbance regimes, and anthropogenic changes: What do these mean for resilience of Pacific salmon populations? Ecol Soc 14:3

Warner MD, Mass CF, Salathé EP (2015) Changes in winter atmospheric rivers along the North American west coast in CMIP5 climate models. J Hydrometeorol 16:118–128

Yvon-Durocher G, Montoya JM, Trimmer M, Woodward G (2011) Warming alters the size spectrum and shifts the distribution of biomass in freshwater ecosystems. Glob Change Biol 17:1681–1694

Zabel RW, Scheuerell MD, McClure MM, Williams JG (2006) The interplay between climate variability and density dependence in the population viability of Chinook salmon. Conserv Biol 20:190–200

2

Validation of the WRF regional climate model over the subregions of Southeast Asia: climatology and interannual variability

Satyaban B. Ratna[1,*], J. V. Ratnam[1], S. K. Behera[1], Fredolin T. Tangang[2], T. Yamagata[1]

[1]Application Laboratory, JAMSTEC, 3173-25 Showa-machi, Kanazawa-ku, Yokohama, Kanagawa, 236-0001, Japan
[2]School of Environmental and Natural Resource Sciences, Faculty of Science and Technology, the National University of Malaysia, Malaysia

ABSTRACT: This study investigates the capability of a regional climate model in simulating the climate variability over Southeast Asia (SE Asia). The present-day climate, covering the period 1991 to 2015, was dynamically downscaled using the Weather Research and Forecasting (WRF) model with a horizontal resolution of 27 km. The initial and boundary conditions for the WRF model is provided with the European Centre for medium-range weather forecasting (ECMWF) reanalysis (ERA-Interim) data. The model reproduced the mean precipitation climatology as well as the annual cycle. Nevertheless, the model overestimated the boreal summer precipitation over the SE Asian mainland, and underestimated the boreal winter precipitation over the Indonesian region. Model biases are associated with the bias in simulating the vertically integrated moisture fluxes. At an interannual scale, the model shows good performance over the SE Asian mainland and the Philippines in all seasons except for the boreal summer. The influence of El Niño/Southern Oscillation (ENSO) on rainfall over mainland SE Asia and the Philippines during JJA is weak, and the model successfully simulated the weak relationship realistically. In contrast, model interannual variability over the Indonesia region is good only in boreal summer and autumn seasons. This is because the model successfully simulated the significant negative correlation between rainfall and ENSO. The influence of the Indian Ocean Dipole (IOD) is seen only in the boreal autumn over the Indonesian region, and the model reproduced it reasonably well. The improvement in the representation of precipitation anomaly associated with ENSO/IOD is due to reasonably accurate simulation of large-scale circulation over SE Asia.

KEY WORDS: Downscaling · Regional climate model · WRF · Southeast Asia · ENSO · IOD

1. INTRODUCTION

Southeast Asia (hereafter SE Asia) is characterized by complex terrain and land–water contrasts with major river systems, tropical forests, and many islands. It rests between the waters of the Indian Ocean in the west and the Pacific Ocean in the east. It contains a mainland section to the north (Myanmar, Cambodia, Laos, Thailand, Peninsular Malaysia, and Vietnam – known as the mainland SE Asia) and a maritime section to the south (Brunei, the Philippines, Singapore, East Malaysia, East Timor, Papua New Guinea and Indonesia — known as the Maritime SE Asia) (Fig. 1a).

The climate of SE Asia, and especially maritime SE Asia is mainly tropical: hot and humid all year round with abundant rainfall. The SE Asian region lies in the range of the classical tropical monsoons (Zeng & Zhang 1998, Li & Zeng 2005), which is surrounded by the positions of the Intertropical Convergence Zone

*Corresponding author: satyaban@jamstec.go.jp

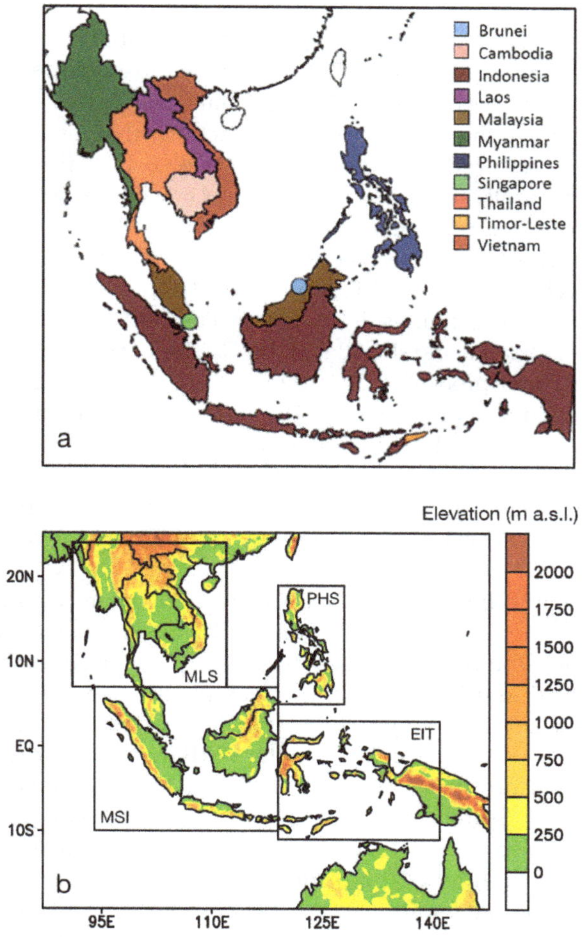

Fig. 1. (a) Southeast Asia with country boundaries; (b) WRF model domain with a 27 km horizontal grid resolution used in this study. Topography (expressed in meters above mean sea level) is shaded. The 4 boxes in (b) are the sub-regions of Southeast Asia considered in this study for the model validation (see Fig. 2 for definitions)

(ITCZ) in summer and winter. The majority of the SE Asia region is influenced by seasonal shifts in winds or the monsoon, and much of the region is annually affected by extreme weather events, particularly tropical cyclones, droughts, and floods (Yusuf & Francisco 2009). During some years, rainfall over SE Asia region is altered by large-scale conditions associated with El Niño/Southern Oscillation (ENSO) and Indian Ocean dipole (IOD) events (e.g. Kripalani & Kulkarni 1997, McBride et al. 2003, Tangang & Juneng 2004, Naylor et al. 2007, Behera et al. 2008, Ummenhofer et al. 2013, Salimun et al. 2014). The influence of large-scale climate phenomena varies across the region due to orography and ocean–atmosphere interactions. Variability in the climate has a great impact on the socio-economic conditions of the region, and accurate forecasting of the weather and climate would be beneficial. However, the climate of SE Asia has not been sufficiently studied to improve the seasonal forecasts over the region.

Despite the development of many global general circulation model (GCM) systems for seasonal forecasts, the skill of the models in forecasting precipitation is still a challenge. Recenly, Salimun et al. (2014) have suggested that GCMs tend to have lower skill for seasonal rainfall predictions over SE Asia. The poor skill of a GCM in forecasting precipitation is often attributed to low horizontal resolution of the GCM. An important aspect of improving precipitation in GCM simulations is to resolve the regional heterogeneity (Giorgi & Mearns 1999). The technique of dynamical downscaling, wherein a high-resolution regional climate model (RCM) is embedded into a low-resolution GCM output, is often used to improve the simulation of precipitation (e.g. Dickinson et al. 1989, Giorgi 1990, Giorgi et al. 2004). It is assumed that the RCM, due to better representation of regional processes, will improve the GCM simulated precipitation.

Before using a RCM for the purpose of seasonal forecast using global climate model (GCM) boundary conditions, the fidelity of the RCM in reproducing the observed regional climate is assessed in order to understand the systematic biases in the model. Often, the assessment is carried out by forcing a RCM with observed/reanalysis data (Giorgi & Mearns 1999). In recent years, many attempts have been made to understand regional climate variability over SE Asia using regional climate models (e.g. Aldrian et al. 2004, Francisco et al. 2006, Im et al. 2008, Phan et al. 2009, Takahashi et al. 2009, 2010, Chotamonsak et al. 2011, 2012, Ngo-Duc et al. 2014, Phan et al. 2014, Raktham et al. 2015, Raghavan et al. 2016, Juneng et al. 2016). To the authors' knowledge, there are no comprehensive studies available evaluating the performance of an RCM over SE Asia and its subregions for different seasons of the year. Also, RCM performance has not been fully evaluated with regards to mean climatology, interannual variability, and the relationship of the latter with ENSO and IOD. Past studies are limited either to study of a specific event or a particular season, and over a particular sub-region of SE Asia. In this study, we try to fill the gap by performing a regional model simulation over a relatively long period of 25 yr at a horizontal resolution of 27 km, and evaluating its performance in simulating the precipitation over 4 subregions of SE Asia during boreal summer (June-July-August; JJA), boreal autumn (September-October-November; SON), boreal winter (December-January-February;

DJF), and boreal spring (March-April-May; MAM). The performance of the model is evaluated at seasonal, annual, and interannual time scales.

2. MODEL, METHODOLOGY AND DATA

Advanced Research Weather research and forecasting (WRF) model (ARW) version 3.6.1 (Skamarock et al. 2008), developed by the National Centre for Atmospheric Research (NCAR) and designed to serve both operational forecasting and atmospheric research, is used in this study. The WRF model is a non-hydrostatic, fully compressible, and terrain-following sigma coordinate model. In this study, WRF simulations over SE Asia are performed using a domain with horizontal resolutions of 27 km (Fig. 1b). The model has 32 levels in the vertical with the upper boundary at 10 hPa. The WRF domain covers the SE Asia land mass, as well as parts of the surrounding Pacific and Indian Oceans (19.12° S to 25.15° N, 86.96° to 147.77° E), with 251 grid points in the east–west and 188 grid points in north–south directions. The model analysis is carried out by dividing the SE Asia region into 4 subregions (Fig. 1b): (1) mainland SE Asia (MLS), (2) Malaysia, Singapore, Brunei, western Indonesia (MSI), (3) the Philippines (PHS), and (4) eastern Indonesia (EIT). The regions of MLS and PHS that lie to the north of 5° N receive most of their annual rainfall during the boreal summer. However, the MSI and EIT regions located in the equatorial belt receive rainfall throughout the year except for a reduced amount during the boreal summer. The subregions were chosen based on rainfall variability, and in such a way that the major landmasses can be accommodated in the chosen boxes (Fig. 1b).

Physical parameterization schemes considered in this study include cumulus parameterization schemes of Betts-Miller-Janjic (Betts & Miller 1986, Janjic 1994), the microphysics scheme of the WSM 3-class simple ice scheme (Hong et al. 2004), the Unified NOAH scheme for land surface processes (Chen & Dudhia 2001), the Yonsei University scheme for the planetary boundary layer (Noh et al. 2003), the rapid radiative transfer model (RRTM) scheme for long waves (Mlawer et al. 1997), and the Dudhia scheme for short waves (Dudhia 1989). The first author in an earlier study (Ratna et al. 2014) for the South Africa region also used a similar configuration, and the model shows good skill simulating climatology and interannual variability of rainfall. Recently, Ratna et al. (2016b) evaluated 4 different convection parameterization schemes over SE Asia for a shorter period of 3 yr, 1990–1993. They found that all convection schemes could reproduce the spatial distribution of rainfall over SE Asia, but the area-averaged rainfall bias with the BMJ scheme was lower compared to that of the other 3 convection schemes. Also, they showed that the annual cycle of precipitation was well simulated by the BMJ scheme with a higher correlation coefficient compared to the other schemes. So, based on Ratna et al. (2014, 2016b), we have used the combination of most suitable parameterization schemes in this study. The WRF model was initialized using the 00:00 h UTC 1 January 1990 data and was integrated up to 00:00 h UTC 1 January 2016 considered in this study. The first year of simulations was regarded as model spin-up and excluded from analysis. The model output data were saved in 6 h intervals, but converted into the monthly mean to study the seasonal climatology and interannual variability.

The 6 hourly, 0.75° × 0.75° grid European Centre for medium-range weather (ECMWF) reanalysis (ERA-Interim) data (Dee et al. 2011) was used as the initial and boundary conditions for the simulations. Sea surface temperatures (SST) from ERA-Interim fields were interpolated to the WRF model grid resolution and used as slowly varying lower boundary input. Surface topography data and 24 category land-use index data based on climatological averages, both at a 30 s resolution, were obtained from the United States Geological Survey (USGS). The NOAA optimum interpolation SST version 2 data (Reynolds et al. 2002) is used to identify the El Niño/La Niña and IOD events. We note that the El Niño and La Niña events in this study for all the seasons are based on the SST anomaly over Niño 3.4 region (5° S to 5° N, 170° to 120° W). The El Niño (La Niña) events are identified when the area-averaged SST anomaly over Niño 3.4 region is above (below) 1 standard deviation for a season during the study period 1991 to 2015. Similarly, the positive (negative) IOD events are defined when an anomalous SST difference between the western equatorial Indian Ocean (10° S to 10° N, 50 to 70° E) and the southeastern equatorial Indian Ocean (10° S to 0° N, 90° to 110°E) is above (below) 1 standard deviation for a season. The mean, standard deviation, and anomaly values for each seasons are calculated with respect to the baseline period 1991—2015. The years identified with ENSO and IOD for different seasons of the year are presented in Table 1.

The large-scale model-simulated parameters in this study are compared with ERA-interim data by interpolating the model data to ERA-Interim grid. To

Table 1. ENSO and IOD years considered here in this study. The years in the DJF correspond to January and February of the given year and December of the previous year

	MAM	JJA	SON	DJF
El Niño	1992, 1993, 1998, 2015	1997, 2002, 2015	1997, 2002, 2015	1992, 1998, 2003, 2010
La Niña	1999, 2000, 2008, 2011	1998, 1999, 2010	1998, 1999, 2007, 2010	1999, 2000, 2008, 2011
+ve IOD	1991, 1994, 2000, 2007, 2009	1994, 1997, 2003, 2008, 2012	1994, 1997, 2006, 2015	1996, 1998, 2001, 2010
−ve IOD	1992, 1996, 2002, 2013	1992, 1996, 1998, 2013, 2014	1992, 1996, 1998, 2010	2005, 2006, 2015

validate the model simulated rainfall with the observed rainfall, the 0.5° Climate Research Unit version 3.24 dataset (CRU; Harris et al. 2014) was used. In a recent study, Ratna et al. (2016b) evaluated various monthly mean precipitation data over Southeast Asia region, such as APHORODITE (Yatagai et al. 2012), GPCP (Adler et al. 2003), and CRU precipitation data for the common period of 1991 to 2007. GPCP and CRU are closer to each other, but APHRODITE shows lower precipitation values over SE Asia. Ratna et al. (2016b) also compared the CRU, GPCP, and TRMM (Huffman et al. 2007) precipitation data for the common period 1998 to 2012 and found that CRU and TRMM are closer to each other and the GPCP data show slightly lower values. Based on the above analysis and also due to the availability of data for the whole period of 1991 to 2015, we have used CRU data in this study. Also, CRU data are widely used for the evaluation of regional climate modelling studies over Southeast Asia (Phan et al. 2009, Chotamonsak et al. 2011, Liew et al. 2014, Raghavan et al. 2016 and others).

3. RESULTS AND DISCUSSION

3.1. Mean rainfall climatology

The annual cycle (based on 25 yr climatology) of monthly mean rainfall for 4 subregions are compared with the WRF simulated rainfall (Fig. 2). Due to seasonal migration of the ITCZ, the peak season of the average monthly rainfall is different among subregions. Precipitation in the MLS region peaks in the months of boreal summer, whereas the MSI region has a peak mostly in the months of boreal winter. The PHS region has a slightly longer rainy season compared to the MLS region, though the peak in both regions are in the boreal summer period. The seasonality is relatively weak in the EIT region, with slightly higher precipitation in the boreal winter and boreal spring season. Overall, the model simulated a realistic annual cycle in all the subregions. However, the model tends to overestimate (underestimate) the precipitation in the period of high (low) rainfall over different subregions (Fig. 2).

Further analysis of the WRF simulated 25 yr climatology for the spatial distribution of precipitation was carried out at both annual and seasonal scales. For the seasonal scale evaluation, we averaged the precipitation over all the months within a season. The seasonal movement of the ITCZ controls the precipitation over

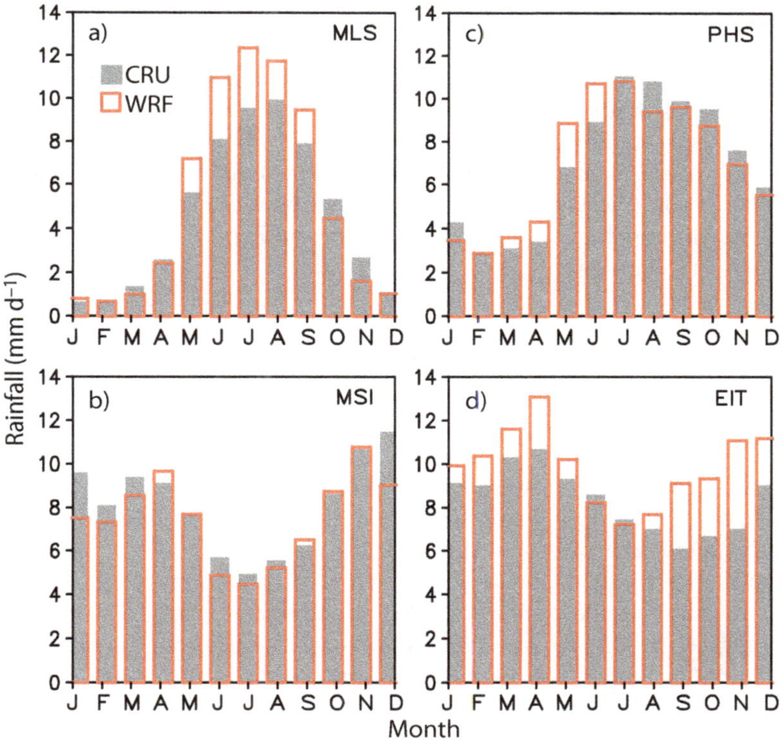

Fig. 2. Annual cycle of the area-averaged rainfall climatology (mm d^{-1}) of the 25 yr (1991–2015) over 4 sub-regions of the Southeast Asia: (a) mainland SE Asia (MLS), (b) Malaysia, Singapore, Brunei, western Indonesia (MSI), (c) the Philippines (PHS) and (d) eastern Indonesia (EIT) both for CRU and WRF

this region. The ITCZ is at the southernmost position in January just near south of the equator; during May, it moves northward to ~5° N, and during July–August, it moves further northward to ~20° N (Matsumoto & Murakami 2002, Lau & Yang 1997).

The observed annual precipitation averaged over the 25 yr period (1991–2015) has high precipitation over the equatorial region (Fig. 3a). A region of high precipitation is also evident over the orographic region of the western coast of the MLS region and in parts of Laos, Vietnam, and the Philippines. The WRF model was able to simulate the north–south rainfall gradient realistically over SE Asia (Fig. 3f). However, the WRF model underestimated precipitation over the western coast of Myanmar, Malaysia, and parts of Kalimantan Island, and overestimated precipitation over eastern parts of MLS, western Sumatra, and the Philippines (Fig. 3k). The model seems to have underestimated the orographic rainfall over the MLS region and overestimated the orographic rainfall over the Indonesian region. The model was also able to capture the spatial variability of the precipitation (Fig. 3g–j) for the different seasons of the year. The model has a dry bias over the region from the equator to 8° N during DJF when maximum annual rainfall occurs (Fig. 3l) During JJA, the maximum annual rainfall occurred over the MLS region when the ITCZ is at its northernmost point, but model simulates large positive bias there (Fig. 3n). The large bias in the precipitation over the MLS region during JJA is also reflected in the annual precipitation bias over the region (Fig. 3k). In the following section, we discuss the mechanism for the simulated precipitation bias in the model.

3.2. Thermodynamic characteristics

To understand the processes that caused the rainfall bias in the WRF simulation, we analyzed vertically integrated moisture convergence and moisture fluxes for the boreal summer and boreal winter season when the biases in the simulated precipitation are largest (Fig. 4). The region north of the equator is a moisture convergence zone during the boreal summer, when the moisture transport occurs from the Bay of Bengal to the MLS region and reaches up to the Philippines as part of the summer monsoon flow (Fig. 4a). The model could simulate the moisture convergence zone, but the simulated moisture convergence is very strong compared to the observation (Fig. 4c). There is a cyclonic circulation of the moisture flux, and the positive bias of moisture convergence (Fig. 4e) extending from eastern MLS to the Philippines contributed to the positive bias of rainfall (Fig. 3n). The southwesterly monsoon wind from the Indian Ocean contributes to the seasonal rainfall over the Myanmar and neighboring zones (Tsai et al. 2015, Ratna et al. 2016a). However, the model simulates a weaker moisture flux transport toward the Myanmar coast compared to the observation, and this has caused the dry bias of rainfall (Fig. 3n). During the DJF season, a moisture convergence region is observed over the region south of 10° N, and moisture transport occurs from the Pacific Ocean (Fig. 4b). However, the model simulates a weaker moisture convergence compared to the observation (Fig. 4d). The weaker moisture convergence is clearly seen in the difference between the model and observations (Fig. 4f). The negative bias of moisture convergence over the region coincides with the dry bias of rainfall (Fig. 3l), especially over Malaysia, eastern Sumatra, and north Kalimantan. In general, biases in the moisture fluxes and their convergence (divergence) as seen in Fig. 4 explain the biases in the spatial distribution of the rainfall (Fig. 3).

To further understand the reasons for the rainfall biases for different seasons and over different subregions of SE Asia, the mean climatology bias of model-simulated vertical profiles of specific humidity and the vertical velocity were analyzed over the MLS and MSI regions by calculating the area-averaged values over the land grid point. The vertical profile analysis is calculated only for the MLS and MSI regions because the model-simulated rainfall shows maximum bias over these 2 regions compared to the other regions (Fig. 3), and this would help to understand the physical processes associated with the model bias. It can be seen from Fig. 3 that only the MLS region has a dominant wet bias during the JJA season, whereas only the MSI region has dominant dry bias during DJF when the maximum annual rainfall occurs over the chosen regions. The vertical profile biases are calculated with respect to the ERA-Interim data. Fig. 5a shows that there is a positive bias of moisture extending from the surface to 300 hPa level during the JJA season over the MLS region. At the same time, there is a positive bias of vertical velocity from the surface to 600 hPa level over MLS region (Fig. 5c). The strong moist atmosphere with strong upward motion in the model during the JJA caused a positive bias of rainfall over MLS region, as seen in Fig. 3n. However, the same region during the DJF season shows low moisture and weak vertical motion compared to the JJA season but still suffers from positive rainfall bias (Fig. 5a,c). The MLS region receives

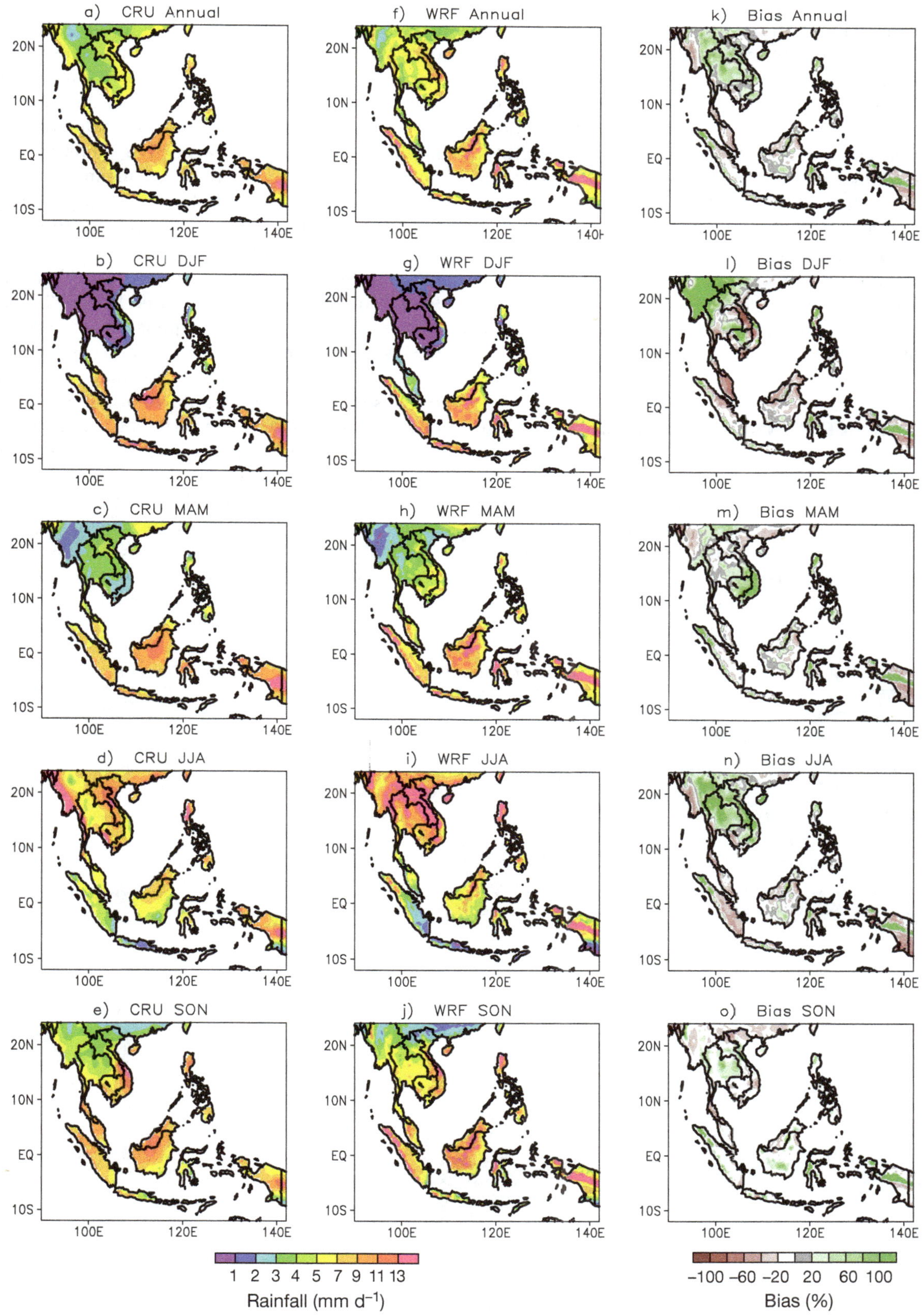

Fig. 3. (a–j) 25 yr (1991–2015) annual and seasonal mean rainfall climatology ([a–e] CRU; [f–j] WRF. (k,l,m,n,o) Same as (a–j), but for WRF bias (%) with respect to CRU

very scanty rainfall during the DJF season, known to be a dry season of the year. In the MSI region, the model simulates a slightly negative bias of moisture from the surface to the 300 hPa level during the DJF season (Fig. 5b). At the same time, the model-simulated vertical velocity is weaker than the ERA-Interim data from the 600 hPa to 150 hPa level (Fig. 5d). These low moisture and weak vertical velocity biases during the DJF season have led to the dry bias of rainfall (Fig. 3l) over the MSI region. However, during the JJA season, the biases of low moisture and weak vertical velocity are smaller compared to those during the DJF season, leading to a small precipitation bias (Fig. 3n).

4. INTERANNUAL VARIABILITY

To employ a regional climate model for short- and longer-term predictions, it is important to determine if the RCM is capable of capturing the observed interannual variability. In this section, we examine the interannual variability simulated by the WRF model for different seasons and for different subregions. The interannual variability of rainfall in terms of standardized rainfall anomaly for the 4 subregions of MLS, MSI, PHS, and EIT with the seasonal mean rainfall for MAM, JJA, SON, and DJF is analyzed and compared with WRF simulated rainfall (Fig. 6). The observed and WRF seasonal anomalies are calculated based on their respective climatology. The observed and simulated MAM mean rainfall have good correlations for both the MLS and PHS regions, but the correlations are very weak for the MSI and EIT regions (Table 2). During the MAM season, very little rainfall (2 to 5 mm d^{-1}) occurs over the MLS and PHS regions, but more rainfall (6 to 12 mm d^{-1}) occurs over MSI and EIT regions. The correlation for MSI is low, as the model has failed to capture the correct sign of rainfall anomaly for many years where the observed rainfall was above or below 1 standard deviation (Fig. 6a). During JJA, although the model has difficulties in simulating the anomalies in most of the years in the MLS and PHS regions, it does a good job simulating the anomalies in the MSI and EIT regions (Fig. 6b), and this can be seen in terms of correlation value (Table 2). In the SON season, all the regions showed the correct sign of the precipitation anomalies (Fig. 6c) with significant correlations of 0.64, 0.45, 0.69, and 0.80 for the MLS, MSI, PHI, and

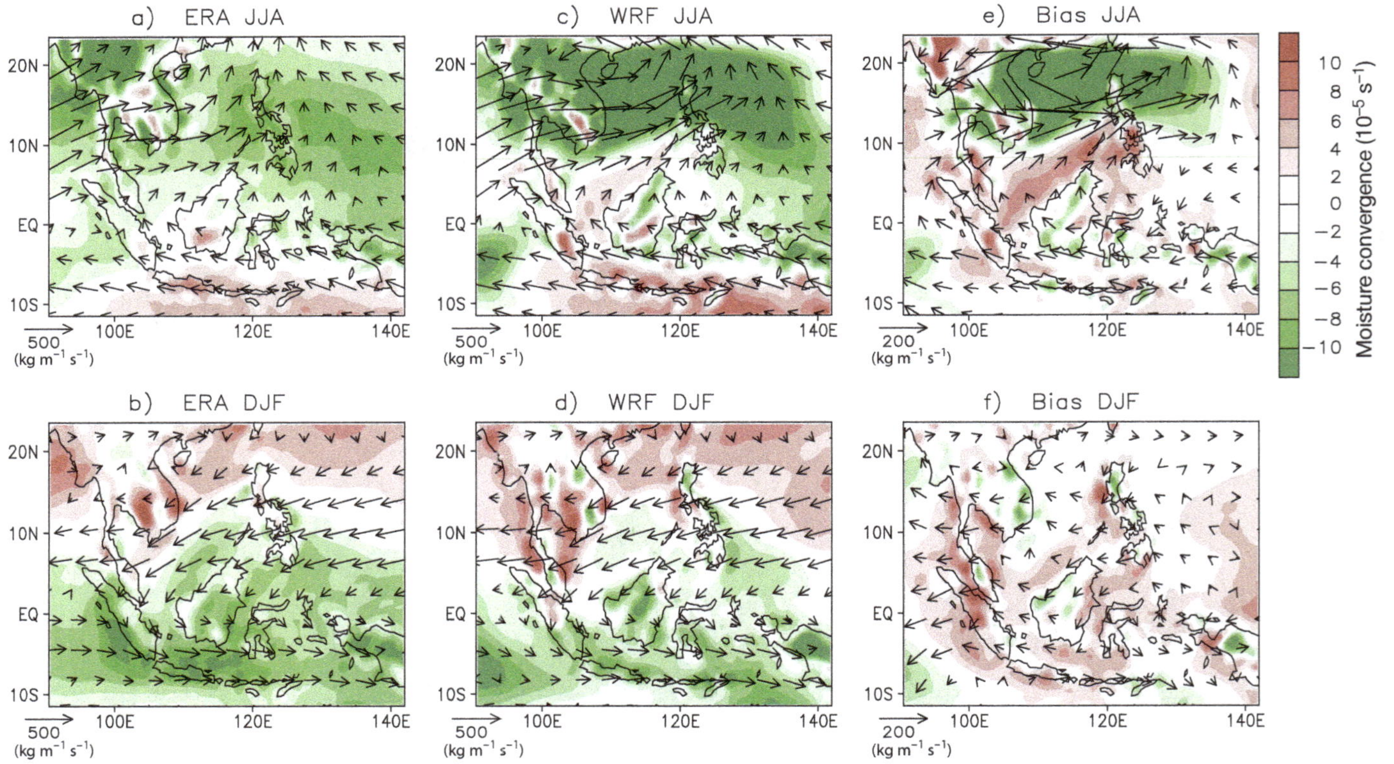

Fig. 4. (a–d) Twenty-five yr (1991–2015) mean climatology of vertically integrated (from 1000 to 300 hPa) moisture convergence and moisture fluxes (arrows) for the JJA and DJF seasons using (a,b) ERA and (c,d) WRF data. (e, f) Same as (a–d), but for bias as calculated from the difference between WRF and ERA

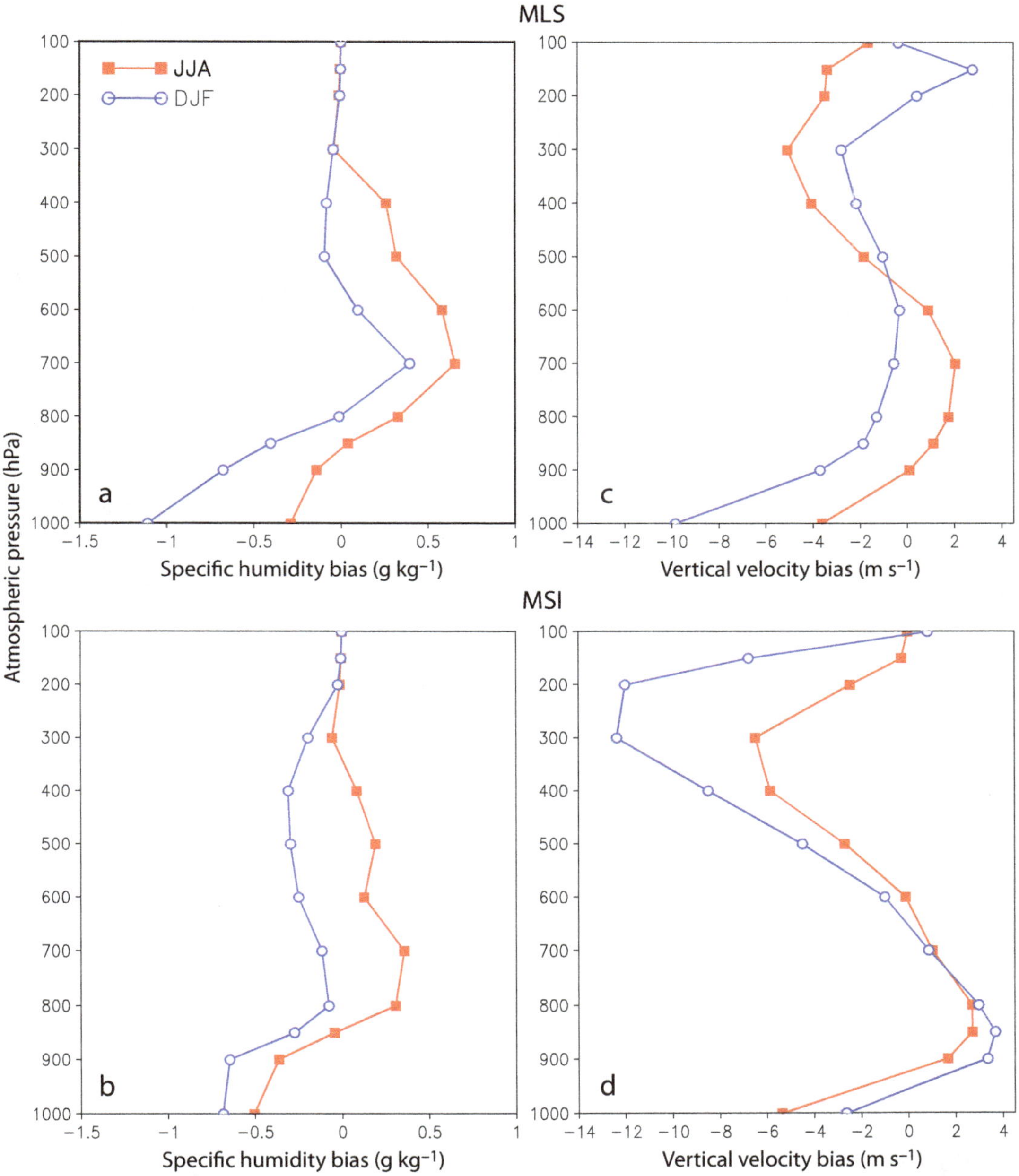

Fig. 5. Twenty-five yr (1991–2015) mean climatology bias for the vertical profiles of area-averaged specific humidity and vertical velocity for the JJA and DJF seasons, for (a,c) the MLS and (b,d) MSI regions (see Fig. 2 for definitions). Area averages are calculated over the land grid points, and the model bias is generated relative to ERA interim data

EIT regions, respectively. This is the only season where all the regions have relatively good model performance in terms of interannual variability. During the DJF season, maximum rainfall occurs over the MSI and EIT regions, and there was very little rainfall over the MLS and PHS regions. Here again, we see the performance of the model for the interannual variability in terms of correlation is poor for the MSI and EIT regions but is good for the MLS and PHS (Table 2) regions.

From the above analysis, it is clear that the model has a systematic precipitation bias during the peak

Fig 6. Interannual variability of standardized precipitation anomaly over 4 sub regions of SE Asia, namely MLS, MSI, PHS and EIT regions (see Fig. 2 for definitions), for the seasons (a) MAM, (b) JJA, (c) SON and d) DJF

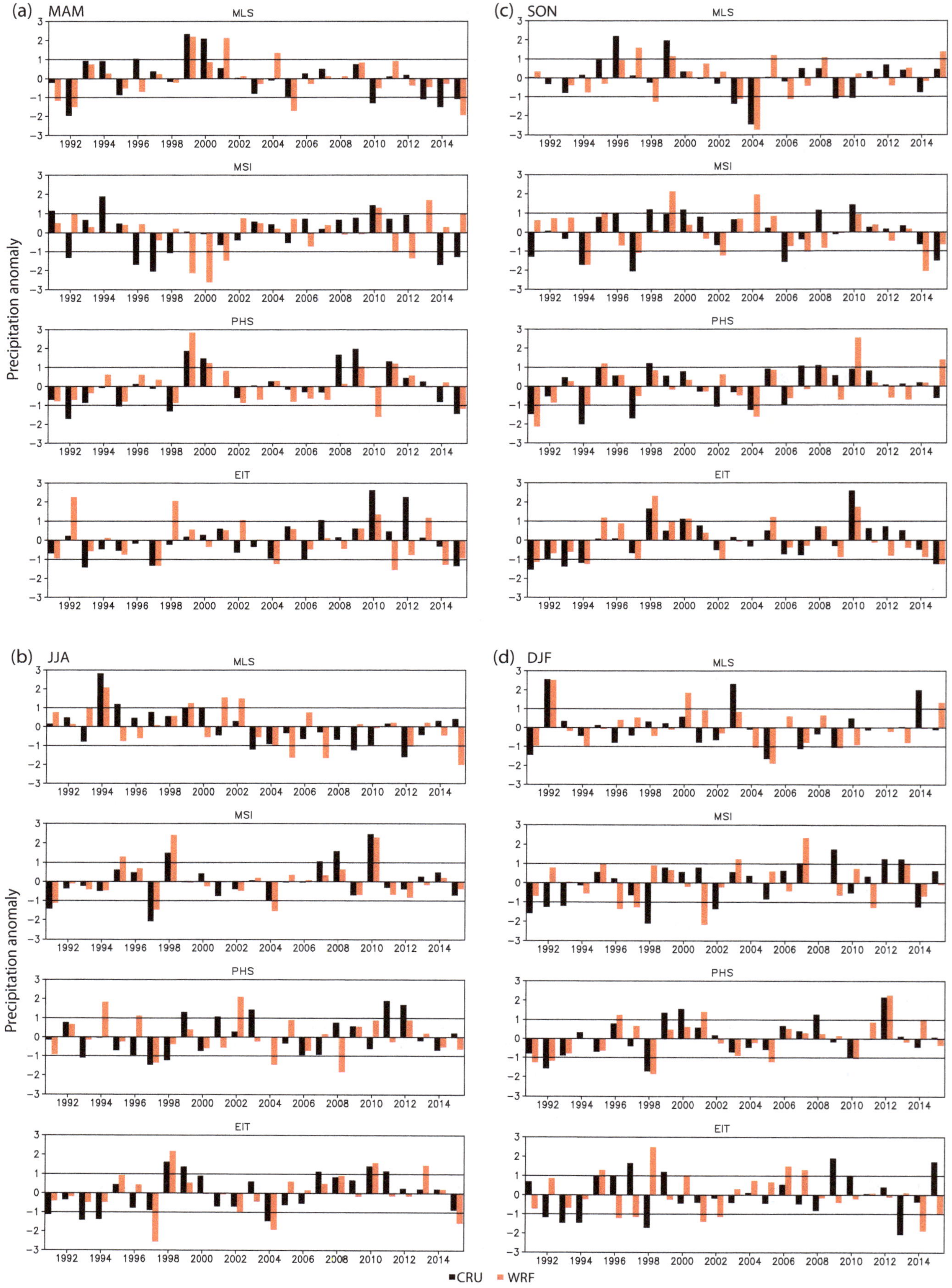
(a) MAM
(b) JJA
(c) SON
(d) DJF
MLS
MSI
PHS
EIT
Precipitation anomaly
1992 1994 1996 1998 2000 2002 2004 2006 2008 2010 2012 2014
CRU
WRF

Table 2. Time correlation coefficient between observed and simulated rainfall. *Statistically significant at the 95% level with 2-tailed Student's *t*-test

	MAM	JJA	SON	DJF
MLS	0.68*	0.39	0.64*	0.66*
MSI	−0.01	0.88*	0.45*	−0.06
PHS	0.74*	0.12	0.69*	0.84*
EIT	0.31	0.65*	0.80*	−0.33

rainfall seasons. For example, the MLS region receives substantial precipitation during the JJA season, and the model shows a large bias in mean precipitation, while doing a poor job in simulating the interannual variability. After verifying the performance of interannual variability, we tried to check whether the WRF model can simulate the rainfall anomalies associated with large-scale climate modes, such as ENSO and IOD. The tropical climate modes are known to have better predictability owing to the intrinsic nature of ocean-atmosphere coupling. The predictability of the associated regional climate variations also increases during ENSO and IOD years. Therefore, it is interesting to see whether the model can simulate precipitation anomalies correctly during these events. In the following sections, we evaluate the fidelity of the WRF model in simulating the effects of these climate modes by computing the anomaly correlation and event composites. IOD and ENSO events co-occur in some years, and we have included them in both IOD and ENSO composites, as our interest is to see the overall impact rather than the unique behavior of each climate mode in a regional climate model. In an earlier study, Ratna et al. (2014) have already shown the capability of WRF model in simulating rainfall and ENSO relationship over South Africa, and here, we check whether the model performs equally well for the SE Asia region.

4.1. SE Asia rainfall variability and ENSO

To evaluate the effect of ENSO on the subregions of SE Asia, we computed the correlation between the Niño 3.4 index and the precipitation anomalies over different subregions for different seasons. The correlation was computed for both the observed and the model simulated precipitation (Fig. 7, Table 3). In addition, we prepared composites of the precipitation anomaly from the identified years based on the phase of the ENSO (Fig. 7).

During MAM, the precipitation over the MLS and PHS regions is significant and is negatively correlated with ENSO. However, no significant correlation is seen over the MSI and EIT regions (Fig. 7a) in the observations. The WRF model-simulated precipitation also shows similar negative correlation over the MLS and PHS regions like in the observations, but has spurious positive correlations over the MSI and EIT regions (Fig. 7d). A similar pattern emerges when the events are composited based on the identified events (Table 1). In the JJA season, the MSI and EIT regions are significantly negatively correlated to the ENSO index with no significant correlation over the MSL and PHS regions (Fig. 7g). This is also reflected in the composite of the El Niño and La Niña events (Fig. 7 h,i). During El Niño (La Niña), the Indonesia regions receive lower (higher) precipitation, which is in agreement with Juneng & Tangang (2005). The simulated rainfall anomaly over MLS in JJA is not accurate with respect to ENSO events. This may be the reason the model reproduces weak interannual rainfall variability compared to the observed variability. The influence of ENSO on the MLS region during SON is similar to that of the JJA season, which does not have significant correlation with the Niño 3.4 SST (Fig. 7m). However, the other 3 regions—PHS, MSI, and EIT—have negative correlations with ENSO (Fig. 7m). However, the model-simulated rainfall correlation with ENSO over these 3 regions is low and less significant compared to the observations (Fig. 7p). So, the model produces a weaker intensity of rainfall anomaly compared to the observation in the composite of the El Niño/La Niña events. (Fig. 7q,r). During DJF, the SE Asia region has a weaker ENSO connection compared to that of the other 3 seasons (Fig. 7c). Observations do not show any significant correlation except over the Philippines and northeast region of Kalimantan Island. The model is able to simulate these correlations with ENSO but with weak intensity. The correlation map (Fig. 7s,v) reflects in the composite rainfall anomaly of El Niño and La Niña years, but the model erroneously simulates excess rainfall over southern Sumatra and southwest Kalimantan.

The above discussion indicates that the model is able to simulate the spatio-temporal variability of rainfall in relation to ENSO, although there are some biases. For example, the model captures the negative ENSO correlation over the Indonesian region during JJA and SON, though model has weaker intensity during SON. The model also captures the ENSO connection over the smaller region of SE Asia during the DJF season. The only exception is that the model

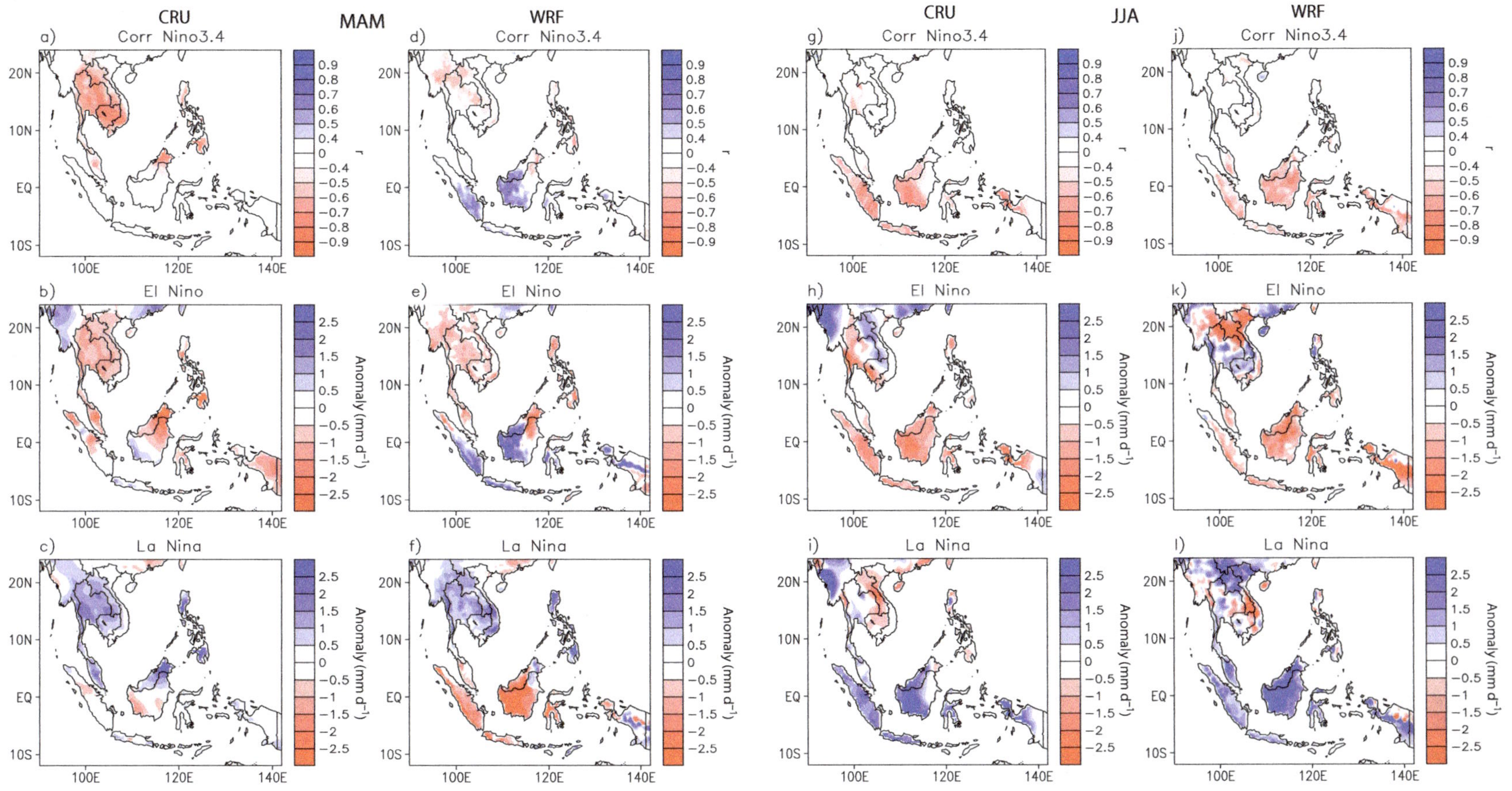

Fig 7. (a,d) Correlation coefficient between MAM mean Niño 3.4 index and MAM mean rainfall anomalies for the 25 yr (1991–2015) period both for CRU and WRF data. (b,c) CRU composite rainfall anomaly (mm d^{-1}) for the MAM season during El Niño and La Niña respectively. (e,f) Same as (b,c) but using WRF simulated composite anomaly. (g–l) Same as (a–f) but for JJA season; (m–r) same but for SON season; (s–x) same but for DJF season. Correlation coefficient values shown are statistically significant at 95% level with 2-tailed Student's *t*-test

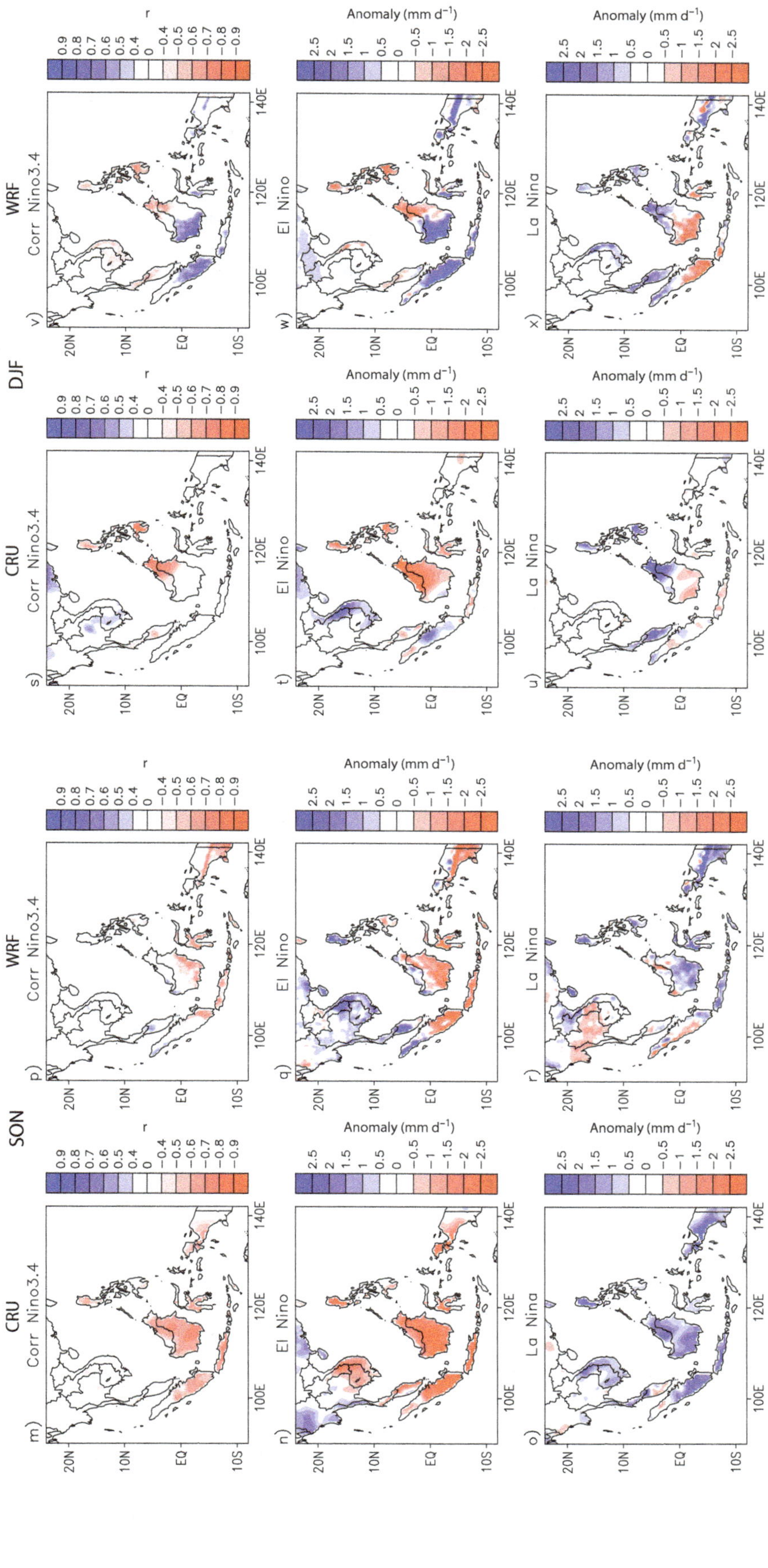

Fig 7. (continued)

incorrectly simulates a positive ENSO correlation over Indonesia region during MAM and DJF. This incorrect correlation with ENSO over the Indonesian region in these particular seasons is responsible for the weak correlation (Table 2) between observations and modelled seasonal rainfall at an interannual scale during the study period of 25 yr.

To further check if the model simulates the divergence and circulation processes realistically during the El Niño and La Niña, we analyzed the composite anomaly of divergence and wind vectors at 850 hPa during the El Niño and La Niña years (Fig. 8). In this SE Asia sector, the evolution of anomalous 850 hPa winds is strongly linked to the evolution of SST during ENSO and IOD events, as discussed in the rich literature (Wang et al. 2003, Juneng & Tangang 2005, Ummenhofer et al. 2013, Tsai et al. 2015). The divergence and circulation patterns were analyzed for all the seasons of the year, but we here show those only for JJA and SON because of the strong influence of ENSO on the seasonal rainfall. The composite anomaly of divergence analysis in JJA shows divergence over the Indonesian region during El Niño years and convergence during La Niña years (Fig. 8a,b). The model successfully simulated the divergence over Indonesia during El Niño years, but the simulated convergence during La Niña is stronger than the observed convergence (Fig. 8c,d), which can be seen from the bias of divergence (Fig. 8e,f). The stronger convergence of wind over the Indonesian region during La Niña (Fig. 8d) contributes to excess rainfall (Fig. 7l) compared to the observations. Also, the difference between the model and observation (Fig. 8e) shows an anomalous cyclonic circulation over the Philip-

Table 3. Correlation of ENSO and IOD with the area averaged rainfall for different subregions of SE Asia. *Statistically significant at 95% level with 2-tailed Student's *t*-test

		MAM	JJA	SON	DJF
ENSO					
MLS	CRU	−0.62*	−0.03	−0.24	0.29
	WRF	−0.55*	−0.13	0.09	−0.10
MSI	CRU	−0.27	−0.76*	−0.79*	−0.48*
	WRF	0.64*	−0.70*	−0.40*	0.54*
PHS	CRU	−0.86*	−0.02	−0.74*	−0.81*
	WRF	−0.70*	−0.12	−0.33	−0.77*
EIT	CRU	−0.23	−0.71*	−0.64*	−0.23
	WRF	0.35	−0.83*	−0.76*	0.39*
IOD					
MLS	CRU	0.42*	0.14	−0.00	−0.07
	WRF	0.24	0.03	0.04	−0.13
MSI	CRU	0.54*	−0.23	−0.84*	−0.23
	WRF	−0.37	−0.35	−0.47*	0.04
PHS	CRU	0.37*	0.22	−0.69*	−0.21
	WRF	0.19	−0.13	−0.39*	−0.10
EIT	CRU	0.05	−0.06	−0.52*	−0.25
	WRF	−0.43*	−0.38	−0.66*	0.22

pines during El Niño years, which generates a positive rainfall anomaly compared to the negative rainfall anomaly observed. Similar characteristics of divergence and convergence are also observed during the SON season (Fig. 8g,h), and the model successfully simulated these features (Fig. 8i,j). The simulated anomalous convergence over the MLS and PHS regions during El Niño years (Fig. 8k) contributed to excess rainfall compared to the observations (Fig. 7n,q).

4.2. SE Asia rainfall variability and IOD

The IOD is also known to influence the climate of SE Asia because this region is located near the eastern pole of the IOD. To understand the influence of the IOD with rainfall over SE Asia in a regional climate model, we calculated the correlation coefficient between the IOD index and rainfall for the 4 seasons MAM, JJA, SON, and DJF for the period 1991 to 2015 using the observed and model simulated data (Table 3b). Except for the MLS region, IOD has the strongest influence on the SE Asia rainfall variability during the SON season compared to the other 3 seasons, clearly due to the IOD peaking in the boreal autumn season (Saji et al. 1999). Therefore, further analysis with IOD is carried out only for the SON season. The model successfully simulates this IOD and rainfall relationship, but the simulated correlation is generally weaker compared to that observed (Table 3b).

The observed rainfall shows the strongest negative correlation with IOD during the SON season over the Indonesian region, and a weak negative correlation over the PHS region (Fig. 9). The IOD is known to peak during the SON season (Saji et al. 1999), and Indonesia receives deficient (excess) rainfall anomaly during positive (negative) IOD years. The model also simulates these rainfall anomalies with respect to the positive and negative IOD, but with a somewhat weaker amplitude. The model successfully simulates the negative rainfall anomaly over the MSI region, associated with positive IOD years in 1994 and 1997, but fails to simulate the excess rainfall anomaly during the negative IOD years of 1996 and 1998. However, the model is successful in simulating the positive rainfall anomaly for the 2010 negative IOD event (see Fig. 6c).

The performance of the WRF model was also checked to see whether it can simulate the 850 hPa divergence and circulation patterns over SE Asia associated with IOD events during SON season. There is strong divergence over SE Asia, with strong easterly winds over southwestern region of Indonesia during positive IOD events (Fig. 10a). These winds diverge over the Indonesian region with a cold SST anomaly, and converge over the western Indian Ocean due to a warm SST anomaly there (not shown). This divergence over the Indonesian region causes a negative rainfall anomaly during the positive IOD events. During the negative IOD years, an anomalous convergence zone is observed over SE Asia, and this causes a anomalous excess rainfall during negative IOD years. The model-simulated divergence is weaker than that observed during positive IOD events (Fig. 10c,e), but the convergence of winds from west and east is well simulated by the model during the negative IOD years (Fig. 10d,f). The model simulations have an anomalous cyclonic circulation bias over southern MLS, Malaysia, and north Sumatra (Fig. 10e) during positive IOD years, and hence, the model simulates a wet rainfall anomaly compared to that observed.

5. SUMMARY

To verify the performance of the WRF model in reproducing the regional climate variability over SE Asia, we conducted a 25 yr (1991–2015) model simulation at 27 km horizontal resolution using the European Centre for Medium-Range Weather Forecasts

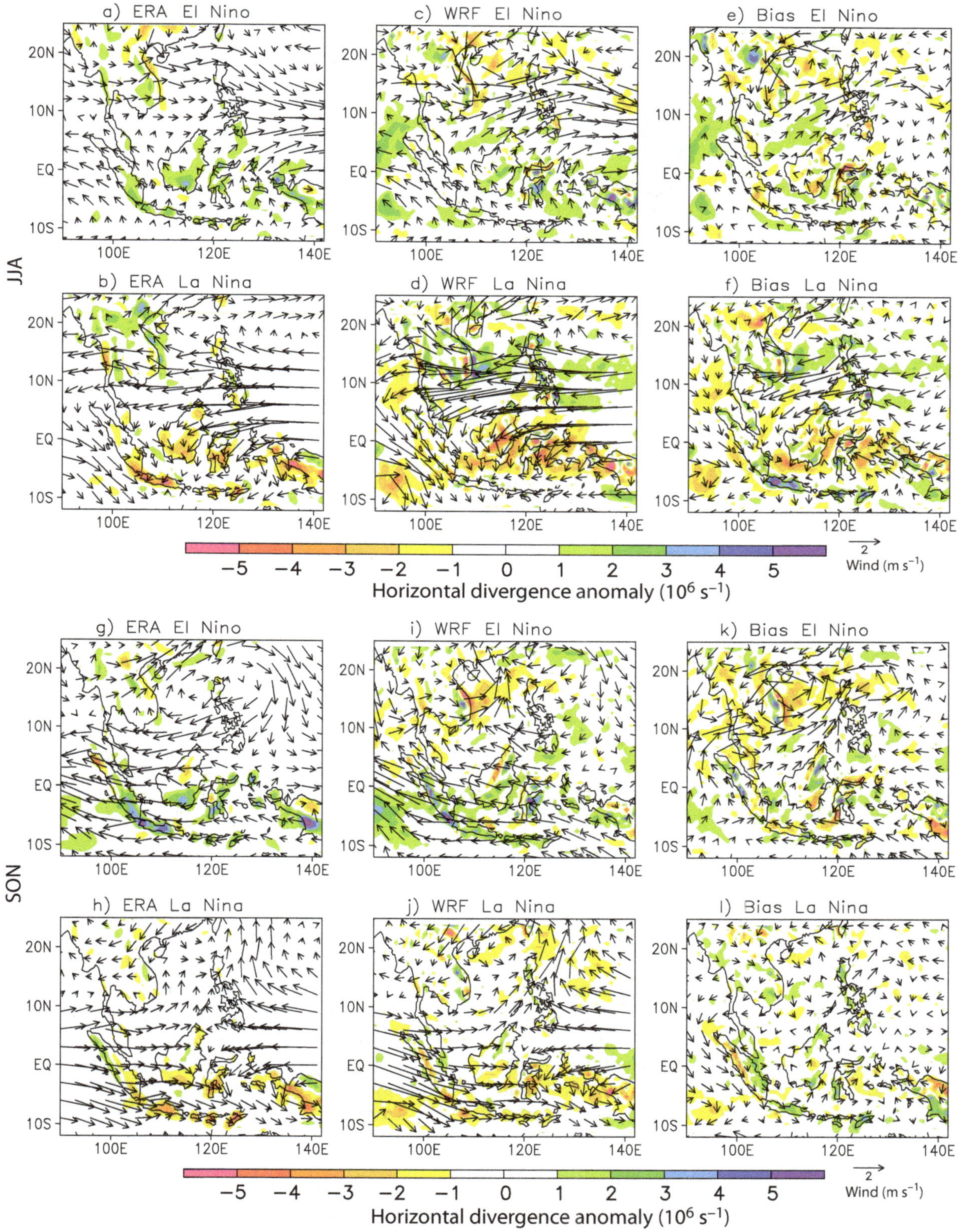

Fig. 8. (a–d) JJA mean composite anomaly of horizontal divergence (color) and winds (vectors) at 850 hPa for El Niño and La Niña events using (a,b) ERA and (c,d) WRF data. (e,f) Bias as calculated from the difference between WRF and ERA data for the El Niño and La Niña events, respectively. (g–l) same as (a–f) but for SON composite anomaly

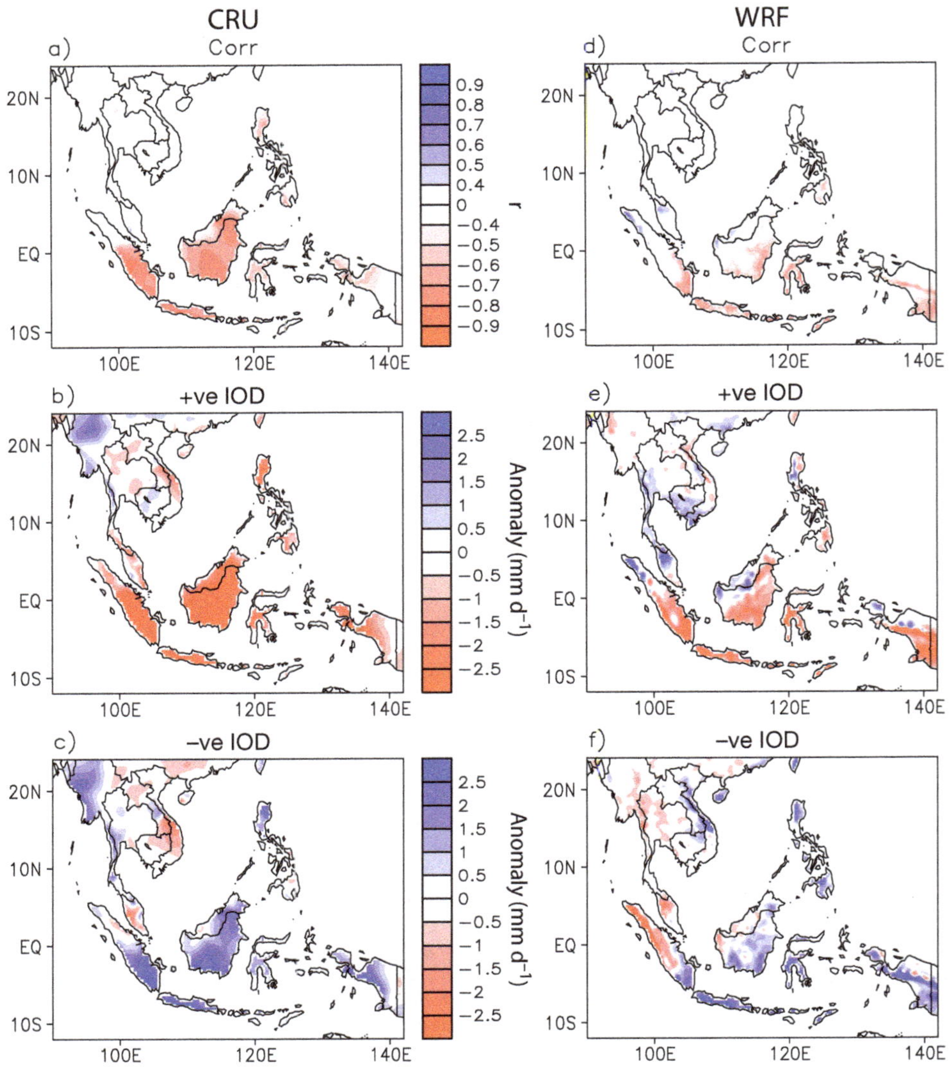

Fig 9. (a,d) Correlation coefficient between SON mean IOD index and SON mean rainfall for the 25 yr (1991–2015) period for CRU (left column) and WRF (right column) data. (b,c,e,f) Composite rainfall anomaly for the MAM season during positive and negative IOD events. (e,f) Same as (b,c) but using WRF simulated composite anomaly. Correlation coefficient values shown are statistically significant at 95 % level with 2-tailed Student's *t*-test

(ECMWF) reanalysis (ERA-Interim) data as the forcing boundary data. We checked the fidelity of the model in reproducing the climatology, annual cycle, and interannual variability of rainfall for the different subregions of SE Asia during 4 seasons (MAM, JJA, SON, and DJF). The rainfall variability over SE Asia with relation to large-scale climate modes such as ENSO and IOD was also analyzed.

The results indicate that the model simulates the spatial variability and annual rainfall climatology with maximum rainfall over Malaysia and Indonesia and minimum rainfall over mainland SE Asia and the Philippines. However, we found a dry bias over the western coast of Myanmar, Malaysia, and parts of the Kalimantan Island and a wet bias over eastern part of MLS, western Sumatra, and the Philippines. Based on analyzing the individual seasons, the model simulates a wet bias over MLS region (except Myanmar) during JJA. The analysis of the climatology of the vertically integrated moisture flux indicates that excess moisture convergence over the MLS and PHS during JJA season contributes to the excess rainfall. Compared to observations, the model simulated a weaker moisture flux along the Myanmar coast, and hence generated a dry bias in orographic rainfall. The dry bias simulated over Malaysia, northern Sumatra, and parts of Kalimantan Island in the annual climatology comes from the strong bias during DJF.

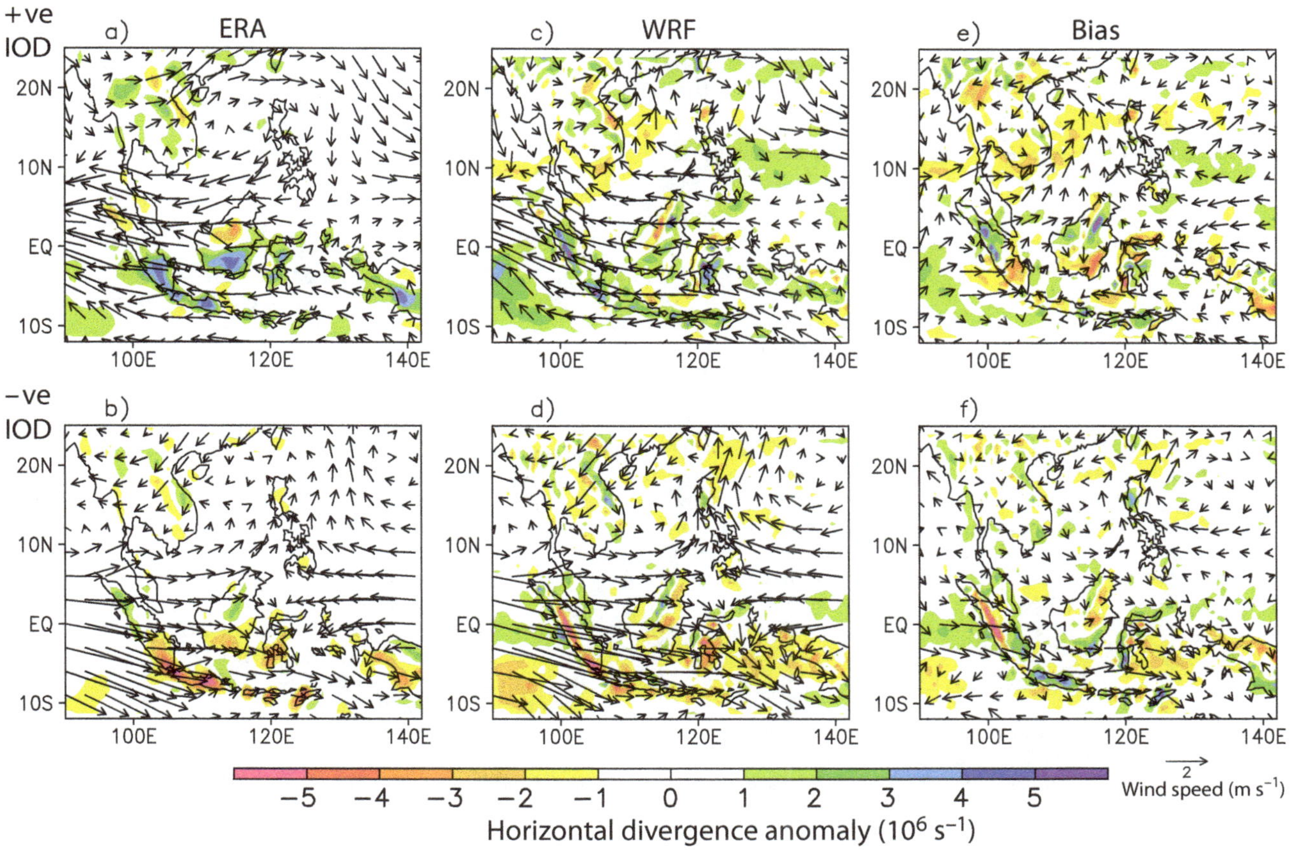

Fig. 10. (a–d) SON mean composite anomaly of horizontal divergence (color) and winds (vectors) at 850 hPa for positive IOD and negative IOD events using (a,b) ERA and (c,d) WRF data. (e,f) Bias, calculated as the difference between WRF and ERA data for the positive IOD and negative IOD events, respectively

The moisture convergence zones shift over the Indonesia region during the DJF season with the movement of the ITCZ southward. The model simulates weaker moisture convergence compared to the observation results in dry bias during DJF. The vertical profile of moisture and vertical velocity bias is also reflected in the wet (dry) bias over the MLS (MSI) region during JJA (DJF).

This study also evaluates the WRF model's ability to simulate the interannual variability of regional rainfall over SE Asia, and its association with the large-scale climate modes, such as ENSO and IOD. Due to its location, the maritime continent is influenced by both the ENSO and IOD. On an interannual scale, the model performs well over mainland SE Asia and the Philippines for all seasons of the year except for the boreal summer (JJA). In general, this is the region where ENSO has very little correlation during JJA, and the model successfully simulated the low correlation between rainfall and the Niño 3.4 index. In contrast, the model has good skill over the MSI region during the boreal summer compared to the other 3 seasons. The Indonesian region has a strong ENSO influence during the boreal summer season, and the model accurately reproduces the rainfall and Niño 3.4 correlation. Also, the model shows good performance in the EIT region during both the boreal summer and fall seasons. The model is able to realistically capture SON rainfall over the Indonesian region during IOD events. During the SON season, both IOD and ENSO have stronger influences over the Indonesia region, but the model reproduces only a weak but significant negative correlation (rainfall–ENSO and rainfall–IOD) compared to the strong correlation in the observation. The accurate representation of the relationship between rainfall and ENSO/IOD is due to the model's ability to realistically simulate the large-scale circulation patterns. This indicates that the WRF model could reproduce the spatio-temporal variability of rainfall with relation to ENSO and IOD.

The WRF model simulates the seasonal climatology and annual cycle, and captures the interannual variability reasonably well over all the subregions of the

SE Asia. However, the model suffers from severe systematic biases of overestimating (underestimating) precipitation during peak rainfall seasons over the MLS (MSI) region. Further experiments with various combination of physical parameterization schemes together with air–sea interactions (e.g. Wei et al. 2014) are needed to reduce these biases. We intend to carry out such experiments in the future.

Acknowledgements. This research was supported by the Environment Research and Technology Development Fund (2-1405) of the Ministry of the Environment, Japan. The authors are thankful to the ECMWF for making available the ERA-Interim reanalysis datasets used in this research. NOAA OI SST data were provided by the NOAA/OAR/ESRL PSD, Boulder, Colorado, USA, at www.esrl.noaa.gov/psd/. The availability of CRU rainfall data from University of East Anglia is highly appreciated. We acknowledge mapchart.net for the free web application to create Fig. 1a. We thank the 3 reviewers for their constructive comments which helped us to improve the quality of this manuscript.

LITERATURE CITED

Adler RF, Huffman GJ, Chang A, Ferraro R and others (2003) The version 2 global precipitation climatology project (GPCP) monthly precipitation analysis (1979-present). J Hydrometeorol 4:1147–1167

Aldrian E, Dumenil-Gates L, Jacob D, Podzun R, Gunawan D (2004) Long-term simulation of Indonesian rainfall with the MPI regional model. Clim Dyn 22:795–814

Behera SK, Luo JJ, Yamagata T (2008) The unusual IOD event of 2007. Geophys Res Lett 35:L14S11

Betts AK, Miller MJ (1986) A new convective adjustment scheme. II. Single column tests using GATE wave, BOMEX, and arctic air-mass data sets. QJR Meteorol Soc 112:693–709

Chen F, Dudhia J (2001) Coupling an advanced land-surface hydrology model with the Penn State-NCAR MM5 modeling system. I. Model implementation and sensitivity. Mon Weather Rev 129:569–585

Chotamonsak C, Salathe EP Jr, Kreasuwan J, Chantara S, Siriwitayakorn K (2011) Projected climate change over Southeast Asia simulated using a WRF regional climate model. Atmos Sci Lett 12:213–219

Chotamonsak C, Salathe EP Jr, Kreasuwan J, Chantara S (2012) Evaluation of precipitation simulations over Thailand using a WRF regional climate model. Warasan Khana Witthayasat Maha Witthayalai Chiang Mai 39: 623–638

Dee DP, Uppala SM, Simmons AJ, Berrisford P and others (2011) The ERA-interim reanalysis: configuration and performance of the data assimilation system. QJR Meteorol Soc 137:553–597

Dickinson RE, Errico RM, Giorgi F, Bates GT (1989) A regional climate model for the western United States. Clim Change 15:383–422

Dudhia J (1989) Numerical study of convection observed during the winter monsoon experiment using a mesoscale two-dimensional model. J Atmos Sci 46: 3077–3107

Francisco RV, Argete J, Giorgi F, Pal J, Bi X, Gutowski W (2006) Regional model simulation of summer rainfall over the Philippines: effect if choice of driving fields and ocean flux schemes. Theor Appl Climatol 86:215–227

Giorgi F (1990) Simulation of regional climate using a limited area model nested in a general circulation model. J Clim 3:941–963

Giorgi F, Mearns LO (1999) Introduction to special section: regional climate modelling revisited. J Geophys Res 104: 6335–6352

Giorgi F, Bi X, Pal JS (2004) Mean, interannual variability and trends in a regional climate change experiment over Europe. I. Present-day climate (1961–1990). Clim Dyn 22:733–756

Harris I, Jones PD, Osborn TJ, Lister DH (2014) Updated high-resolution grids of monthly climatic observations–the CRU TS3.10 dataset. Int J Climatol 34:623–642

Hong SY, Dudhia J, Chen SH (2004) A revised approach to ice microphysical processes for the bulk parameterization of clouds and precipitation. Mon Weather Rev 132: 103–120

Huffman GJ, Adler RF, Bolvin DT, Gu GJ and others (2007) The TRMM Multisatellite Precipitation Analysis (TMPA): quasi-global, multiyear, combined-sensor precipitation estimates at fine scales. J Hydrometeorol 8:38–55

Im ES, Ahn JB, Remedio AB, Kwon WT (2008) Sensitivity of the regional climate of East/Southeas Asia to convective parameterizations in the RegCM3 modelling system. 1. Focus on the Korean peninsula. Int J Climatol 28: 1861–1877

Janjic ZI (1994) The step-mountain eta coordinate model: further developments of the convection, viscous sublayer and turbulence closure schemes. Mon Weather Rev 122: 927–945

Juneng L, Tangang FT (2005) Evolution of ENSO-related rainfall anomalies in Southeast Asia region and its relationship with atmosphere–ocean variations in Indo-Pacific sector. Clim Dyn 25:337–350

Juneng L, Tangang F, Chung JX, Ngai ST and others (2016) Sensitivity of the Southeast Asia rainfall simulations to cumulus and ocean flux parameterization in RegCM4. Clim Res 69:59–77

Kripalani RH, Kulkarni A (1997) Rainfall variability over South-east Asia: connections with Indian monsoon and ENSO extremes: new perspective. Int J Climatol 17: 1155–1168

Lau KM, Yang S (1997) Climatology and interannual variability of the southeast Asian summer monsoon. Adv Atmos Sci 14:141–162

Li JP, Zeng QC (2005) A new monsoon index, its interannual variability and relation with monsoon precipitation. Clim Environ Res 10:351–365

Liew SC, Raghavan SV, Liong SY (2014) Development of Intensity-Duration-Frequency curves at ungauged sites: risk management under changing climate. Geosci Lett 1:8

Matsumoto J, Murakami T (2002) Seasonal migration of monsoons between the northern and southern hemisphere as revealed from equatorially symmetric and asymmetric OLR data. J Meteorol Soc Jpn 80:419–437

McBride J, Haylock M, Nicholls N (2003) Relationship between the maritime continent heat source and the El Niño–Southern oscillation phenomenon. J Clim 16: 2905–2914

Mlawer E, Taubman S, Brown P, Iacono M, Clough S (1997) Radiative transfer for inhomogeneous atmosphere: RRTM, a validated correlated-k model for the long-wave. J Geophys Res 102:16663–16682

Naylor RL, Battisti DS, Vimont DJ, Falcon WP, Burke MB (2007) Assessing risks of climate variability and climate change for Indonesian rice agriculture. Proc Natl Acad Sci USA 104:7752–7757

Ngo-Duc T, Kieu C, Thatcher M, Nguyen-Le D, Phan-Van T (2014) Climate projections for Vietnam based on regional climate models. Clim Res 60:199–213

Noh Y, Cheon WG, Hong SY, Raasch S (2003) The improvement of the K-profile model for the PBL using LES. Boundary-Layer Meteorol 107:401–427

Phan VT, Ngo-Duc T, Ho TMH (2009) Seasonal and interannual variations of surface climate elements over Vietnam. Clim Res 40:49–60

Phan VT, Nguyen HV, Tuan LT, Quang TN, Ngo-Duc T, Laux P, Xuan TN (2014) Seasonal prediction of surface air temperature across Vietnam using the regional climate model version 4.2 (RegCM4.2). Adv Meteorol 2014:245104

Raghavan SV, Vu MT, Liong SY (2016) Regional climate simulations over Vietnam using the WRF model. Theor Appl Climatol 126:161–182

Raktham C, Bruyère C, Kreasuwun J, Done J, Thongbai C, Promnopas W (2015) Simulation sensitivities of the major weather regimes of the Southeast Asia region. Clim Dyn 44:1403–1417

Ratna SB, Ratnam JV, Behera SK, Rautenbach CJ, Ndarana T, Takahashi K, Yamagata T (2014) Performance assessment of three convective parameterization schemes in WRF for downscaling summer rainfall over South Africa. Clim Dyn 42:2931–2953

Ratna SB, Cherchi A, Joseph PV, Behera SK, Abish B, Masina S (2016a) Moisture variability over the Indo-Pacific region and its influence on the Indian summer monsoon rainfall. Clim Dyn 46:949–965

Ratna SB, Ratnam JV, Behera SK, Doi T, Yamagata T (2016b) Downscaled prediction of extreme seasonal climate over Southeast Asia using a regional climate model. Techno-Ocean 2016, IEEE Kobe, p 375–380

Reynolds RW, Rayner NA, Smith TM, Stokes DC, Wang W (2002) An improved in situ and satellite SST analysis for climate. J Clim 15:1609–1625

Saji NH, Goswami BN, Vinayachandran PN, Yamagata T (1999) A dipole mode in the tropical Indian Ocean. Nature 401:360–363

Salimun E, Tangang F, Juneng L, Behera SK, Yu W (2014) Differential impacts of conventional El Niño versus El Niño Modoki on Malaysian rainfall anomaly during winter monsoon. Int J Climatol 34:2763–2774

Skamarock WC, Klemp JB, Dudhia J, Gill DO and others (2008) A description of the advanced research WRF version 3. NCAR technical note, NCAR/TN\u2013475?STR, Boulder, CO

Takahashi HG, Yoshikane T, Hara M, Yasunari T (2009) High resolution regional climate simulations of the long-term decrease in September rainfall over Indochina. Atmos Sci Lett 10:14–18

Takahashi HG, Yoshikane T, Hara M, Takata K, Yasunari T (2010) High resolution modeling of the potential impact of land surface conditions on regional climate over Indochina associated with diurnal precipitation cycle. Int J Climatol 30:2004–2020

Tangang F, Juneng L (2004) Mechanisms of Malaysian rainfall anomalies. J Clim 17:3616–3622

Tsai C, Behera SK, Waseda T (2015) Indo-China monsoon indices. Sci Rep 5:8107

Ummenhofer CC, D'Arrigo RD, Anchukaitis KJ, Buckley BM, Cook ER (2013) Links between Indo-Pacific climate variability and drought in the Monsoon Asia Drought Atlas. Clim Dyn 40:1319–1334

Wang B, Wu R, Li T (2003) Atmosphere–warm ocean interaction and its impacts on Asian–Australian monsoon variation. J Clim 16:1195–1211

Wei J, Malanotte-Rizzoli P, Eltahir EAB, Xue P, Xu D (2014) Coupling of a regional atmospheric model (RegCM3) and a regional ocean model (FVCOM) over the Maritime Continent. Clim Dyn 43:1575–1594

Yatagai A, Kamiguchi K, Arakawa O, Hamada A, Yasutomi N, Kitoh A (2012) APHRODITE: constructing a long-term daily gridded precipitation dataset for Asia based on a dense network of rain gauges. Bull Am Meteorol Soc 93:1401–1415

Yusuf AA, Francisco HA (2009) Climate change vulnerability mapping for Southeast Asia. Economy and Environment Program for Southeast Asia (EEPSEA), Singapore

Zeng Q, Zhang B (1998) On the seasonal variation of atmospheric general circulation and the monsoon. Chinese J Atmos Sci 22:805–813 (in Chinese)

3

Projection of future hot weather events and potential population exposure to this in South Korea

Changsub Shim[1,*], Jihyun Seo[1], Jihyun Han[1], Jongsik Ha[1], Tae Ho Ro[1], Yun Seop Hwang[2], Jung Jin Oh[3]

[1]Korea Environment Institute, 30147 Sejong, ROK
[2]Department of International Business and Trade, Kyung Hee University, 02453 Seoul, ROK
[3]Department of Chemistry, Sookmyung Women's University, 04312 Seoul, ROK

ABSTRACT: Heat waves, often caused by consecutive severe hot weather events, are responsible for the majority of medical costs associated with climate change in South Korea. In this study, we obtained a regional climate change scenario (RCP4.5) for South Korea, with 7.5 × 7.5 km horizontal resolution and extending up to 2100, by dynamically downscaling from results of the Community Earth System Model (CESM) with the Weather Research and Forecasting (WRF) model. We analyzed hot weather events (daily maximum temperature >33°C) in summer (June–August), focusing on changes in extent and frequency. According to our analysis, the area exposed to hot weather events in August will expand to cover ~70% of the nation in the middle of this century, with a rate of increase of 0.24% yr^{-1}. We calculated the population exposed to hot weather events in Korea, considering both spatial coverage and number of event days. Population exposure was projected to increase almost 3-fold, from 26% of the national population during the 2010s to 72% during the 2090s. In particular, exposure of the elderly population (>65 yr old), who are particularly vulnerable, was expected to rapidly increase, with ~22% of the national population (~10.4 million people aged >65 yr) affected in the middle of this century when we considered the future projection of rapid aging of the South Korean population structure. Our projection of extensive hot weather events starting from the middle of the 21st century suggests the need for urgent government long-term measures and enforcement to ensure an early response to extreme weather events in Korea.

KEY WORDS: Climate change · Climate vulnerability · Hot weather · Korea · Population exposure · Representative Concentration Pathways · RCP scenarios

1. INTRODUCTION

Heat waves, which are generally defined as extreme weather conditions due to an extended period of unusually high air temperature, often lead to adverse health consequences (Robinson 2001) and great economic loss (Meehl et al. 2000, Beniston & Stephenson 2004, Kysely & Huth 2004, Knowlton et al. 2011). In the USA, 10 major droughts/heat wave events in 1980–2003 accounted for the largest economic loss (US$ 144 billion) among weather-related disasters (Ross & Lott 2003). In South Korea, 3384 people died due to an unusual heat wave in 1994 (KMA 2012, 2013). There has been a warming trend since the 20th century in Korea, as indicated by the increase in daily summer maximum temperature (Choi et al. 2007, Na 2015). Specifically, maximum temperature has increased over the major cities of South Korea from 1961 to 2000, with a rate of increase of more than twice that of the global average (0.028°C yr^{-1} for Seoul), which includes the impact of urbanization (Choi et al. 2007).

The number of heat wave days in Korea in the 1990s and 2000s has ranged from 5 to 20 days yr^{-1}

*Corresponding author: marchell@gmail.com

under different heat wave definitions, and the occurrence of heat waves in Korea has increased over this period (Park et al. 2008, Son et al. 2012). In particular, heat waves have been more intense over southern provinces (including the Yong-Nam and Ho-Nam regions) (B. C. Choi et al. 2007, G. Y. Choi et al. 2008, Jeung et al. 2013, KMA 2013, Kim et al. 2014, Na 2015), and heat wave warnings or alerts were issued for most regions of Korea in summer 2013 (KMA 2013).

Heat waves in South Korea are most intense in late July through early August and the mortality rate is highest for people aged >65 yr due to their impaired ability to adapt to heat (Woodward & Scheraga 2003, Kim et al. 2006, Lee 2015). Not including the extreme heat wave disaster in 1994, the death toll due to heat waves has continued to rise since the 1990s (Park et al. 2008, Kim et al. 2014, Lee 2015).

Knowledge of potential extreme events in the future climate is necessary for the development of national plans for adaptation. According to the IPCC, global average surface temperature is projected to rise by 1.1–6.4°C by the 2090s (compared to 1980–1999; IPCC 2007). The frequency, intensity, and duration of heat waves will likely increase in the future (IPCC 2014). In addition, it has been reported that the adverse impacts on health of increased heat wave frequency in the future greatly outweigh the benefits of reduced cold spells associated with climate change (Kinney et al. 2012, Ebi & Mills 2013), due in part to the larger and more immediate rises in mortality caused by heat waves (Deschênes & Moretti 2009). For the USA, the Representative Concentration Pathway (RCP) scenarios suggest that mortality risks at the end of this century are expected to increase by an order of magnitude compared with the current risk (2002–2004) (Wu et al. 2014).

A number of studies have been conducted regarding future hot weather patterns and the responses required for the vulnerable population in Korea. According to RCP-based climate scenarios, future hot weather is likely to occur in the metropolitan area over the central regions (e.g. Seoul Metropolitan Area [SMA] with a population of ~20 million) and the southern regions of Korea (Jeung et al. 2013, Park et al. 2013). Park et al. (2013) concluded that most regions of South Korea, except for the mountainous regions, will have >30 d of heat waves in 2070, based on RCP8.5. Mortality due to heat waves is also expected to increase accordingly, with the current rate of 0.7 deaths per 100 000 population in 2001–2010 increasing to 1.5 deaths per 100 000 population in 2040 (Yang & Ha 2013).

The present study used the RCP4.5 scenario, which represents relatively moderate greenhouse gas mitigation to stabilize radiative forcing at 4.5 W m^{-2} in this century, to analyze the future projection of hot weather events in Korea. Our RCP4.5 scenario for Korea is based on the Community Earth System Model (CESM) developed by the US National Center for Atmospheric Research (NCAR), and the global model results were downscaled by a regional-scale meteorological model (Weather Research and Forecasting [WRF]) to obtain a climate scenario with a horizontal resolution of 7.5 × 7.5 km over Korea.

With the estimated regional climate scenario for Korea, we investigated the projection of hot weather events regarding the spatial distribution and the cumulative number of days where a hot weather event is experienced in the 21st century. Additionally, this study investigated changes in the population exposed to hot weather events in Korea by estimating the population of affected administrative districts with corresponding model grids. Estimation of the exposed population can be used to gauge the potential health effects of hot weather events in the future, which is critical information for developing national adaptation measures to climate change in Korea.

The main objective of this study was to estimate the future population exposed to hot weather events in Korea based on the national-level climate scenario in order to support a long-term national policy.

2. DATA AND METHODOLOGY

2.1. Dynamical downscaling of global climate model results

We used the RCP4.5 scenario over Korea, which is based on results of the CESM model (www.cesm.ucar.edu/models) with a spatial resolution of 1.9° × 2.5°. CESM is a climate model that simulates the Earth's climate system and consists of submodels of the atmosphere (CAM), ocean (OCN), land (CLM), sea ice (CICE), and land ice (GLC). These submodels are connected through a coupler (CPL) that exchanges the feedbacks. We used CESM (v.1.0), developed by the NCAR and supported by the National Science Foundation (NSF) and Department of Energy (DOE), USA. WRF is a mesoscale numerical weather prediction model designed to support atmospheric forecasting and research by providing a variety of dynamic cores, with data assimilation and parallel computing (www.wrf-model.org/index.php/).

To dynamically downscale the CESM results, meteorological variables from the CESM model were used as the initial and boundary conditions for the WRF (v.3.3) model at 6 h intervals to produce regional-scale scenarios with a 7.5 × 7.5 km spatial resolution for the Korean Peninsula for 2006–2100. Detailed methodology for the dynamical downscaling has been reported previously by Seo et al. (2013). Fig. 1 shows the model domain for the CESM and the boundary of the WRF model for dynamic downscaling.

2.2. Bias correction based on ground observations

The climate model results inherently contain uncertainties, which can be amplified during the downscaling (Ho et al. 2011). It is agreed that downscaled surface variables, such as precipitation and temperature, should be comparable to observations to ascertain credibility in the projected scenarios (Lee et al. 2014).

Bias correction was estimated from the difference between simulated and observation data on a monthly basis from 2006 to 2014. The calculated bias correction was applied to the future climate projection, as in previous studies (delta change approach) (e.g. Graham et al. 2007, Sperna Weiland et al. 2010, Ho et al. 2012).

Observation data from ~400 nationwide ground stations were used for bias correction over South Korea. To undertake bias correction for temperature, we mapped the observational points with the corresponding model grid (7.5 × 7.5 km) considering geographical elevation. If the model grid contained a measurement station, bias was calculated directly. If the model grid did not contain a measurement station, data from the closest adjacent station were used to calculate bias over the shortest distance, which is similar to the gap-filling process for missing satellite observations where the interpolation of surrounding good-quality data is used to cover gaps (Zhao et al. 2011). This grid-based bias correction was applied to monthly temperatures from 2006 to 2014. The estimated bias for each grid was applied to the climate scenarios for the future projection of temperature.

Although there are many definitions of heat waves, we adopted the current official definition used by the Korea Meteorological Administration (KMA), i.e. a daily maximum temperature >33°C, based on regional information for health impacts (KMA 2013, Park et al. 2013). Because this definition simply classifies single days based on a predefined temperature threshold, we adopted the term 'hot weather events' for clarity. Hot weather events in Korea mostly occur in July and August, with the number of hot weather days being highest in August (Park et al. 2008, Kim et al. 2014). Based on the frequency of national hot weather events for the last 20 yr, we analyzed hot weather events in the Korean summer months (June–August).

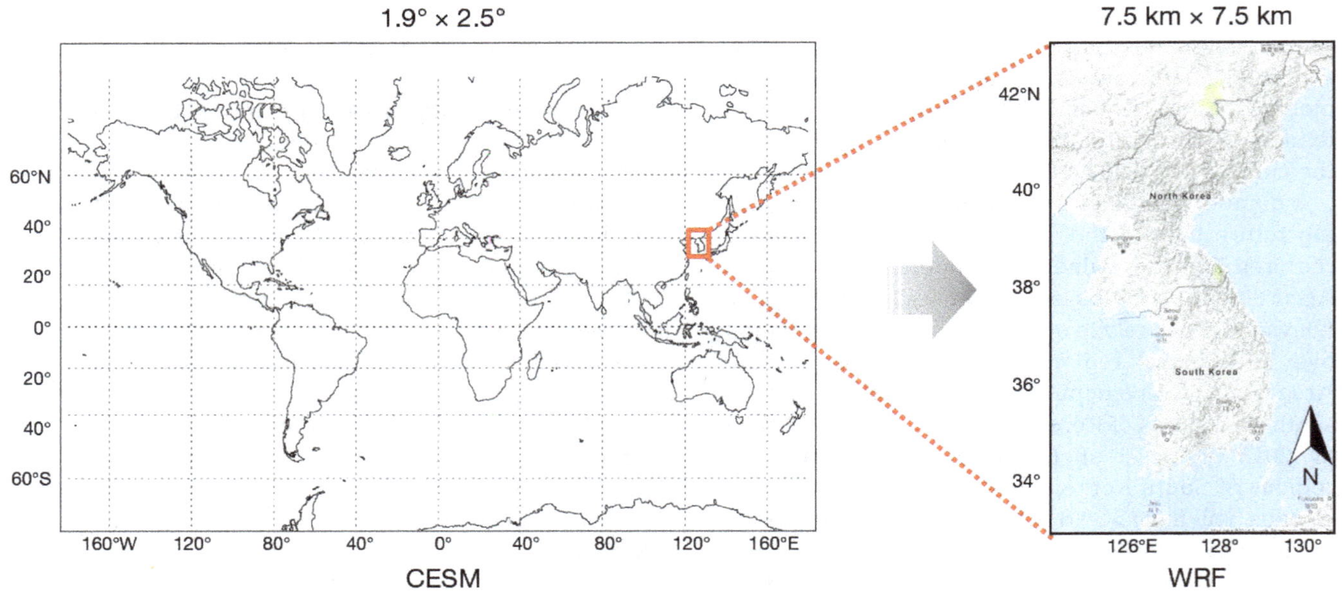

Fig. 1. Model domain for dynamic downscaling of climate scenarios from the global scale climate model Community Earth System Model (CESM; 1.9° × 2.5° spatial resolution) (left) by the Weather Research and Forecasting model (WRF; 7.5 × 7.5 km horizontal resolution) (right)

2.3. Considering both area and frequency of hot weather events

We evaluate the national impact of hot weather events in terms of both area and frequency by integrating the area and frequency (IAF, %) of hot weather events. IAF is defined here to evaluate the spatiotemporal extent of hot weather events. IAF is obtained by the integration of each unit of area with a hot weather event multiplied by the frequency (number of days), which is then divided by the area of the entire nation multiplied by the total number of days in the month, as follows:

$$\mathrm{IAF}(\%) = \frac{\sum_{i=1}^{h} A_i d_i \times 100}{\sum_{i=1}^{t} A_i D} \quad (1)$$

where A is the area of grid i with a hot weather event, h is the total number of grids with hot weather events, d is the number of heat wave days in grid i, t is the total number of grids over South Korea, and D is the total number of days per month (~30 d). For example, an IAF of 100% indicates that the entire area of South Korea is exposed to hot weather events every day during that month.

2.4. Estimating population exposed to hot weather events

One of the main objectives of this study was to estimate the future population that will be exposed to hot weather events in Korea. Estimation of the exposed population was undertaken by mapping the 228 administrative districts (vector data) containing population information to the geographical information on the corresponding model grid (7.5 × 7.5 km, raster data). The area used for sampling was the model grid with total number of hot weather days per year >11 d, which was the average number of hot weather days over a 30 yr period (1981–2010; Na 2015). Population data, with age groups, for each administrative district, was based on 2010 data from the Korea Statistical Information Service (KOSIS) (http://kosis.kr/statHtml/statHtml.do?orgId=101&tblId=DT_1IN0001_ENG&conn_path=I3).

For mapping, we used ArcGIS (v.10.1) software, with the zonal statistics method. If one administrative district matched multiple grids, we selected the median value of temperature, which would show a more realistic current hot weather distribution than the mean value (not shown). Our estimation of total population exposure is based on data from 2010, which is not expected to change substantially until the middle of this century due to the low national birthrate and the limited residential area in Korea (~65% of the nation is mountainous) (KOSIS 2015). However, changes in population structure are considered by 2060, reflecting a rapidly aging society (KOSIS 2015).

3. RESULTS

3.1. Distribution and frequency of hot weather events

Fig. 2 displays the spatial distribution of the projection of daily maximum temperature (T_{max}) in summer with the RCP4.5 climate scenario on a monthly mean basis. Differences of future projection of T_{max} compared to the 2010s are presented in Fig. 3. While T_{max} was not expected to increase greatly over northwestern Korea in June until the 2050s, T_{max} over southeastern Korea increased markedly in the 2050s (Figs. 2 & 3). T_{max} was projected to increase over most regions of South Korea in the last decade of the 21st century, which was a similar pattern to the summer months over Korea (Figs. 2 & 3). An increase in T_{max} over the western coast and southeastern Korea in July in the 2050s and 2090s was noticeable (Figs. 2 & 3), which was also reported previously by Choi et al. (2008), based on observations made over several decades. In particular, a generally higher T_{max} in August (>35°C) was expected in the 2090s (Fig. 2), which is fairly remarkable considering that the projection was made with the RCP4.5 scenario that represents relatively moderate climate change.

Projection of the distribution of hot weather frequency (total number of hot weather days) over South Korea is shown in Fig. 4, and differences in the projected frequency compared to the 2010s are in Fig. 5. Similar to T_{max}, the area affected by hot weather events was expected to expand over southeastern Korea (Young-Nam region) in the 2090s (Fig. 4). In addition to expansion of the area affected by hot weather events in July and August, the higher number of hot weather days (>15 d mo^{-1}) over the western and southeastern regions of Korea are expected in the last decade of the 21st century (in July and August; Figs. 4 & 5). Similar projections were also made by Jeung et al. (2013) and Park et al. (2013), although with different extreme weather definitions and using different climate scenarios.

To investigate hot weather event projection, we focused on 2 factors: expansion of the area affected by hot weather events, and frequency (number of days).

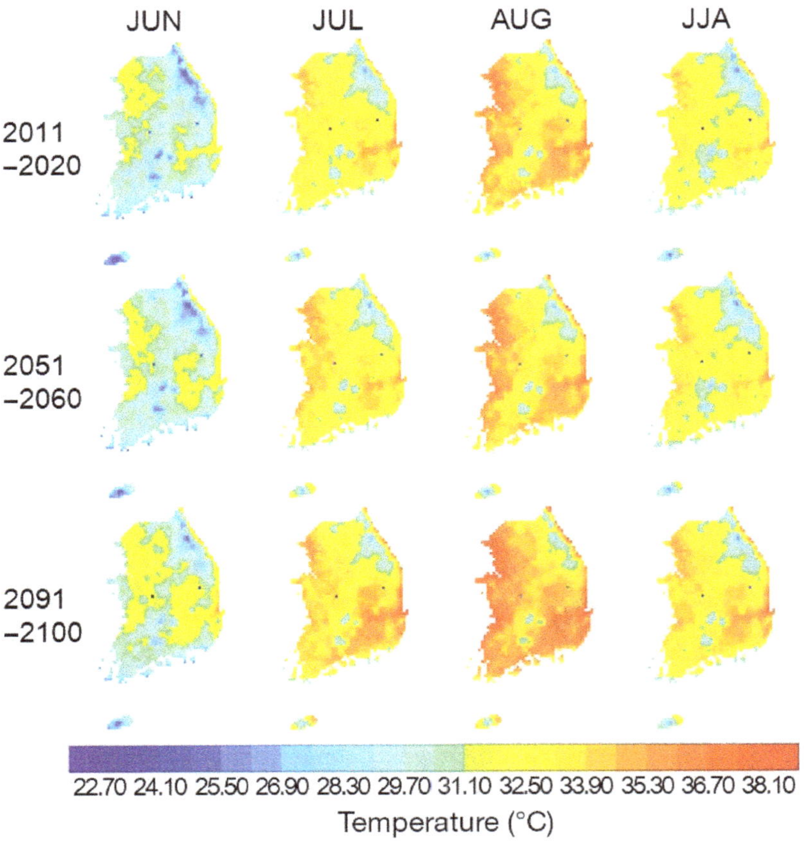

Fig. 2. Distribution of daily maximum temperature (T_{max}) of South Korea in June, July, August, and the summer mean (average of JJA), for 3 decadal bins (2011–2020, 2051–2060, and 2091–2100)

We estimated expansion of the area affected by hot weather events by defining it as the area exposed to hot weather events more than once in a month. Fig. 6 shows the decadal trend of expansion of the area affected by hot weather events for each summer month, with the area affected by hot weather events expressed as a fraction (%) of the total area of the country.

Although the area currently affected by hot weather events was not extensive in June (<10%), it was projected to increase to almost 25% before 2040, which is similar to the current area affected by hot weather events in July (~30%) (Fig. 6). The fastest expansion of the area affected by hot weather events occurred in July (Fig. 6). The current extent of the area affected by hot weather events (30%) almost doubled by the end of this century, and the rate of increase was 0.29% yr^{-1} (Fig. 6). In particular, a large expansion in the area affected by hot weather events in the 2060s was noticeable (Fig. 6). The area affected by hot weather events was largest in August and expanded at an average rate of 0.24% yr^{-1}, resulting in 70% of South Korea being exposed to hot weather events in the 2060s (Fig. 6).

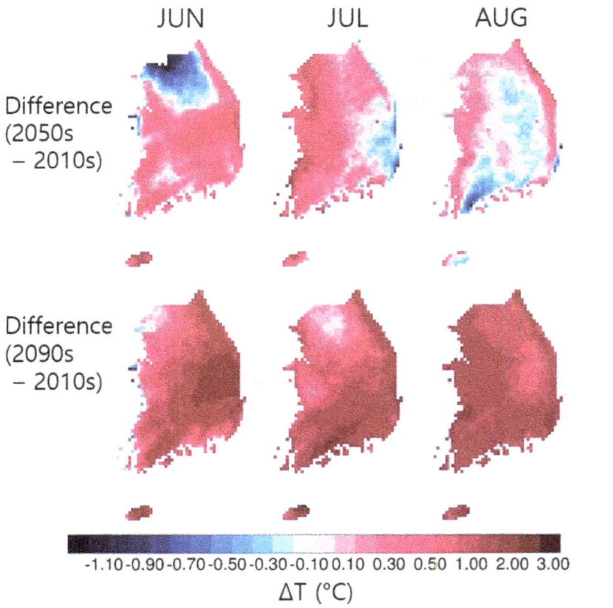

Fig. 3. Distribution of daily maximum temperature (T_{max}) of South Korea in June, July, and August, showing the differences between the 2050s and 2010s and the 2090s and 2010s

Expansion of hot weather events in the 21st century is not linear with time, with instant drops predicted in the 2050s and 2080s (Figs. 6 & 7). This is likely due to the complex geophysical response to external forcing and internal climate variability in the simulation (Manabe et al. 1990, Hall 2004, Seneviratne et al. 2010, Deser et al. 2012, Ishizaki et al. 2012, Good et al. 2015).

In addition to the area affected by hot weather events, we also considered the frequency of hot weather events by integrating the area and frequency of hot weather events (IAF, %) (defined in Section 2.3). The nationwide IAF is currently <5% in July, but was projected to increase to 10% in the 2060s (Fig. 7). IAF in August is a more serious concern. It was projected to increase sharply and reach ~20% in the 2040s (Fig. 7). In the 2040s, IAF will reach 25%, with an average rate of increase of 0.15% yr^{-1} (Fig. 7). Our analysis indicated that the national frequency of hot weather events and the population's exposure to them will be severe in August, with a large potential mortality associated with the increased hot weather intensity in Korea.

3.2. Population exposed to hot weather events

As explained in Section 2.4, we estimated the population exposed to hot weather events in the summer months based on the mapping of local administrative districts and the model grid. Currently (2010), a quarter of the Korean population (~12.5 million people) is exposed to hot weather events (Fig. 8). However, the exposed population was projected to rapidly increase to more than half of the total population (58%, 27.9 million) in the middle of this century (2050; Fig. 8), which is closely associated with the expansion of the area affected by hot weather events into the SMA and southeastern region, where there is higher population density (Figs. 4 & 5). In the last decade of this century, the population exposed to hot weather events was projected to reach 72% of the total population, indicating that the majority of the nation's population will be exposed to hot weather events by the end of this century. Furthermore, the population structure in Korea is expected to change significantly. The Korean population is continuously aging and has been considered an 'aging society' since 2000 (Gong & Jang 2010). The elderly population is projected to increase rapidly in the future (KOSIS 2015), and this portion of the population is particularly vulnerable to heat waves (McGeehin & Mirabelli 2001, Schifano et al. 2009, Huang et al. 2010, Son et al. 2012), having significantly higher mortality from heat waves in South Korea, especially those >65 yr of age (Lee 2015). Thus, we estimated the population exposure for the elderly (>65 yr old) to hot weather events. Currently (2010), the population of >65 yr olds is about 5.5 million, accounting for ~11% of the total population of South Korea.

If we apply the future changes in national population structure with the same spatial distribution of the population in 2010, our estimation shows that the ~1.4 million elderly exposed to hot weather events (~3% of the national population) in 2010 could rapidly increase to ~10.4 million (~22% of the national population) in 2050 (Fig. 8).

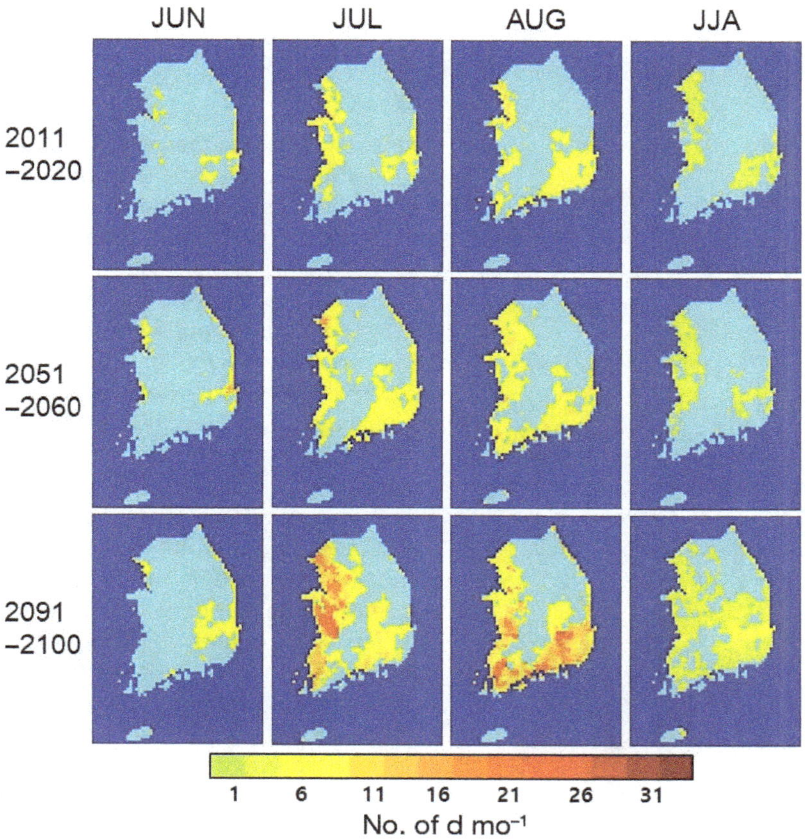

Fig. 4. Projection of hot weather frequency in South Korea in June, July, August, and the summer mean (JJA), for 3 decadal bins (2011–2020, 2051–2060, and 2091–2100)

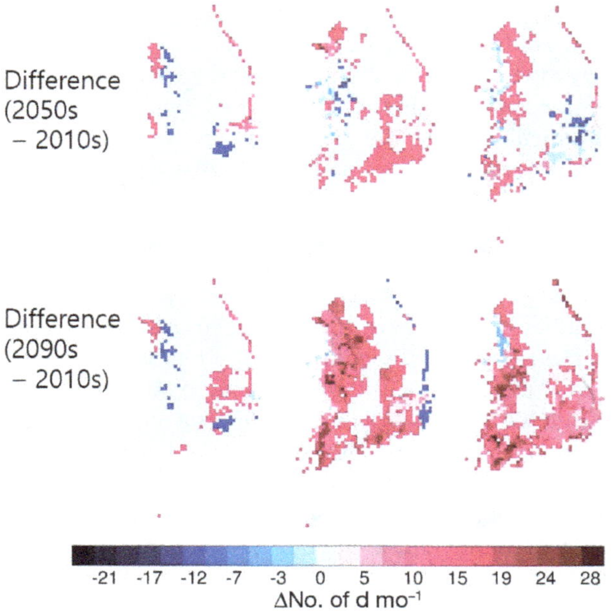

Fig. 5. Projection of hot weather frequency (number of days per month) in South Korea in June, July, and August, for the difference from the 2050s to 2010s and the 2090s to 2010s

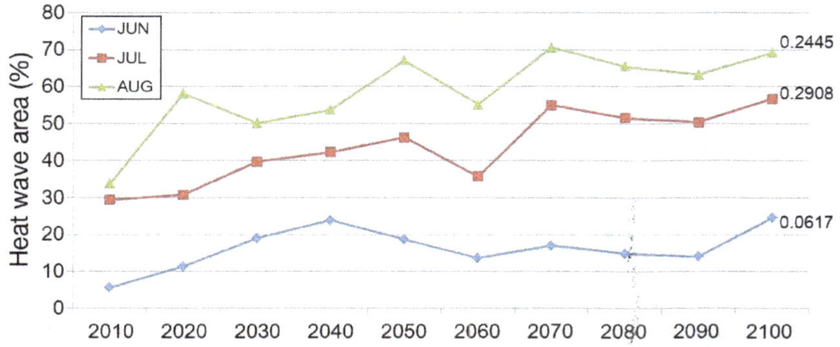

Fig. 6. Projection of area affected by hot weather events (%) in South Korea, defined as the sum of the area experiencing ≥1 hot weather event in each summer month (June–August), divided by the area of the entire nation. Numbers on the right side are the rate of increase (% area per year) for each month

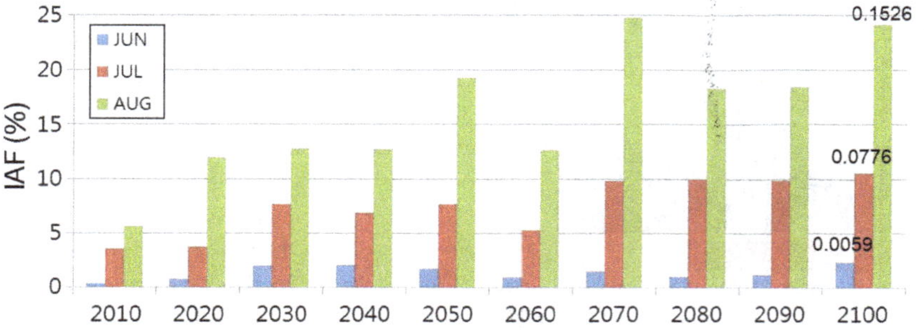

Fig. 7. Projection of integration of area and frequency (IAF) of hot weather events in South Korea (see Section 2.3 and Eq. 1 for definition). Numbers above bars for 2100 are the rate of increase of IAF (% per year) for each month

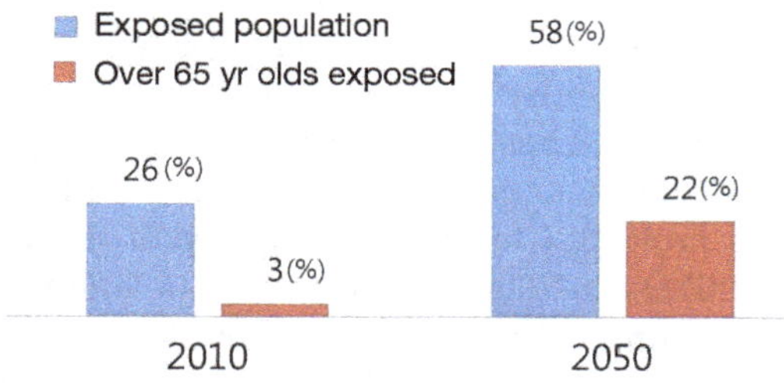

Fig. 8. Projection of population (overall and elderly) exposed to summer hot weather events in South Korea. A projection of the national population structure was used. Note that the population of >65 yr olds is considered to have significantly higher heat wave mortality in Korea

Although our estimation contains uncertainty from the assumption of a static population distribution due to lack of future data available at the local administrative district level, the uncertainty would be small, because the changes in the population structure at the regional administrative district level are small (KOSIS 2015). Our results therefore have important implications for future national security associated with climate change.

4. DISCUSSION AND CONCLUSIONS

Here, we discuss the projections of hot weather events and the potential population exposure over South Korea in the 21st century. The regional RCP4.5 scenario, with a 7.5 × 7.5 km horizontal resolution, was calculated by dynamical downscaling of CESM model results, in which the estimated temperature was validated by ground observations, and the mean bias was applied for future temperature projection.

This study first presented quantitative information on future projections of hot weather events in Korea with information on changes in the area affected and frequency for the entire nation. This study also estimated future nationwide population exposure to hot weather events, which is essential information for preparing national climate change adaptation policy.

Our analysis showed that the area affected by hot weather events will increase to cover ~70 % of the nation in August in the middle of this century, with an overall rate of increase of 0.24 % yr^{-1}. In terms of frequency, the larger number of hot weather days (>15 d mo^{-1}) in the western and southeastern regions of Korea in the last decade of the 21st century is critical because the majority of the Korean population, nearly 30 million people, reside in these regions, which includes the SMA and Busan.

We introduced the index of integration of area and frequency (IAF, %), which is a summation of the area affected by hot weather events multiplied by the number of heat wave days for each area. According to the IAF results, a quarter of the nation will be exposed to daily hot weather events in August during the 2060s and 2090s, which is almost a 500 % increase compared with the 2000s. An IAF of 25 % can cause a great impact on the nation since it is statistically equivalent to a quarter of the nation being exposed to hot weather events every day of a month, or all of the nation being exposed to hot weather events for about 7.5 d mo^{-1}. IAF contains frequency information associated with hot weather intensity and duration, which is closely related to mortality (Son et al. 2012). Ho et al. (2011) also estimated the projection of heat waves.

Although they used a different definition and the Special Report on Emissions Scenarios (SRES), they concluded that heat wave frequency will increase by 583% in the A1FI scenario, implying that our analysis with RCP4.5 was almost as serious as the previous analysis using the A1FI scenario.

In addition, we estimated the population exposed to hot weather events by mapping population data from the nation's administrative districts with the model grid. Despite the limitations and uncertainty in the projection of future population, the exposed population was projected to increase almost 3-fold, from 26% during the 2010s to 72% during the 2090s. In particular, the exposure of the elderly population (>65 yr old), who are particularly vulnerable to heat waves, was expected to reach ~22% of the national population (~10.4 million) in the middle of this century if we apply the projection of an aging population to our results. It is important to note that those results are only based on the RCP4.5 scenario, implying that even more severe hot weather events could be expected under a possible business-as-usual scenario (e.g. RCP8.5).

The Korean government has taken actions against heat waves with an annual national plan for heat waves coordinated by the Ministry of Public Safety and Security (MPSS) and a national adaptation plan to climate change coordinated by the Ministry of Environment (MOE). However, cross-sectoral collaboration and engagement for regional-specific plans need to be improved (Ha et al. 2014). In particular, our results strongly suggest the need for long-term measures to mitigate the effects of heat waves, for which there are no current provisions (Ha et al. 2014).

This study may contain uncertainties for the following reasons. (1) Bias correction over a relatively short period (9 yr) can ignore long-term interannual climate variability. We further estimated that the difference in monthly bias between the 9 yr period 2006–2014 and 30 yr period 1985–2014 ranges from −0.04 to 0.45°C (national mean). Applying the bias correction over a 30 yr period could reduce the number of total grids with hot weather events in summer by 16% (2015–2100), leading to reductions in population exposure of 50% (total population) and 19% (elderly population) in 2050 (corresponding to Fig. 8). Although the application of the same approaches to bias correction with a 30 yr period was not possible due to the much smaller number of ground stations (~70), these facts imply the possibility of the overestimated projections.

(2) There are uncertainties associated with the climate model schemes and downscaling method, which requires further study, including comparisons with multi-model scenarios, including those from the KMA (Jeung et al. 2013, Park et al. 2013), because the projection of heat wave mortality could be highly variable, depending on the exact climate model used (Ho et al. 2011, Peng et al. 2011).

(3) Heat waves earlier in the year can produce higher mortality than later heat waves (Son et al. 2012), and the Korean government has already issued heat wave alerts several times in May 2016 (Kim 2016), while the occurrence of heat waves in Korea used to be rare (<10 in the last 20 yr; Kim et al. 2014). Thus, studies of early heat waves and their impacts on health are necessary.

Acknowledgements. This work was jointly supported by the Korea Meteorological Administration Research and Development Program under Grant KMIPA (No. 2015-2022) and the National Research Foundation of Korea (Nos. NRF-2013 S1A5B6043772 and NRF-2013R1A2A1A03070600) funded by the Ministry of Science, ICT and Future Planning (MSIFP). The system for producing regional climate scenarios was supported partly by the Korea Environmental Institute (KEI) and partly by the Korean Ministry of Environment (KMOE)'s Climate Change Correspondence Program. We thank Dr. Sung-Dae Kang and Ms. Jiyoun Hong for supporting our analysis of the CESM climate model. The CESM project is supported by the National Science Foundation and the Office of Science (BER) of the US Department of Energy.

LITERATURE CITED

Beniston M, Stephenson DB (2004) Extreme climate events and their evolution under changing climate conditions. Glob Planet Change 44:1–9

Choi BC, Kim JY, Lee DG, Kysely J (2007) Long-term trends of daily maximum and minimum temperatures for the major cities of South Korea and their implications on human health. Atmosphere (Korean Meteorol Soc) 17: 171–183

Choi GY, Kwon WT, Boo KO, Cha YM (2008) Recent spatial and temporal changes in means and extreme events of temperature and precipitation across the Republic of Korea. J Korean Geogr Soc 43:681–700

Deschênes O, Moretti E (2009) Extreme weather events, mortality, and migration. Rev Econ Stat 90:659–681

Deser C, Knutti R, Solomon S, Phillips AS (2012) Communication of the role of natural variability in future North American climate. Nat Clim Change 2:775–779

Ebi KL, Mills D (2013) Winter mortality in a warming climate: a reassessment. WIREs Clim Change 4:203–212

Gong S, Jang HJ (2010) A literature survey on environmental issues and life pattern changes of an aging society. Working Paper of Korea Environment Institute, Seoul (in Korean)

Good O, Lowe JA, Andrews T, Wiltshire A and others (2015) Nonlinear regional warming with increasing CO_2 concentrations. Nat Clim Change 5:138–142

Graham LP, Andréasson J, Carlsson B (2007) Assessing climate change impacts on hydrology from an ensemble of regional climate models, model scales and linking methods — a case study on the Lule River basin. Clim Change 81(Suppl 1):293–307

Ha J, Jung HC, Lee JH, Kim DH, Choi J (2014) A study on establishment and management of a long-term heatwave plan addressing climate change. Research Report of Korea Environment Institute, Sejong (in Korean)

Hall A (2004) The role of surface albedo feedback in climate. J Clim 17:1550–1568

Ho CH, Park TW, Jun SY, Lee MH and others (2011) A projection of extreme climate events in the 21st century over East Asia using the Community Climate System Model 3. Asia-Pac J Atmos Sci 47:329–344

Ho CK, Stephenson DB, Collins M, Ferro CAT, Brown SJ (2012) Calibration strategies: a source of additional uncertainty in climate change projections. Bull Am Meteorol Soc 93:21–26

Huang W, Kan H, Kovats S (2010) The impact of the 2003 heat wave on mortality in Shanghai, China. Sci Total Environ 408:2418–2420

IPCC (2007) Climate change 2007: synthesis report. Summary for policymakers. Cambridge University Press, Cambridge

IPCC (2014) Climate change 2014: impacts, adaptation, and vulnerability. Contribution of Working Group II to the Fifth Assessment Report of the Intergovernmental Panel on Climate Change. Cambridge University Press, Cambridge

Ishizaki Y, Shiogama H, Emori S, Yokohata T and others (2012) Temperature scaling pattern dependence on representative concentration pathway emission scenarios. Clim Change 112:535–546

Jeung SJ, Sung JH, Kim BS (2013) Change projection of extreme indices using RCP climate change scenario. J Korea Water Resour Assoc 46:1089–1101

Kim Y (2016) The Korean government takes action against heat waves in 2016. Environmental News of Korea. www.hkbs.co.kr/?m=bbs&bid=envnews7&uid=394186 (accessed on May 23) (in Korean)

Kim H, Ha J, Park J (2006) High temperature, heat index, and mortality in 6 major cities in South Korea. Arch Environ Occup Health 61:265–270

Kim EB, Park JK, Jung WS (2014) A study on the occurrence characteristics of tropical night day and extreme heat day in the metropolitan city, Korea. J Environ Sci Int 23: 873–885

Kinney PL, Pascal M, Vautard R, Laaid K (2012) Winter mortality in a changing climate: Will it go down? Bull Epidemiol Hebd (Paris) 12–13:148–151

KMA (Korea Meteorological Administration) (2012) The annual report of extreme climate in 2012. KMA, Seoul (in Korean)

KMA (Korea Meteorological Administration) (2013) The annual report of extreme climate in 2013. KMA, Seoul (in Korean)

Knowlton K, Rotkin-Ellman M, Geballe L, Max W, Solomon GM (2011) Six climate change-related events in the United States accounted for about $14 billion in lost lives and health costs. Health Aff 30:2167–2176

KOSIS (Korea Statistical Information Service) (2015). The projection of population structure. http://kosis.kr/statisticsList/statisticsList_01List.jsp?vwcd=MT_ZTITLE&parentId=A. (In Korean) (accessed October 19, 2015)

Kysely J, Huth R (2004) Heat-related mortality in the Czech Republic examined through synoptic and 'traditional' approaches. Clim Res 25:265–274

Lee J (2015) The damages of heat waves and responding strategy. Forum for responding to heat waves, Seoul, July 30, 2015. Korea Meteorological Administration, Seoul, p 67–69 (in Korean)

Lee JW, Hong SY, Chang EC, Suh MS, Kang HS (2014) Assessment of future climate change over East Asia due to the RCP scenarios downscaled by GRIMs-RMP. Clim Dyn 42:733–747

Manabe S, Bryan K, Spelman MJ (1990) Transient response of a global ocean-atmosphere model to a doubling of atmospheric carbon dioxide. J Phys Oceanogr 20:722–749

McGeehin MA, Mirabelli M (2001) The potential impacts of climate variability and change on temperature-related morbidity and mortality in the United States. Environ Health Perspect 109(Suppl 2):185–189

Meehl GA, Karl T, Easterling DR, Changnon S and others (2000) An introduction to trends in extreme weather and climate events: observations, socioeconomic impacts, terrestrial ecological impacts, and model projections. Bull Am Meteorol Soc 81:413–416

Na D (2015) The meteorological characteristics of Korean summer in 2015 and the warning of heat waves. Forum for responding to heat waves, Seoul, July 30, 2015. Korea Meteorological Administration, Seoul, p 25–38 (in Korean)

Park JK, Jung WS, Kim EB (2008) A study on development of the extreme heat standard in Korea. J Environ Sci (China) 17:657–669

Park CY, Choi YE, Kwon YA, Kwon JI, Lee HS (2013) Studies on changes and future projections of subtropical climate zones and extreme temperature events over South Korea using high resolution climate change scenario based on PRIDE model. J Korean Assoc Reg Geogr 19: 600–614

Peng RD, Bobb JF, Tebaldi C, McDaniel L, Bell ML, Dominici F (2011) Toward a quantitative estimate of future heat wave mortality under global climate change. Environ Health Perspect 119:701–705

Robinson PJ (2001) On the definition of a heat wave. J Appl Meteorol 40:762–775

Ross T, Lott N (2003) A climatology of 1980–2003 extreme weather and climate events. Technical Report (2003-01) of National Climate Data Center, Asheville, NC

Schifano P, Cappai G, De Sario M, Michelozzi P and others (2009) Susceptibility to heat wave-related mortality: a follow-up study of a cohort of elderly in Rome. Environ Health 8:50

Seneviratne SI, Corti T, Davin EL, Hirschi M and others (2010) Investigating soil moisture-climate interactions in a changing climate: a review. Earth-Sci Rev 99:125–161

Seo J, Shim C, Hong J, Kang S, Moon N, Hwang YS (2013) Application of the WRF model for dynamical downscaling of climate projections from the Community Earth System Model (CESM). Atmosphere (Korean Meteorol Soc) 23:347–356

Son J, Lee JT, Anderson GB, Bell ML (2012) The impact of heat waves on mortality in seven major cities in Korea. Environ Health Perspect 120:566–571

Sperna Weiland FC, van Beek LPH, Kwadijk JCJ, Bierkens MFP (2010) The ability of a GCM-forced hydrological model to reproduce global discharge variability. Hydrol Earth Syst Sci 14:1595–1621

Woodward A, Scheraga JD (2003) Looking to the future: challenges for scientists studying climate change and health. In: McMichael AJ, Campbell-Lendrum DH, Corvalan CF, Ebi KL and others (eds) Climate change and

human health: risks and responses. World Health Organization, Geneva, p 61–78
Wu J, Zhou Y, Gao Y, Fu JS and others (2014) Estimation and uncertainty analysis of impacts of future heat waves on mortality in the eastern United States. Environ Health Perspect 122:10–16
Yang J, Ha J (2013) Estimation of future death burden of high temperatures from climate change. Korean J Environ Health Sci 39:19–31
Zhao M, Running SW, Heinsch FA, Nemani RR (2011) MODIS-derived terrestrial primary production. In: Ramachandran B, Justice CO, Abrams MJ (eds) Land remote sensing and global environmental change. Springer, New York, NY, p 635–660

4

Seasonal patterns and consistency of extreme precipitation trends in Europe, December 1950 to February 2008

Ewa B. Łupikasza*

Department of Climatology, Faculty of Earth Sciences, University of Silesia in Katowice, 41-200 Sosnowiec, Poland

ABSTRACT: Seasonal trends in extreme precipitation indices were investigated for 30 yr moving periods between December 1950 and February 2008. To update the 2008 to 2015 data, supplementary calculations were performed for >120 meteorological stations. A linear regression of the least squares method was used to calculate trend magnitudes. Trend significance was tested using the Mann-Kendall method. Changes in short-term trend frequency and temporal coherence were assessed. Extreme precipitation was defined as a daily amount exceeding the 95th percentile, calculated separately for each month and station using daily totals ≥1 mm. The spatial pattern of extreme precipitation trends varied by season. Significant extreme precipitation trends were rare, constituting approximately 25 to 30% of all analysed trends, and were seldom temporally coherent. Most of these significant trends were upward, except in summer, when a nearly equal frequency of positive and negative trends was found. Increases in the frequency and the total were a characteristic feature of extreme precipitation changes, particularly in winter. Seasonal variations in the spatial patterns of extreme precipitation trends may have resulted from seasonal changes in the prominence of the driving factors of precipitation. In spring, upward trends in Central and Western Europe were twice as frequent as the downward trends found primarily in Southern Europe. In summer, the percentages of significant downward trends in Western Europe and upward trends in Eastern Europe were similar. In autumn, a coherent decrease in extreme precipitation was clear in Central Europe. The spatial distribution of trend directions was the most consistent in winter.

KEY WORDS: Precipitation trends · Heavy precipitation · Climate change · Trend variability and coherence · Europe

1. INTRODUCTION

Extreme precipitation events are known to trigger many hydrological and geomorphological processes (e.g. floods, run-off, landslides, and transport of sediments) and directly influence the living conditions of humans and other species. Due to their large inherent variability, the unequivocal identification of precipitation trends is very difficult. Spatial and seasonal changes in extreme precipitation, particularly on a continental scale, are a manifestation of changes in processes that lead to the formation of precipitation (e.g. convection, processes inside clouds). Locally, precipitation is modified by the availability of moisture and is affected by circulation systems that are subject to local energetic constraints (Hartmann et al. 2013). Therefore, the sign of extreme precipitation trends can indicate the direction of possible future changes in precipitation formation processes.

Extreme precipitation trends in Europe have been analysed on continental and regional scales. Selected results of this research are presented as a schematic map in Fig. 1. Europe has experienced a significant increase in extreme winter precipitation, although the trends were not spatially coherent (Klein-Tank & Können 2003). In summer, when conditions that

*Corresponding author: ewa.lupikasza@us.edu.pl

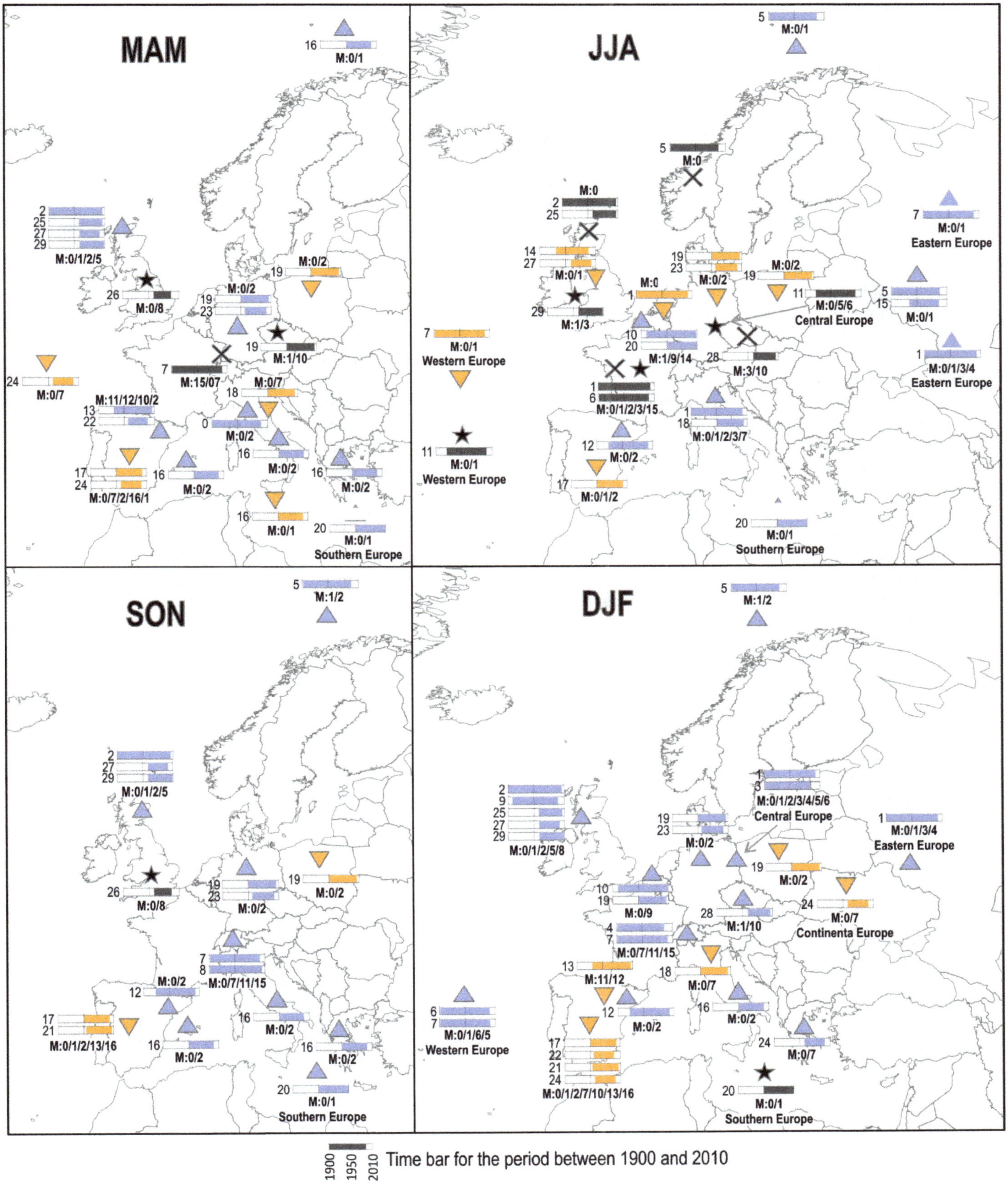
MAM
JJA
SON
DJF
1900 1950 2010 Time bar for the period between 1900 and 2010

favour the development of convection are the most common, extreme precipitation trends are spatially variable in direction and magnitude (e.g. Moberg & Jones 2005). Positive trends were found in Eastern Europe, whereas decreases were predominant in Western Europe (Moberg & Jones 2005, Zolina et al. 2005, Moberg at al. 2006, Burt & Horton 2007). Besselaar et al. (2013) reported stronger regularity of the trends in the recurrence period of the highest 1 d and 5 d precipitation totals in the second half of the 20th century and the following decade (1951–2010; Besselaar et al. 2013). Northern Europe exhibited predominantly growing trends in these indices, whereas growing but weaker trends were confirmed in Southern Europe in spring, summer and autumn (Besselaar et al. 2013). The trends in heavy cyclonic precipitation (≥60 mm) were positive in Northern Europe and negative in Southern Europe (1956 to 2000). However, their statistical significance depended on the period and season: Negative trends were significant in the last or last 2 decades of the 20th century, whereas no trends reached statistical significance during the entire period (Karagiannidis et al. 2012). The direction of trends in extreme precipitation also varied depending on the study period in Northern Italy (Pavan et al. 2008).

Research at a regional scale has demonstrated broad spatial and seasonal differences in trend directions. Persistent increasing trends in various characteristics of extreme precipitation were detected in the Netherlands in both warm and cool seasons during 1951–2009 (Buishand et al. 2013) and in the Hungarian region of the Carpathian Basin in the last quarter of the 20th century (Bartholy & Pongrácz 2007). Gajić-Čapka & Cindrić (2011) detected no evidence of major secular changes in the amount and frequency of extreme precipitation since the beginning of the 20th century (1901 to 2008) over most of Croatia. In other parts of Europe, the direction and statistical significance of trends varied depending on the season and research period, as discussed below.

In summer, extreme precipitation significantly increased in Eastern Europe (1936 to 1994) (Easterling et al. 2000) and the Czech Republic (Kyselý 2009). In Norway, precipitation in summer displayed no statistically significant change over nearly a century (i.e. 1901 to 1996) (Easterling et al. 2000). Similarly, no trends were found for intense daily precipitation in summer in the Alpine region of Switzerland (Frei & Schär 2001, Schmidli & Frei 2005). In contrast, extreme summer precipitation decreased in the UK (Osborn et al. 2000, Osborn & Hulme 2002, Maraun et al. 2008, Burt & Ferranti 2012, Jones et al. 2013) and western Germany (Hundecha & Bárdossy 2005).

In winter, increasing trends in various characteristics of extreme precipitation were predominant in Central and Western Europe including the UK (Osborn et al. 2000, Osborn & Hulme 2002, Maraun et al. 2008, Burt & Ferranti 2012, Jones et al. 2013), Germany (Hundecha & Bárdossy 2005, Łupikasza et al. 2011), western Czech Republic (Kyselý 2009), and Switzerland (the Alpine region) during the 20th century (Frei et al. 2000, Frei & Schär 2001). Opposite trends prevailed in southern Poland (Lupikasza 2010, Łupikasza et al. 2011). In transitional seasons (spring and autumn), extreme precipitation increased in western Germany (Hundecha & Bárdossy 2005), decreased in the Czech Republic (Kyselý 2009) and varied in direction in Poland (Lupikasza 2010) and the UK during the second half of the 20th century (1961 to 2000) (Osborn & Hulme 2002). However, when the UK data series was extended to 2006, an upward trend in extreme precipitation was identified (Maraun et al. 2008, Jones et al. 2013).

Fig. 1. Review of research on extreme precipitation trends in Europe, based on selected literature. Blue triangles: significant upward trends; orange triangles: significant downward trends; black stars: various trend directions in a time frame; crosses: no significant trends; M: method of trend analysis; 1/2/: number corresponding to the method of trend analysis (see below); filled parts of time bars: time frame of trend analysis; numbers to the left of time bars: exact period of trend analysis (see below). Numbering of trend analysis methods: (0) Linear least-squares; (1) *t*-test; (2) Mann-Kendall test; (3) Wilcoxon test; (4) Hayashi reliability ratio; (5) block bootstrap technique; (6) resistant methods; (7) Kendall-tau test; (8) *F*-test; (9) Log-linear model; (10) Teil-Sen estimator; (11) Log-likelihood ratio; (12) linear non-stationary model; (13) Spearmann correlation (SC); (14) field significance; (15) logistic regression; (16) differences between averages. Numbering of time bars: (0) 1880–1996, (1) 1900–2002, (2) 1900–2006, (3) 1901–1991, (4) 1901–1994, (5) 1901–1996, (6) 1901–1999, (7) 1901–2000, (8) 1901–2004, (9) 1908–1995, (10) 1910–2009, (11) 1921–1999, (12) 1923–2002, (13) 1930–2006, (14) 1931–1995, (15) 1936–1994, (16) 1951–2000, (17) 1951–2002, (18) 1951–2004, (19) 1951–2006, (20) 1951–2010, (21) 1955–2006, (22) 1958–1997, (23) 1958–2001, (24) 1960–2000, (25) 1960–2006, (26) 1961–1995, (27) 1961–2001, (28) 1961–2005, (29) 1961–2009. Sources: Klein-Tank & Können (2003), Moberg & Jones (2005), Zolina et al. (2005), Moberg et al. (2006), Besselaar et al. (2013), Easterling et al. (2000), Karagiannidis et al. (2012), Easterling et al. (2000), Osborn et al. (2000), Burt & Ferranti (2012), Osborn & Hulme (2002), Maraun et al. (2008), Jones et al. (2013), Hundecha & Bárdossy (2005), Łupikasza et al. (2011), Buishand et al. (2013), Lupikasza (2010), Kyselý (2009), Frei et al. (2000), Frei & Schär (2001), Schmidli & Frei (2005), Bartholy & Pongrácz (2007), Norrant & Douguédroit (2006), Brunetti et al. (2000), Pavan et al. (2008), Gajić-Čapka & Cindrić (2011), Ramos & Martinez-Casanovas (2006), Rodrigo & Trigo (2007), Beguería et al. (2011), López-Moreno et al. (2010), Rodrigo (2010), García et al. (2007)

In Southern Europe, significant changes were found in parts of Greece, Italy, and the Gulf of Valencia (Norrant & Douguédroit 2006). In northern Italy, the frequency of intense and extreme precipitation events during 1951 to 2004 decreased significantly in winter and spring and increased in summer in mountainous regions (Piervitali et al. 1998, Brunetti et al. 2000, Pavan et al. 2008). Studies of the entire Iberian Peninsula indicated a great variety of extreme precipitation trend directions, depending on the research period (e.g. Norrant & Douguédroit 2006, Ramos & Martinez-Casanovas 2006, Rodrigo & Trigo 2007, Beguería et al. 2011, López-Moreno et al. 2010). In general, the Iberian Peninsula was characterized by an increased probability of dry conditions, but no increased probability of extreme wet conditions was detected (Rodrigo 2010). However, predominant decreasing extreme precipitation trends in winter in the Iberian Peninsula were reported by García et al. (2007).

According to the above literature review, extreme precipitation trends tend to be infrequently statistically significant, strongly dependent on the time-frame studied, and not spatially coherent. These attributes may partly result from the variety of methods, definitions, and periods adapted for extreme precipitation research, but may also reflect the inherent variability in precipitation extremes. Willems (2013) found that rainfall extremes in Europe exhibit oscillatory behaviour on multidecadal time scales. He reported that recent upward trends in precipitation extremes in parts of Europe are partly related to a positive phase of this oscillation. Moreover, questions regarding the variability of trends over time have emerged from the work by Casanueva et al. (2014). The above studies provided us the motivation to perform trend analyses covering various time frames to understand changes in precipitation extremes.

In this study, a trend analysis using moving 30 yr periods is proposed that could serve as a supplementary method for long-term trend analysis. This short-term trend analysis provides information on predominant trend directions and accounts for the variability and oscillatory behaviour of precipitation. Additionally, the method allows for the assessment of the temporal coherency of precipitation trends. The 30 yr interval was selected based on World Meteorological Organization standards.

This study aims to evaluate the spatial, seasonal, and long-term variabilities in extreme precipitation trends in Europe, and to assess the temporal consistency in trend direction from December 1950 to February 2008. This study continues and expands on methodologically similar analyses of Poland and Germany during 1951 to 2006 (Lupikasza 2010, Łupikasza et al. 2011). A continent-wide context for the regional variability of trends in extreme precipitation is, thus, provided. The results presented in this paper cover the long-term period defined by the project on precipitation extremes (NN 306 243939) funded by the National Research Centre of the Ministry of Science and Higher Education of Poland. The results obtained for the project time frame (December 1950 to February 2008) are based on >350 meteorological stations. To update the data from 2008 to 2015, supplementary calculations were performed for >120 meteorological stations (126 stations for December-January-February [DJF], 127 stations for March-April-May [MAM] and September-October-November [SON], and 128 stations for June-July-August [JJA]) with available and complete series of daily precipitation, selected to be as evenly distributed throughout the continent as possible. The updated results for temporal changes in significant trend frequencies are presented in Fig. S1 in the Supplement (www.int-res.com/articles/suppl/c072p217_supp.pdf).

2. MATERIALS AND METHODS

2.1. Data

Extreme precipitation events during 1951 to 2008 were identified based on chronological daily precipitation records provided by the European Climate Assessment and Database (ECA&D), which was the primary source (Klein Tank et al. 2002, Klok & Klein Tank 2009; data and metadata available at www.ecad.eu), supplemented with data from e-klima (database no longer available; e-klima can be found at www.eklima.no/), CliWare (http://cliware.meteo.ru/meteo) and the website of the German weather service (www.dwd.de/). The Polish data were from the archive of the Institute of Meteorology and Water Management (IMGW) and include meteorological yearbooks and precipitation yearbooks. Data from after 28 September 1999 were taken from the OGIMET online weather database (www.ogimet.com) and used to verify and fill in minor gaps in the daily precipitation series. This database contains information on weather phenomena (current and past) and was useful in verifying doubtful daily precipitation totals. This study used rigorous completeness criteria to assess the data. The missing data did not exceed (1) 2 days per month, (2) 5% of days in the entire research period (December 1950 to February 2008), and (3) 2%

of days in the reference period (1961 to 1990). When calculating seasonal series of indices, series that lacked >2 annual values were omitted.

Assessing the homogeneity of daily weather data is difficult because no reliable statistical method is available, particularly for precipitation. In this study, substantial effort was made to apply quality control to the data. Homogeneity of the ECA&D data was controlled for by using 4 homogeneity tests: the standard normal homogeneity test, the Buishand range test, the Pettitt test, and the von Neumann ratio test (Von Neumann 1941, Pettitt 1979, Buishand 1982, Alexandersson 1986, Klein Tank et al. 2002, Klok & Klein Tank 2009). The highest quality series (classified according to the test results as 'useful') were selected. Detailed information on the data homogenization procedures is available on the project website (www.ecad.eu). The days with daily precipitation outliers and extremes (Eqs. 1 and 2) were identified, and the precipitation amounts on these days were compared among different databases. The outliers and extremes were calculated according to the following formulas (Hill & Lewicki 2007):

$$\text{outlier} > (\text{avg} + \text{SE}) + 1.5[(\text{avg} + \text{SE}) - (\text{avg} - \text{SE})] \quad (1)$$

$$\text{extreme} > (\text{avg} + \text{SE}) + 3[(\text{avg} + \text{SE}) - (\text{avg} - \text{SE})] \quad (2)$$

where avg is the average daily precipitation total from days with precipitation ≥1 mm, and SE is the standard error.

Most of the values were correct. However, the suspected precipitation totals were manually verified using synoptic charts (due to the possibility of a high-precipitation event), daily precipitation maps created using data from OGIMET (for data from 1999 only) or information from various academic books and papers (e.g. Arléry 1970, Escardó 1970, Johannessen 1970, Manley 1970, Furlan 1977, Lydolph 1977, Schüepp & Schirmer 1977, Kundzewicz 2005, Alfnes & Førland 2006). For example, the very high daily precipitation total of 520 mm recorded on 24 February 1964 at the Mont-Aigoual station proved to be true, whereas a 42 mm thunderstorm-related precipitation event recorded on 27 February 2004 at Lerwic station was incorrectly 'homogenized' to 5 mm in the ECA&D.

In total, 363 stations in autumn and summer, 362 stations in spring, and 346 stations in winter met the completeness and homogeneity criteria (described above). These stations are shown in Fig. 2. The stations are not evenly distributed across the research area. However, due to the large spatial variability in precipitation, particularly extreme precipitation, every station with high-quality data delivers valuable information about changes in extreme precipitation characteristics. The densest set of stations covers Central Europe. Southern Europe, with its variable topography, is strongly underrepresented. The density of stations in Eastern Europe is also low, but due to this region's more uniform landscape, the spatial variability of precipitation in this region is less than that in southern localities. Moreover, the stations in Eastern Europe are evenly distributed. The updated daily precipitation series are relatively evenly distributed in Western, Central, and most of Southern Europe (Fig. 2).

Here, Western Europe refers to Great Britain, Ireland, the Netherlands, central and northern France, Luxembourg, Belgium, the western region of Germany, and Switzerland. Southern Europe refers to the Iberian Peninsula, southern France, the Italian Peninsula, the islands in the Mediterranean Sea, and the Balkan Peninsula, including Slovenia, Croatia, Bulgaria, and Romania. Central Europe refers to

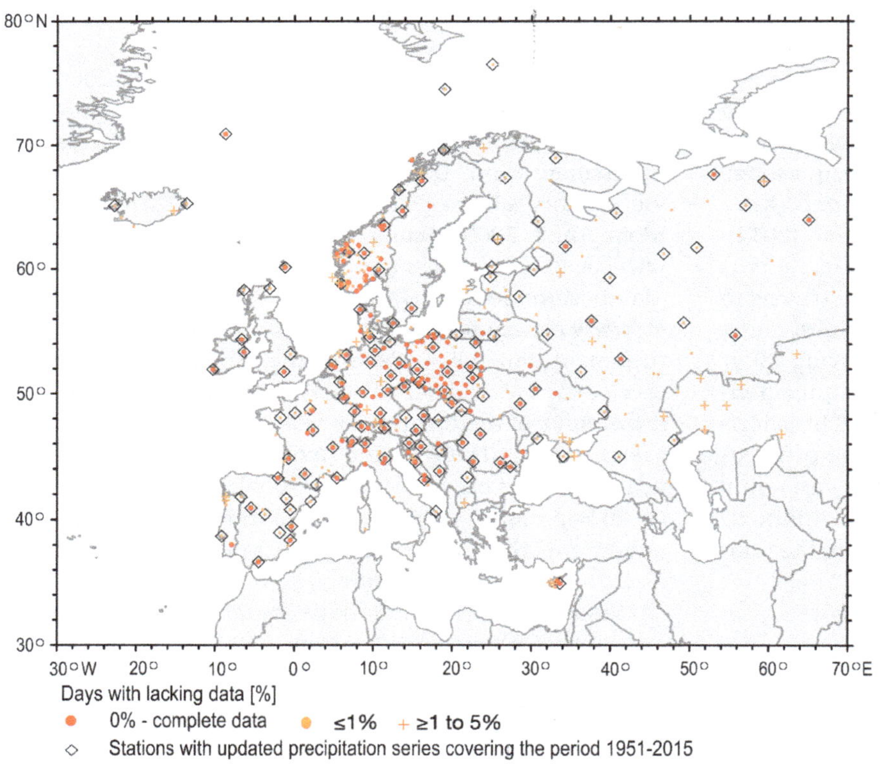

Fig. 2. Study region and the network of stations used for trend detection

Eastern Germany, Poland, the Czech Republic, Slovakia, Austria, and Hungary. Northern Europe refers to Norway, Sweden, Finland and Iceland, and Eastern Europe was defined to cover the remaining areas beyond the eastern borders of Finland and Poland and the northern and eastern borders of Romania.

2.2. Criteria for identifying extreme precipitation events

In the various climatic types of Europe, precipitation is characterized by large differences in the total, frequency, and annual peak timing. The scale of these differences between precipitation regimes influences daily precipitation extremes, and thus, they must be defined on a regional scale. One method to accomplish this is the use of Intergovernmental Panel on Climate Change (IPCC)-recommended percentile-based extreme precipitation indices (IPCC 2001, 2007), which are more useful than absolute values for regional spatial comparisons. This study defined extreme precipitation as a daily amount exceeding the 95th percentile (95p). The threshold value equivalent to 95p was calculated separately for each month and station using an empirical distribution of daily precipitation ≥1 mm for the standard period (1961 to 1990). Precipitation extremes thus defined do not always cause serious material loss, but their frequency is high enough to obtain reliable results from statistical analyses. The percentile-based indices are less suitable for direct assessment of the impacts, but may provide useful indirect information that is relevant to studying the impacts and adaptation. Trends in the days with precipitation exceeding 95p of the daily amount are relevant for assessing changes in the demands on drainage and sewage systems at different locations.

Using the 95p threshold values, 3 precipitation extreme indices were calculated: the days with extreme precipitation (95pNoD), the extreme precipitation total (95pT), and the average daily extreme precipitation amount (95pINT) (Table 1). These indices characterize the frequency, amount and intensity of extreme precipitation events. A seasonal approach was adopted using spring (MAM), summer (JJA), autumn (SON), and winter (DJF). The results obtained by this study based on a single criterion (i.e. the 95p) provide general information about the spatial variability in the trends of moderate and the highest precipitation extremes. Research on Poland and Germany suggests similarity in the spatial distributions of the extreme precipitation index trends based on various percentile thresholds (Łupikasza 2010, Łupikasza et al. 2011). Besselaar et al. (2013) also found that trends in the highest precipitation extremes were consistent with those in more common events identified using the descriptive indices of extremes that occur, on average, several times per year.

2.3. Trend estimation

The trend magnitude was calculated using a linear regression of the least squares method and expressed as a percentage change during the 30 yr period relative to the mean value of the index for 1961 to 1990. The statistical significance of the trends was tested using the non-parametric Mann-Kendall method (Mann 1945, Kendall 1970), which is a robust, rank-based test that, unlike the conventional Student's *t*-test, does not depend on the statistical distribution of the analysed parameter (Schmidli & Frei 2005). A combination of linear regression and the Mann-Kendall test is commonly used for extreme precipitation trend analysis (Brunetti et al. 2000,

Table 1. Definitions of extreme precipitation indices

Symbol	Name	Definition
95pNoD	Count of very wet days	Number of days with RRij ≥ RRwn95, where: *RRij* is the daily precipitation amount on day*i* in period *j*, *RRwn95* is the 95th percentile of precipitation on wet days in the base period *n* (1961–1990)
95pTOT	Precipitation total due to very wet days	95pTOT = sum (RRwj), where RRwj >RRwn95, where: *RRwj* is the daily precipitation amount on a wet day *w* (RR >1 mm) in period *j*, *RRwn95* is the 95th percentile of precipitation on wet days in the base period *n* (1961–1990)
95pINT	Simple daily intensity index: mean precipitation amount on a wet day	95pINT = sum (RRwj) / W, where: RRwj is the daily precipitation amount on wet day w (RR >1 mm) in period *j*, W is the number of wet days in period *j*

Hundecha & Bárdossy 2005, Schmidli & Frei 2005, Norrant & Douguédroit 2006, Ramos & Martinez-Casanovas 2006, Rodrigo & Trigo 2007, Pavan et al. 2008, Gajić-Čapka & Cindrić 2011, Karagiannidis et al. 2012, Jones et al. 2013) and was adopted in this manuscript. Using the most commonly applied methods of trend analysis allows for comparison of the results on continental and regional scales. The significance information from the Mann–Kendall test can be used interchangeably with the *t*-test in practical applications, as both methods generally deliver almost identical results (Önöz & Bayazit 2003, Soro et al. 2016). Weaker criteria of trend significance were applied because of the high variability in precipitation, particularly extreme precipitation (Rapp 2000, Hänsel 2009, Lupikasza 2010, Łupikasza et al. 2011). Three significance levels were considered: $\alpha = 0.2$ (weakly significant trends), $\alpha = 0.1$ (moderately significant trends), and $\alpha = 0.05$ (significant trends).

Extreme precipitation trends were calculated for each moving 30 yr period, producing 27 to 28 short-term trends for each station (or 36 short-term trends for stations with updated series), depending on the season. This approach allowed for the assessment of the temporal consistency of trend directions and the long-term variability in the frequency of significant trends. Moreover, the results of the trend analysis are very sensitive to the beginning and end dates and do not generally reflect long-term climate trends (Hartmann et al. 2013). Therefore, the derived trend information is not strongly dependent on extreme high or low values at the beginning or end of the time series (Łupikasza et al. 2011). Short-term trend analysis provides information on the general direction of change while accounting for its fundamental feature of great variability and is a good complement to long-term analysis. The trends for every 30 yr period for 126 to 128 stations with updated data are included in Fig. S1.

The long-term changes in the frequency of extreme precipitation trends were assessed using the stations with statistically significant trends in each of the 30 yr periods. These changes were expressed as percentages of all stations analysed in a season and in the 30 yr period. Additionally, for each 30 yr period, the field significance of extreme precipitation trends was tested using a global test based on the false discovery rate (FDR) method described by Wilks (2006). In this method, a network is declared significant if at least 1 local null hypothesis satisfies the bracket condition in Eq. (1). The parameter $p_{(j)}$ denotes the *j*th smallest of the *p* local values, and *K* is the number of local tests. The values of the local test were rejected when $p_{(j)}$ did not exceed the following:

$$p_{\mathrm{FDR}} = \max_{j=1,\ldots,k} \left[p_{(j)} : p_{(j)} \leq \alpha_{\mathrm{global}}(j/K) \right] \quad (3)$$

The critical value *p* for the significant field was calculated using $\alpha_{\mathrm{global}} = 0.2$, which was the highest significance level for local precipitation trends adopted in this study. The FDR approach produces good results even when the local tests results are correlated with each other (Wilks 2006). The field significance was tested using all stations and separately for the groups of stations with positive and negative trends.

The temporal consistency in the extreme precipitation trend direction was determined using the statistical significance and direction of the trends for all moving 30 yr periods. This parameter is expressed as a percentage of the 30 yr periods (T) during which these trends were statistically significant ($\alpha \leq 0.2$). The following criteria were used: (1) inconsistent trends: T < 25% or the frequencies of significant positive and negative trends were the same; (2) weakly consistent significant trends: 25% ≤ T < 50% and >60% of significant trends had the same direction; (3) consistent significant trends: 50% ≤ T < 75% and >60% of significant trends had the same direction; and (4) strongly consistent significant trends: T ≥ 75% and >60% of significant trends had the same direction.

Within the categories of weakly consistent, consistent, and strongly consistent trends, the stations with predominantly significant positive trends (i.e. positive trends constituted >60% of the significant trends) were differentiated from those with a majority of significant negative trends (i.e. negative trends constituted >60% of the significant trends) and from those with 'neutral' trends (the frequencies of both positive and negative trends were 40 to 60%). The inconsistent trends category also included stations with T > 25% and similar frequencies of significant positive and negative trends (i.e. between 40 and 60%).

3. RESULTS

3.1. General overview of extreme precipitation trends

The frequencies of statistically significant trends in the seasonal indices of extreme precipitation are shown in Fig. 3. The frequencies were calculated as a ratio of the statistically significant 30 yr trends to the 30 yr trends. The latter were derived by multiplying the stations at which the given season's data were available (i.e. 363 for MAM, 364 for JJA and SON, and 345 for DJF) by the moving 30 yr periods from December 1950 to February 2008 (i.e. 27 MAM, JJA,

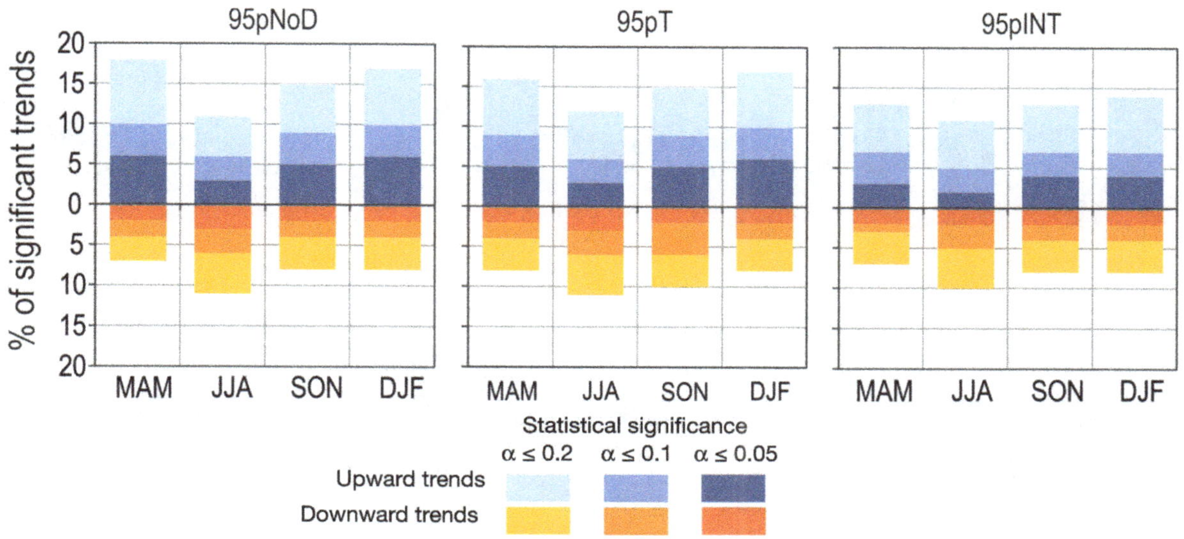

Fig. 3. Frequency of significant trends in extreme precipitation indices in the period December 1950 to February 2008. 95pNoD: number of days with extreme precipitation, 95pT: extreme precipitation totals, 95pINT: extreme precipitation intensity

SON, and 28 DJF). Depending on the season, there were 9660, 9801 or 9828 trends. These frequencies should be interpreted carefully given the temporal correlation between the overlapping 30 yr periods and the spatial correlation among the stations. Henceforth, UTr and DTr denote upward and downward trends, respectively.

Most of the trends for the overlapping 30 yr periods were insignificant at the 0.2 significance level for all of the extreme precipitation indices. Depending on the season, significant trends constituted 20 to 25% of all 30 yr trends. The seasonal variability in their frequency was very low (i.e. ≤3%). In most seasons, except summer, UTr dominated the significant trends. The frequency of significant UTr was the greatest in winter (14 to 17%) and spring (13 to 18%; Fig. 3). In summer, there were almost as many significant UTr as DTr (11 to 12%; Fig. 3). Compared with other seasons, summer had the fewest UTr and the most DTr among significant trends.

For reference, linear trends were also calculated for the entire 1951 to 2008 period. The percentage of significant long-term term trends presented in Table 2 varies from 18% (14% UTr + 4% DTr for 95pINT in JJA) to approximately 40% (33% UTr + 8% DTr for 95pNoD and 33% UTr + 9% DTr for 95pT in DJF), considering a 0.2 significance level. The disproportion between the long-term significant UTr and DTr was larger (e.g. 34% UTr and 4% DTr in the 95pNoD in MAM) than that detected for the short-term trends at every significance level. However, the seasonal patterns of changes were similar (Fig. 3, Table 2).

Table 2. Percent of stations with statistically significant long-term trends (1951–2008) in extreme precipitation indices. 95pT: extreme precipitation totals, 95pNoD: number of days with extreme precipitation; 95pINT: extreme precipitation intensity; MAM: spring; JJA: summer; SON: autumn; DJF: winter; UTr: upward trend; DTr: downward trend

Significance	95pT				95pNoD				95pINT			
level	MAM	JJA	SON	DJF	MAM	JJA	SON	DJF	MAM	JJA	SON	DJF
UTr												
0.2	31	14	26	33	34	13	25	33	19	14	20	19
0.1	21	8	14	25	23	9	14	25	11	7	12	12
0.05	13	4	10	18	15	4	8	18	7	3	7	7
DTr												
0.05	1	1	1	3	1	2	1	3	2	1	2	2
0.1	2	3	2	5	2	4	1	5	4	2	4	4
0.2	5	5	4	8	4	6	5	9	5	4	6	7

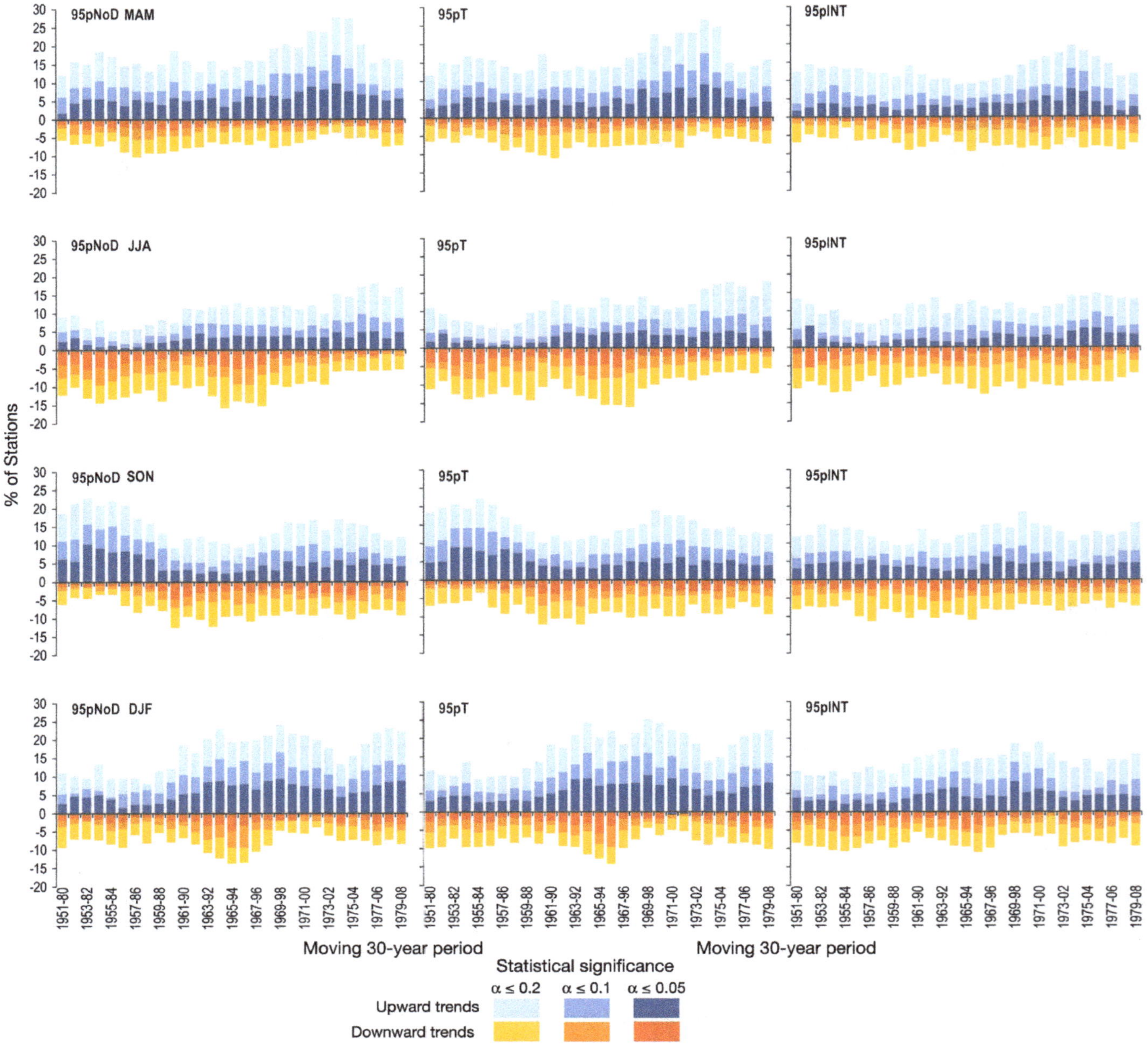

Fig. 4. Percentage of stations with significant seasonal trends in extreme precipitation indices for moving 30 yr periods within the period December 1951 to January 2007. 95pNoD: number of days with extreme precipitation, 95pT: extreme precipitation totals, 95pINT: extreme precipitation intensity

3.2. Long-term variability in the frequency of significant short-term extreme precipitation trends

The percentage of stations with statistically significant seasonal trends in extreme precipitation for the 30 yr moving periods is shown in Fig. 4. Due to the uneven distribution of the stations, these figures reflect more accurately the changes in the trend frequencies in areas with the densest station coverage. For the series updated to 2015, see Fig. S1 in the Supplement. Pearson correlation coefficients were calculated between the updated series (126 to 128 stations) and the series created based on 346 to 363 stations for 1951 to 2007 (Table 3). All the correlations were statistically significant at $\alpha < 0.01$, except for upward trends in extreme precipitation intensity in spring (MAM).

In every season, the frequency of significant trends changed perceptibly during the study period (Fig. 4).

Table 3. Correlation coefficients between percentage of stations with positive (UTr) and negative (DTr) trends in extreme precipitation indices calculated on the base of 366- stations and 127 stations. S: statistics, PC: Pearson correlation coefficient, p: statistical significance

	95pNoD		95pT		95pINT	
	PC	p	PC	p	PC	p
MAM						
UTr	0.859	<0.0001	0.791	<0.0001	0.351	0.067
DTr	0.806	<0.0001	0.723	<0.0001	0.766	<0.0001
JJA						
UTr	0.888	<0.0001	0.885	<0.0001	0.775	<0.0001
DTr	0.774	0.0001	0.798	<0.0001	0.483	0.009
SON						
UTr	0.612	0.001	0.568	0.002	0.671	<0.0001
DTr	0.583	0.001	0.451	0.016	0.783	<0.0001
DJF						
UTr	0.910	<0.0001	0.927	<0.0001	0.770	<0.0001
DTr	0.851	<0.0001	0.764	<0.0001	0.780	<0.0001

The changes in the trend frequencies in all indices were similar over the long term, but the patterns varied seasonally. In most of the 30 yr periods and seasons, UTr were more frequently significant. However, significant 30 yr trends of either sign occurred at a maximum of 30% of the stations. A higher prevalence of UTr than DTr in most seasons was also found in the long-term trends, the frequency of which is included in Table 2. In this section, the first year of the reference period (e.g. 1955 to 2000) is the beginning year of the first 30 yr period (e.g. 1955 to 1984), and the last year is the ending year of the last 30 yr period (e.g. 1971 to 2000).

In spring, significant UTr in the extreme precipitation indices were typically twice as common as DTr (Fig. 4). From 1951 to 1997, UTr occurred at 12 to 16% of stations on average, depending on the index. The percentage of the UTr peaked during 1973 to 2002 (Fig. 4). From the mid-1970s, the frequency of UTr dropped from >25% to approximately 10 to 15% of stations, depending on the index. The frequency of DTr rarely exceeded 10% of stations, with a maximum between 1957 and 1990 and a minimum between 1972 and 2002 (Fig. 4).

In summer, the frequency of DTr was greater than in any other season and was similar to or greater than the UTr frequency for much of the study period. However, from 1967 onward, the percentage of stations with DTr gradually decreased. During the study period, the stations with UTr in the 95pNoD and 95pT increased from 5% in the second half of the 1950s to 18% of stations at the end of the research period. The smallest change was found in the frequency of the 95pINT UTr.

In autumn, UTr occurred at 13 to 15% of stations on average, depending on the index. These numbers were much higher in the 30 yr periods between 1951 and 1985 and much lower between 1960 and 1994 (Fig. 4). A significant DTr of all indices occurred at 8% of stations on average. Between 1951 and 1984, DTr were particularly rare (approximately 4 to 6% of stations) and subsequently continued to oscillate near that frequency until the end of the period (Fig. 4).

The frequency of winter UTr in the 95pNoD and 95pT increased markedly from 10% of stations in the 30 yr periods during 1951 to 1989 to 24 to 25% of stations between 1958 and 1993. No clear changes in the frequency of DTr (8-9% of stations) were found in the 30 yr periods between 1951 and 1990. In subsequent 30 yr periods, quite clear maxima occurred between 1965 and 1995, and minima occurred between 1969 and 2001.

The field significance analysis revealed the most important changes in winter extreme precipitation. The percentage of 30 yr periods with at least 1 local null hypothesis satisfying the condition in Eq. (1), indicating significant fields of precipitation trends, is shown in Table 4. In winter, the fields of precipitation trends in the 95pNoD and 95pT were significant during 69 and 79% of the 30 yr periods, respectively, when considering all local tests; these values were even higher in the group of positive trends: 79 and

Table 4. Percentage of 30 yr periods with significant field trends in extreme precipitation calculated using the False Discovery Rate method (FDR). 95pT: extreme precipitation totals, 95pNoD: number of days with extreme precipitation, 95pINT: extreme precipitation intensity, MAM: spring, JJA: summer, SON: autumn, DJF: winter

Seasons	All stations			Stations with positive trends			Stations with negative trends		
	95pNoD	95pT	95pINT	95pNoD	95pT	95pINT	95pNoD	95pT	95pINT
MAM	46	36	11	57	50	11	11	21	21
JJA	14	4	14	21	14	7	4	4	14
SON	14	21	18	32	54	25	14	4	4
DJF	69	79	28	79	86	41	14	10	28

Table 5. Percent of stations with temporally consistent trends in extreme precipitation indices in Europe (December 1950 to February 2008). UTr: upward trends; DTr: downward trends; INCTr: inconsistent trends; T: percent of 30 yr periods with significant trends in extreme precipitation; 95pT: extreme precipitation totals; 95pNoD: number of days with extreme precipitation; 95pINT: extreme precipitation intensity; MAM: spring; JJA: summer; SON: autumn; DJF: winter

	UTr			INCTr	DTr		
	T ≥ 75%	50% ≤ T < 75%	25% ≤ T < 50%	0% ≤ T < 25%	25% ≤ T < 50%	50% ≤ T < 75%	T ≥ 75%
95pT							
MAM	0.8	6.6	22.9	58.9	8.3	2.5	0
JJA	0	4.4	12.7	64.7	13.8	4.1	0.3
SON	0.5	5.0	23.4	59.0	9.1	3.0	0.3
DJF	0.9	8.1	19.7	61.4	9.0	0.9	0
95pNoD							
MAM	1.1	8.6	22.5	60.0	6.6	2.2	0
JJA	0.3	4.4	12.4	66.1	12.9	3.6	0.3
SON	0.5	4.4	22.6	62.0	8.8	1.4	0.3
DJF	1.5	7.8	19.1	61.4	7.6	2.6	0
95pINT							
MAM	1.1	1.9	22.1	63.8	8.9	2.2	0.1
JJA	0	3.3	14.0	65.9	13.2	3.3	0.3
SON	0.3	4.1	17.9	63.9	11.0	2.8	0
DJF	0.6	3.5	16.8	67.8	9.6	1.7	0

86% of the 30 yr periods, respectively. In the other seasons, the frequency of significant fields was much lower: 46% (95pNoD) and 36% (95pT) in spring and no more than 15% in other seasons. The significant fields of positive precipitation trends were much more frequent than those of negative trends (Table 4).

3.3. Temporal consistency of the significant trends in extreme precipitation indices

The direction of statistically significant short-term trends varied, as demonstrated in Section 4.2. The predominance of short-term trends with the same direction is evidence of their temporal consistency. In contrast, when the amounts of significant short-term trends with opposite directions are similar or when statistically significant trends are rare, the consistency of the trend directions is weak. The temporal consistency of extreme precipitation trend directions is quantified in Table 5 and illustrated in Figs. 5 & 6. In this study, the term 'coherent trends' is used when discussing both consistent trends and strongly consistent trends. The term 'consistent trends' is used when all categories of trend consistency are discussed. When referring to a particular class of trend consistency, the term 'category' is added, or the full name of the class is used (e.g. weakly consistent trends).

The directions of statistically significant short-term trends in extreme precipitation indices were coherent at 5 to 12% of stations, depending on the season and index (Table 5). In most seasons except summer, the consistent UTr in the 95pNoD and 95pT were more than twice as frequent as the DTr. In summer, stations with consistent UTr and DTr had nearly equal frequencies (17 to 18% of stations).

In spring, the trend directions were consistent at 36 to 41% of stations when evaluating all categories of trend consistency, whereas they were coherent at 5 to 12% of stations, depending on the index. Most of these trends were positive and occurred primarily in Western and Central Europe, reaching a latitude of approximately 44° N, and in many Northern European stations. UTr in the 95pNoD were also consistent in the southern area of Eastern Europe (Fig. 5). Weakly consistent DTr were found in all extreme precipitation indices in Southern Europe, particularly in the Iberian Peninsula and west of the Black Sea. In Eastern Europe, most trends were inconsistent, although in the case of the 95pNoD, UTr were predominant amongst the consistent trends (Fig. 5).

In summer, consistent DTr, many of which were coherent, were primarily found in Central Europe, including Germany, Poland, and stations south of Poland and into the Balkan Peninsula (Table 5). Despite the predominance of DTr in Central Europe, many stations had a weakly consistent UTr. Coherent UTr in the 95pT and 95pNoD occurred on the south-

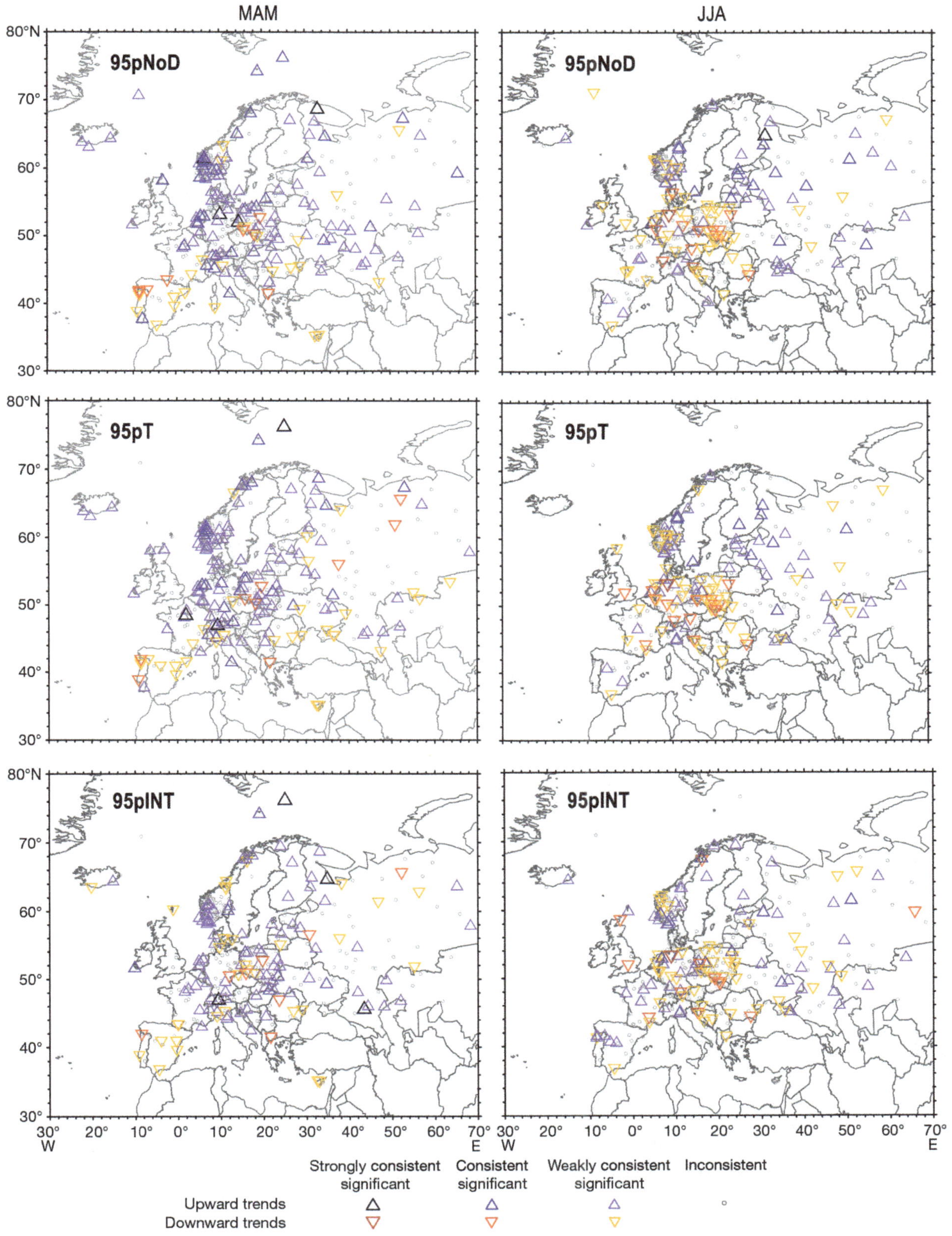

Fig. 5. Temporal consistency of short-term trends in extreme precipitation indices for spring (MAM) and summer (JJA) in Europe in the period 1951 to 2007. Criteria for trends consistency/inconsistency are described in the 'Methods'. 95pNoD: number of days with extreme precipitation, 95pT: extreme precipitation totals, 95pINT: extreme precipitation intensity

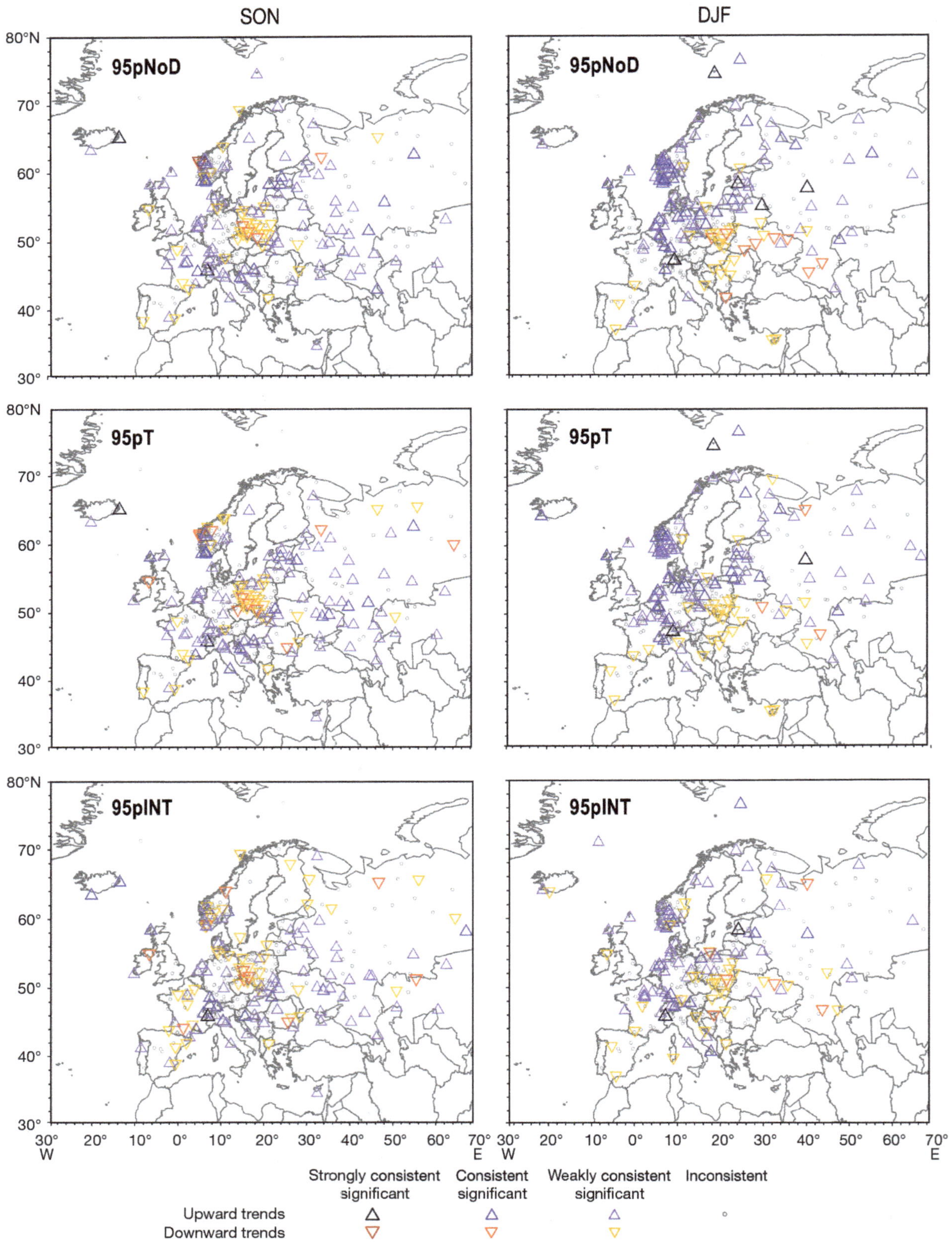

Fig. 6. Temporal consistency of short-term trends in extreme precipitation indices for autumn (SON) and winter (DJF) in Europe in the period 1951 to 2007. See Fig. 5 for more details

ern Scandinavian Peninsula and in the northwest area of Eastern Europe (Fig. 5). Elsewhere in Eastern Europe, both UTr and DTr occurred, but upward trends predominated in the 95pNoD and 95pT.

In autumn, coherent decreases were observed primarily in a small region of Central Europe, and consistent decreases were detected at single stations in France, the Iberian Peninsula, the Balkan Peninsula, and the southern tip of Norway. In other areas of Europe (between France and the Balkan Peninsula, the southwestern tip of Norway and Eastern Europe), weakly coherent UTr, particularly in the 95pNoD and 95pT, prevailed. At certain stations (e.g. Northern Europe), despite weakly coherent increases in the 95pNoD and 95pT, the extreme precipitation intensities declined because of different rates of change in these characteristics at individual stations (Fig. 6).

In winter, the spatial distribution of trend directions followed the clearest pattern of all the seasons: areas with consistent increases were clearly distinguished from areas with consistent decreases in extreme precipitation indices. UTr were predominant among the temporally coherent trends (Table 5) and were found in the eastern region of Western Europe between the eastern border of France and Poland, in north-central Europe, and Eastern Europe and the Scandinavian Peninsula. Consistent and weakly consistent DTr categories appeared in Southern Poland and extended south to the Balkan Peninsula, east to Ukraine, and between the Black and Caspian Seas. Over the Iberian Peninsula and in Southern France, the trends were weak and mostly downward. UTr were consistent in most of Eastern Europe, except in the southern region, where coherent DTr were found, particularly in the 95pNoD (Fig. 6).

3.4. Average magnitude of extreme precipitation trends

The average trend magnitude was calculated for each station as an arithmetic mean of all 30 yr trends. The trend magnitude was then expressed as a percentage of the index value for the standard period (1961 to 1990). The average trend magnitude, which is a weighted trend magnitude, provides information on prevailing short-term extreme precipitation trends (direction and magnitude) during the research period, whereas the trend calculated during the entire period is more sensitive to the values at the end of the series and may change if the research period is extended or changed. As a result, the long-term trend direction may differ from the direction of the average trend. The weighted magnitudes of trends for all stations are shown in Figs. 7 & 8. For magnitudes at stations with coherent trends, see Figs. S2 & S3 in the supplement.

The directions of statistically insignificant trends that were included in analysis of the average trend magnitude fit the seasonal spatial distributions identified during the trend-coherency analysis (compare Fig. 5 with Fig. 7 and Fig. 6 with Fig. 8). In every season, the scale of change was the greatest in the 95pNoD.

In spring, the magnitudes of the relative UTr in the 95pNoD and 95pT exceeded 25% in vast areas of Western, Central and Northern Europe and in the southern region of Eastern Europe (95pNoD). Within this area, strong UTr (>50% of the average) were observed over the Scandinavian Peninsula, between the Frisian Islands and the Gulf of Finland, and between the Alps and the Great Hungarian Plain (Fig. 7). DTr with magnitudes >25% were recorded in Southern Europe, in the Sudeten and Ore Mountains, and in western and northern areas of Eastern Europe (Fig. 7). The magnitudes of these DTr exceeded 50% only in the case of the 95pNoD in the Iberian Peninsula and at single stations in southern Poland and the eastern Mediterranean region.

In summer, the UTr that occurred throughout the continent were mostly weak. Their relative magnitudes were ≤25% in Western Europe. Stronger trends (>25%) were recorded primarily in Eastern Europe. The DTr were concentrated in the western part of the continent (to the 26° meridian [Fig. 7]), and their relative magnitudes exceeded 25% in Poland, Germany and Southern Europe. The most rapid decrease in the 95pNoD was observed in Southern Europe (Fig. 7).

In autumn, the strongest UTr in the 95pNoD (with relative magnitudes >50%) were detected south to the 52° parallel, particularly in southwest Eastern Europe (Fig. 8). The UTr in the 95pT followed a similar distribution pattern, but the magnitudes were lower. An increase in the 95pINT was even weaker. The strongest DTr in the 95pNoD of ≤25% were found in Central Europe and in the western region of Southern Europe (Fig. 8).

In winter, the magnitudes of UTr were >50% in Northern Europe, particularly on west-facing mountain slopes in the southwest Scandinavian Peninsula and between the Frisian Islands and the Gulf of Finland (Fig. 8). DTr with a relative magnitude >50% were rare (5% of stations) and were only found in the 95pNoD. The strongest DTr of the 95pNoD were recorded in southern Poland and continued southward to the Balkan Peninsula (Fig. 8).

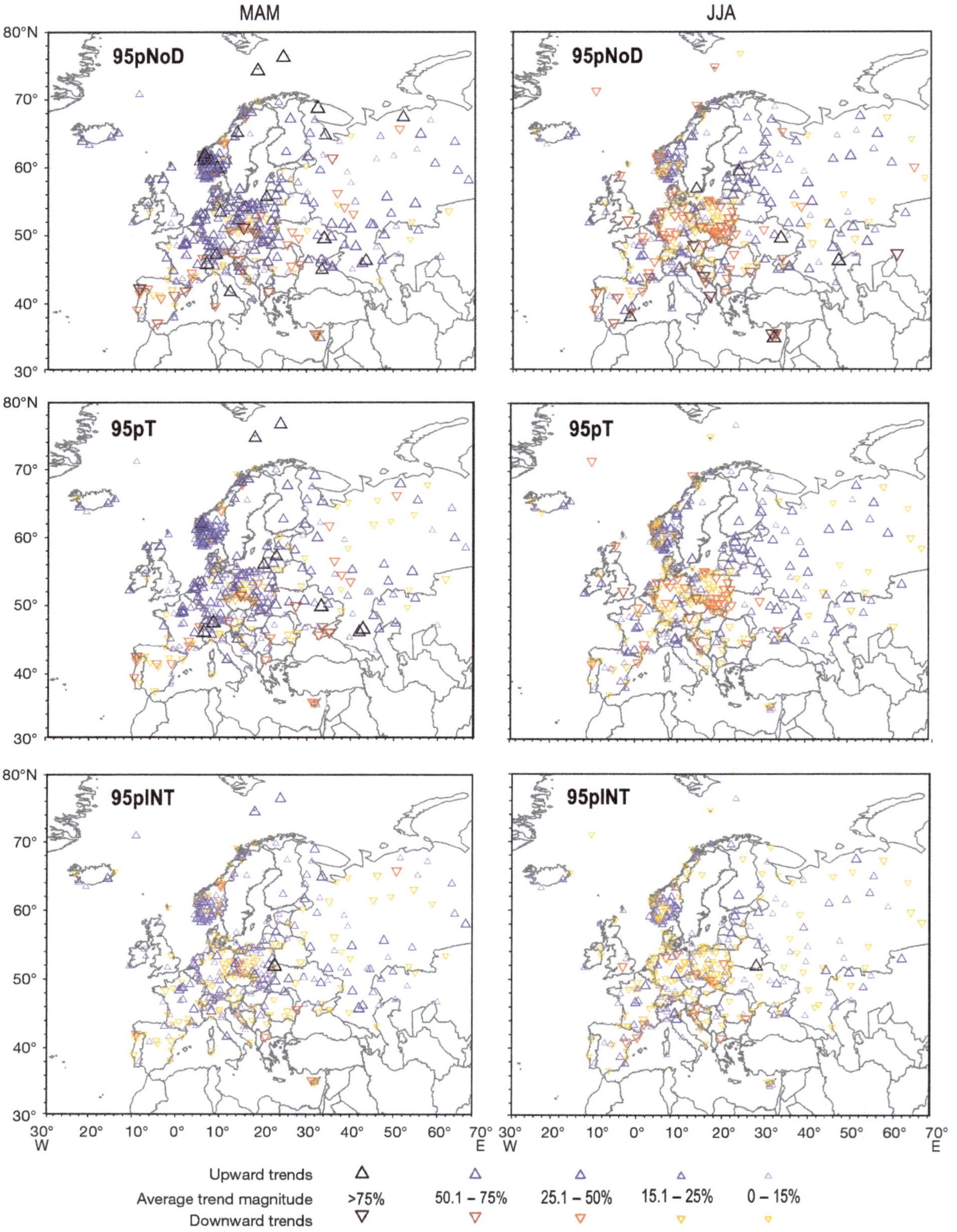

Fig. 7. Average trend magnitudes in extreme precipitation indices for spring (MAM) and summer (JJA), using arithmetical average from 30 yr trend magnitudes within December 1951 to February 2008. Average trend magnitude is expressed as a percentage of average index value in the period 1961 to 1990. 95pNoD: number of days with extreme precipitation; 95pT: extreme precipitation totals; 95pINT: extreme precipitation intensity

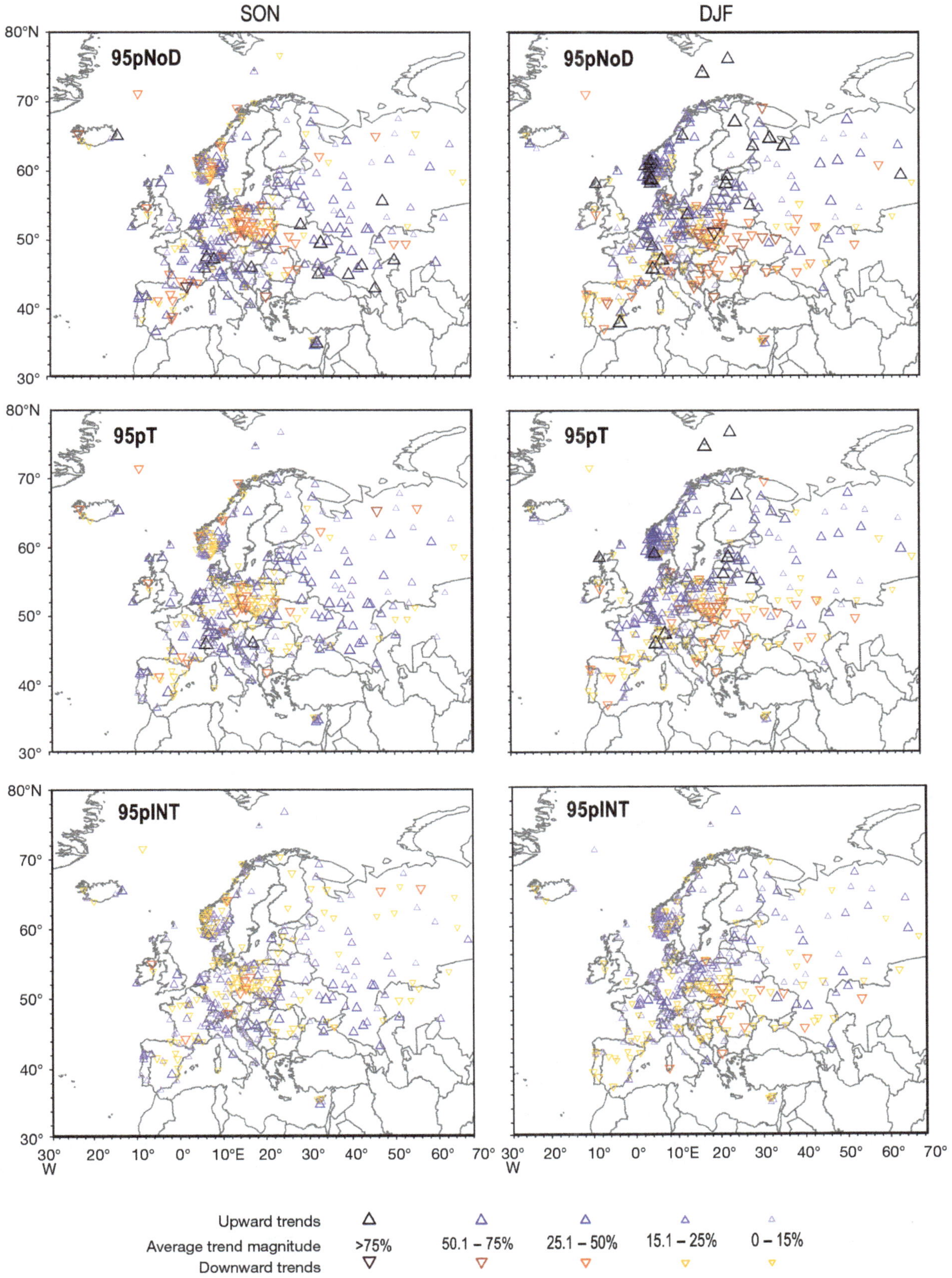

Fig. 8. Average trend magnitudes in extreme precipitation indices for autumn (SON) and winter (DJF). See Fig. 7 for more details

4. DISCUSSION

Short-term trends in extreme precipitation indices were investigated to establish their spatio-seasonal patterns in Europe based on data from 346–366 meteorological stations. The analysis was updated to 2015 for 126–127 select stations (depending on season) using available and good-quality daily precipitation data.

An analysis of precipitation trends should not eliminate statistically insignificant trends. The research on the average magnitudes in this study included insignificant trends, and their directions fit the pattern that was determined using only the significant trends.

In Europe, significant trends in extreme precipitation indices were very rare during the research period, and this may continue into the future because their frequencies did not change noticeably in consecutive 30 yr periods. Among rarely occurring significant trends, upward trends were predominant in all seasons except summer, and thus, an increase can be recognized as a noticeable feature in a long-term extreme precipitation course. Similar frequencies of significant UTr and DTr trends in summer are likely due to the enhanced prominence of local factors (more frequent free convection) that overlap with synoptic-scale circulation influences. However, the predominance of summer UTr at the end of the research period appears to be concordant with near-term projections, which depict increases in both winter and summer precipitation over the mid-latitude regions (Hartmann et al. 2013).

The prominent seasonality and strong regional variations detected in the pattern of long-term trend frequencies and their consistency are caused by the complicated impacts of both local (e.g. topography) and macro- and meso-scale (atmospheric circulation) factors that trigger precipitation formation and cause variability. The impact of synoptic circulation changes on precipitation trends varies among regions and months. The exact location of the strongest influence of synoptic circulation changes varies with the time of year (Fleig et al. 2015). Casanueva et al. (2014) link the seasonal behaviour of trends in extreme precipitation with various atmospheric teleconnection patterns: the Scandinavian and East Atlantic patterns and El Niño/Southern Oscillation events in spring and autumn and the Atlantic Multidecadal Oscillation during the entire year. However, one of the most prominent macroscale factors that impacts variability in European precipitation is the North Atlantic Oscillation (NAO) (Qian et al. 2000). Changes in the zonal air flow associated with swings in the NAO index lead to changes in the transport of humidity and the occurrence of precipitation and its amount across Europe, including extreme precipitation. The relations are the strongest in winter (e.g. Zorita et al. 1992, Hurrell & Van Loon 1997, Dickson et al. 2000, Rodríguez-Puebla et al. 2001, Haylock & Goodess 2004, Trigo et al. 2004, Vicente-Serrano et al. 2009, Casanueva et al. 2014, de Lima et al. 2015). The highest order in the spatial distribution of temporally coherent trends and their magnitudes in winter can be linked to the NAO impact, which is strongest in winter. Hoy et al. (2014) and Fleig et al. (2015) also found a strong linkage between European precipitation and synoptic scale circulation described by Grosswetterlagen circulation types, particularly in the winter months. This finding is confirmed by the pattern of precipitation trends, with an increase of all extreme precipitation indices in the northern part of the continent and a decrease in the south. Modelling studies show that the UTr of extreme precipitation in winter are expected to continue in the future (Hartmann et al. 2013).

Due to the larger sample of stations from central and western regions of the continent analysed in this study, it was possible to complement the results of Moberg et al. (2006) and conclude that significant trends in summer extreme precipitation were negative at stations in Central and Western Europe. Unlike earlier studies that suggested a change in the direction of long-term extreme precipitation trends in Southern Europe (Rodrigo & Trigo 2007, Beguería et al. 2011, Rodrigo 2010), temporally consistent DTr in spring extreme precipitation were found, primarily on the Iberian Peninsula. According to earlier studies (Piervitali et al. 1998, Brunetti et al. 2000, Pavan et al. 2008), the long-term autumn trends in Northern Italy were negative during 1951 to 2004, whereas this study revealed that most significant short-term trends were positive during this period. Differences were also noted in Croatia, where long-term trends were statistically non-significant (Gajić-Čapka & Cindrić 2011), whereas in this study UTr (SON) or DTr (JJA) were predominant among the short-term trends. The enormous scale of precipitation variability, which is linked to the modification of circulation factors by complex landforms in this area of Europe, likely explains these disagreements. The strong fluctuations in precipitation result in a lack of significant long-term trends. This effect is particularly strong in autumn and summer, when precipitation is often linked to free convection.

The seasonal variability in the extreme precipitation trends is a consequence of the seasonality inherent in the processes of precipitation formation. Extremes in precipitation are also associated with the coincidence of particular weather patterns (e.g. Lavers et al. 2011). Currently, adequate understanding of what factors control the return time and persistence of such rare events is lacking (Hartmann et al. 2013). Understanding the effects of these mechanisms could be improved by finding patterns in the spatial distributions of trends or areas or groups of stations with consistent changes in extreme precipitation. Despite the spatial variability in the distribution of trends in each season, such areas can be identified. Their existence can reflect the linkage between long-term variability and changes in extreme precipitation and atmospheric circulation, the latter of which is a leading natural factor in climate variability. As mentioned above, the NAO plays an important role in climate variability over Europe and leaves a significant signature in precipitation. According to Qian et al. (2000), however, the NAO does not appear to be the most important signal of atmospheric variability in precipitation during seasons other than winter over the continent. In contrast, some studies suggested that the strength of the relationships between the NAO and the surface climate depends on the shift in sea-level pressure anomalies associated with this teleconnection pattern (Jung & Hilmer 2001, Jung et al. 2003, Beranová & Huth 2007, Vincente-Serrano et al. 2009). The location of the NAO pressure centres has also been reported to undergo secular changes (Ulbrich & Christoph 1999), and thus, the common assumption of a constant NAO spatial structure has several limitations in explaining the role of the NAO in climate variability (Jung et al. 2003, Beranová & Huth 2007). Further investigations should include shifts in the NAO configuration. Regional studies indicate that, in certain parts of Europe, extreme precipitation events are clearly related to local circulation types (Ustrnul & Czekierda 2001).

Extreme precipitation intensity on daily and subdaily scales was found to be linked to an increase in the atmospheric water-holding capacity associated with a temperature increase under warmer climates (Betts & Harshvardhan 1987, Trenberth et al. 2003, Held & Soden 2006, Scherrer et al. 2016). The applicability of these relations to climate change is not straightforward but can shed light on the physical mechanisms that are important for extreme rainfall (Drobinski et al. 2016). The rate of increase in extreme daily precipitation associated with atmospheric warming may be 7% $°C^{-1}$ globally, as determined from Clausius-Clapeyron relations (Allen & Ingram 2002, Pall et al. 2007, Kharin et al. 2007). However, these relations are seasonal and not straightforward; regional specificities are apparently strong drivers (Berg et al. 2009), the most important of which is the dynamic contribution of orography (Drobinski et al. 2016). Berg et al. (2009) found that extreme daily precipitation intensity increases with the daily surface air temperature in winter, but decreases as the daily surface air temperature increases in summer. In summer, the availability of moisture — not the atmosphere's capacity to hold this moisture — is the predominant factor at the daily timescale.

5. CONCLUSION

This study of extreme precipitation in Europe revealed that significant trends in extreme precipitation constituted approximately 25–30% of all short-term trends, and that this percentage did not change substantially in subsequent 30 yr periods. Furthermore, in the majority of the 30 yr periods, the frequency of UTr was clearly higher than that of DTr, except in summer. However, in some recent years, upward trends were also predominant in summer. Therefore, increases in the frequency and total were recognized as characteristic features of extreme precipitation changes in Europe, particularly in winter. This finding concerns stations with significant trends, and is supported by local trend analysis and field significance of precipitation extremes. To the best of the author's knowledge, this study also documented for the first time that significant trend directions were temporally consistent at 6–13% of stations and perceivably consistent at 28–35% of stations, depending on the season and index. Furthermore, the UTr were more temporally coherent than the DTr.

Clear seasonality was apparent in the spatial distribution of extreme precipitation trend coherency and magnitude. This seasonality — resulting from changes throughout the year in the strength of the impact of local and mesoscale factors on precipitation occurrence and totals — was characterized as follows:

- Winter: The winter season showed the most consistent spatial distribution of trends. The UTr occurred in the northern area of the continent, whereas the DTr occurred in the southern area of the continent.
- Summer: A seasonal peak in the frequency of DTr occurred mainly in Central and Western Europe, with the most coherent DTr in the central region of the continent. Less-frequent UTr were found in Eastern Europe, with the most coherent UTr in western

and southern Eastern Europe and in most of the Iberian Peninsula except the southern region, where coherent DTr were predominant.

• Spring: In most of the 30 yr periods, the frequencies of UTr in Central, Western and Northern Europe were twice as high as those of DTr in western Southern Europe and the central-European Sudeten and Ore mountain ranges.

• Autumn: A clear majority of UTr occurred across the continent, despite the prevalence of coherent and strong DTr in Central Europe, Norway, and western Southern Europe. The strongest UTr were found in the northern region of Southern Europe.

Climatological research on extreme precipitation trends can be a useful tool to verify the assessments of processes that lead to the formation of precipitation, and for forecasting it in the future. Further research could assess the input of every driving mechanism of precipitation into extreme precipitation variability and trends. Indirect assessment of the influence of human activity on precipitation by scaling precipitation vs. air temperature might also include atmospheric circulation as an explanatory variable (e.g. for various circulation patterns).

Acknowledgements. This paper reports on a study carried out under the project 'Effects of underlying processes and the spatial and temporal variability of extreme precipitation in Europe' (NN 306 243939) funded by the National Research Centre, Ministry of Science and Higher Education of Poland. I acknowledge the data providers in the ECA&D project. The publication has been partially financed from the funds of the Leading National Research Centre (KNOW) received by the Centre for Polar Studies of the University of Silesia, Poland.

LITERATURE CITED

Alexandersson H (1986) A homogeneity test applied to precipitation data. J Clim 6:661–675

Alfnes E, Førland EJ (2006) Trends in extreme precipitation and return values in Norway 1900–2004. Report 2/2006, Norwegian Meteorological Institute

Allen MR, Ingram WJ (2002) Constraints on future changes in climate and the hydrological cycle. Nature 419:224–232

Arléry R (1970) The climate of France, Belgium The Netherlands and Luxembourg. In: Wallen CC (ed) Climates of Northern and Western Europe. World survey of climatology, Elsevier, Amsterdam 6:135–193

Bartholy J, Pongrácz R (2007) Regional analysis of extreme temperature and precipitation indices for the Carpathian Basin from 1946 to 2001. Global Planet Change 57:83–95

Beguería S, Angulo-Martínez M, Vicente-Serrano SM, López-Moreno JI, El-Kenawy A (2011) Assessing trends in extreme precipitation events intensity and magnitude using non-stationary peaks-over-threshold analysis: a case study in northeast Spain from 1930 to 2006. Int J Climatol 31:2102–2114

Beranová R, Huth R (2007) Time variations of the relationships between the North Atlantic Oscillation and European winter temperature and precipitation. Stud Geophys Geod 51:575–590

Berg P, Haerter JO, Thejll P, Piani C, Hagemann S, Christensen JH (2009) Seasonal characteristics of the relationship between daily precipitation intensity and surface temperature. J Geophys Res 114:D18102

Betts AK, Harshvardhan (1987) Thermodynamic constraint on the cloud liquid water feedback in climate models. J Geophys Res 92:8483–8485

Brunetti M, Buffoni L, Maugeri M, Nanni T (2000) Precipitation intensity trends in Northern Italy. Int J Climatol 20: 1017–1031

Buishand TA (1982) Some methods for testing the homogeneity of rainfall data. J Hydrol 58:11–27

Buishand TA, Martino GD, Spreeuw JN, Brandsma T (2013) Homogeneity of precipitation series in the Netherlands and their trends in the past century. Int J Climatol 33: 815–833

Burt TP, Ferranti EJS (2012) Changing patterns of heavy rainfall in upland areas: a case study from northern England. Int J Climatol 32:518–532

Burt TP, Horton BP (2007) Inter-decadal variability in daily rainfall at Durham (UK) since the 1850s. Int J Climatol 27:945–956

Casanueva A, Rodríguez-Puebla C, Frías MD, González-Reviriego N (2014) Variability of extreme precipitation over Europe and its relationships with teleconnection patterns. Hydrol Earth Syst Sci 18:709–725

de Lima MIP, Espírito Santo F, Ramos AM, Trigo RM (2015) Trends and correlations in annual extreme precipitation indices for mainland Portugal, 1941–2007. Theor Appl Climatol 119:55–75

Dickson RR, Osborn TJ, Hurrell JW, Meincke J and others (2000) The Arctic Ocean response to the North Atlantic Oscillation. J Clim 13:2671–2696

Drobinski P, Alonzo B, Bastin S, Da Silva N, Muller C (2016) Scaling of precipitation extremes with temperature in the French Mediterranean region: What explains the hook shape? J Geophys Res D 121:3010–3119

Easterling DR, Evans JL, Groisman PYa, Karl TR, Kunkel KE, Ambenje P (2000) Observed variability and trends in extreme climate events: a brief review. Bull Am Meteorol Soc 81:417–425

Escardó AL (1970) The climate of the Iberian Peninsula. In: Wallen CC (ed) Climates of Northern and Western Europe. World survey of climatology. Elsevier, Amsterdam 6:195–239

Fleig AK, Tallaksen LM, James P, Hisdal H, Stahl K (2015) Attribution of European precipitation and temperature trends to changes in synoptic circulation. Hydrol Earth Syst Sci 19:3093–3107

Frei C, Schär C (2001) Detection probability of trends in rare events: theory and application to heavy precipitation in the alpine region. Int J Climatol 14:1568–1584

Frei C, Davies HC, Gurtz J, Schär C (2000) Climate dynamics and extreme precipitation and flood events in Central Europe. Integrated Assess 1:281–299

Furlan D (1977) The climate of Southeast Europe. In: Wallen CC (ed) Climates of Central and Southern Europe. World survey of climatology. Elsevier, Amsterdam 6:185–235

Gajić-Čapka M, Cindrić K (2011) Secular trends in indices of precipitation extremes in Croatia, 1901–2008. Geofizika 28:293–312

García JA, Gallego MC, Serrano A, Vaquero JM (2007) Trends in block-seasonal extreme rainfall over the Iberian Peninsula in the second half of the twentieth century. J Clim 20:113–130

Hänsel S (2009) Changes in Saxon precipitation characteristics. Trends of extreme precipitation and droughts. PhD dissertation, TU Bergakademie Freiberg, Cuvillier Verlag, Göttingen

Hartmann DL, Klein Tank AMG, Rusticucci M, Alexander LV and others (2013) Observations: atmosphere and surface. In: Stocker TF, Qin D, Plattner GK, Tignor M and others (eds) Climate change 2013: the physical science basis. Contribution of Working Group I to the Fifth Assessment Report of the Intergovernmental Panel on Climate Change. Cambridge University Press, Cambridge, p 159–254

Haylock MR, Goodess CM (2004) Interannual variability of European extreme winter rainfall and links with mean large-scale circulation. Int J Climatol 24:759–776

Held IM, Soden BJ (2006) Robust responses of the hydrological cycle to global warming. J Clim 19:5686–5699

Hill T, Lewicki P (2007) Statistics: methods and applications. StatSoft, Tulsa, OK

Hoy A, Schucknecht A, Sepp M, Matschullat J (2014) Large-scale synoptic types and their impact on European precipitation. Theor Appl Climatol 116:19–35

Hundecha Y, Bárdossy A (2005) Trends in daily precipitation and temperature extremes across western Germany in the second half of the 20th century. Int J Climatol 25: 1189–1202

Hurrell JW, Van Loon H (1997) Decadal variations in climate associated with the North Atlantic Oscillation. Clim Change 36:301–326

IPCC (2001) Climate Change 2001: the scientific basis. Contribution of Working Group I to the Third Assessment Report of the Intergovernmental Panel on Climate Change. Cambridge University Press, Cambridge and New York, NY

IPCC (2007) Climate Change 2007: the physical science basis. Contribution of Working Group I to the Fourth Assessment Report of the Intergovernmental Panel on Climate Change. Cambridge University Press, Cambridge

Johannessen TW (1970) The climate of Scandinavia. In: Wallen CC (ed) Climates of Northern and Western Europe. World survey of climatology. Elsevier, Amsterdam 6:23–79

Jones MR, Fowler HJ, Kilsby CG, Blenkinsop S (2013) An assessment of changes in seasonal and annual extreme rainfall in the UK between 1961 and 2009. Int J Climatol 33:1178–1194

Jung T, Hilmer M (2001) The link between the North Atlantic oscillation and Arctic Sea ice export through Fram Strait. J Clim 14:3932–3943

Jung T, Hilmer M, Ruprecht E, Kleppek S, Gulev SK, Zolina O (2003) Characteristics of the recent eastward shift of interannual NAO variability. J Clim 16:3371–3382

Karagiannidis AF, Karacostas T, Maheras P, Makrogiannis T (2012) Climatological aspects of extreme precipitation in Europe, related to mid-latitude cyclonic systems. Theor Appl Climatol 107:165–174

Kendall MG (1970) Rank correlation methods, 4th edn. Griffin, London

Kharin VV, Zwiers FW, Zhang X, Hegerl GC (2007) Changes in temperature and precipitation extremes in the IPCC ensemble of global coupled model simulations. J Clim 20:1419–1444

Klein Tank AMG, Wijngaard JB, Können GP, Böhm R and others (2002) Daily dataset of 20th-century surface air temperature and precipitation series for the European Climate Assessment. Int J Climatol 22:1441–1453

Klein Tank AMG, Können GP (2003) Trends in indices of daily temperature and precipitation extremes in Europe, 1946–99. J Clim 16:3665–3680

Klok EJ, Klein Tank AMG (2009) Updated and extended European dataset of daily climate observations. Int J Climatol 29:1182–1191

Kundzewicz ZW (2005) Intense precipitation and high river flows in Europe—observations and projections. Acta Geophysica Polonica 53(4):385–400

Kyselý J (2009) Trends in heavy precipitation in the Czech Republic over 1961–2005. Int J Climatol 29:1745–1758

Lavers DA, Allan RP, Wood EF, Villarini G, Brayshaw DJ, Wade AJ (2011) Winter floods in Britain are connected to atmospheric rivers. Geophys Res Lett 38:L23803

López-Moreno JI, Vicente-Serrano SM, Angulo-Martínez M, Beguería S (2010) Trends in daily precipitation on the northeastern Iberian Peninsula, 1955–2006. Int J Climatol 30:1026–1041

Lupikasza E (2010) Spatial and temporal variability of extreme precipitation in Poland in the period 1951–2006. Int J Climatol 30:991–1007

Łupikasza EB, Hänsel S, Matschullat J (2011) Regional and seasonal variability of extreme precipitation trends in southern Poland and central-eastern Germany 1951–2006. Int J Climatol 31:2249–2271

Lydolph PE (1977) Climates of the Soviet Union. World survey of climatology, Vol 7. Elsevier, Amsterdam

Manley G (1970) The climate of the British Isles. Wallen CC (ed) Climates of Northern and Western Europe. World survey of climatology, Elsevier, Amsterdam 6:81–133

Mann HB (1945) Nonparametric test against trends. Econometrica 13:245–259

Maraun D, Osborn TJ, Gillett NP (2008) United Kingdom daily precipitation intensity: improved early data, error estimates and an update from 2000 to 2006. Int J Climatol 28:833–842

Moberg A, Jones P (2005) Trends in indices for extremes in daily temperature and precipitation in Central and Western Europe, 1901–99. Int J Climatol 25:1149–1171

Moberg A, Jones PD, Lister D, Walther A and others (2006) Indices for daily temperature and precipitation extremes in Europe analyzed for the period 1901–2000. J Geophys Res 111:D22106

Norrant C, Douguédroit A (2006) Monthly and daily precipitation trends in the Mediterranean (1951–2000). Theor Appl Climatol 83:89–106

Önöz B, Bayazit M (2012) Block bootstrap for Mann-Kendall trend test of serially dependent data. Hydrol Process 26:3552–3560

Osborn TJ, Hulme M (2002) Evidence for trends in heavy rainfall events over the UK. Philos Trans R Soc Lond A 360:1313–1325

Osborn T, Hulme M, Jones PD, Basnett TA (2000) Observed trends in the daily intensity of United Kingdom precipitation. Int J Climatol 20:347–364

Pall P, Allen MR, Stone DA (2007) Testing the Clausius-Clapeyron constraint on changes in extreme precipitation under CO_2 warming. Clim Dyn 28:351–363

Pavan V, Tomozeiu R, Cacciamani C, Lorenzo MD (2008) Daily precipitation observations over Emilia-Romagna: mean values and extremes. Int J Climatol 28:2065–2079

Pettitt AN (1979) A non-parametric approach to the change-point problem. Appl Stat 28:126–135

Piervitali E, Colacino M, Conte M (1998) Rainfall over the central western Mediterranean basin in the period 1951–1995. I. Precipitation trends. Nuovo Cimento Soc Ital Fis, C, Geophys Space Phys 21:331–344

Qian B, Corte-Real J, Xu H (2000) Is the North Atlantic Oscillation the most important atmospheric pattern for precipitation in Europe? J Geophys Res 105:11901–11910

Ramos CM, Martinez-Casanovas JA (2006) Trends in precipitation concentration and extremes in the Mediterranean Penedés-Anoia Region, NE Spain. Clim Change 74: 457–474

Rapp J (2000) Konzeption, Problematik und Ergebnisse klimatologischer Trendanalysen für Europa und Deutschland. Berichte des Deutschen Wetterdienst 212, Offenbach

Rodrigo FS (2010) Changes in the probability of extreme daily precipitation observed from 1951 to 2002 in the Iberian Peninsula. Int J Climatol 30:1512–1525

Rodrigo FS, Trigo RM (2007) Trends in daily rainfall in the Iberian Peninsula from 1951 to 2002. Int J Climatol 27: 513–529

Rodríguez-Puebla C, Encinas AH, Sáenz J (2001) Winter precipitation over the Iberian Peninsula and its relationship to circulation indices. Hydrol Earth Syst Sci 5:233–244

Scherrer SC, Fischer EM, Posselt R, Liniger MA, Croci-Maspoli M, Knutti R (2016) Emerging trends in heavy precipitation and hot temperature extremes in Switzerland. J Geophys Res Atmos 121:2626–2637

Schmidli J, Frei C (2005) Trends of heavy precipitation and wet spells in Switzerland during the 20th century. Int J Climatol 25:753–771

Schüepp M, Schirmer (1977) Climates of Central Europe. In: Wallen CC (ed) Climates of Central and Southern Europe. World survey of climatology, Elsevier, Amsterdam 6:3–73

Soro GE, Noufé D, Bi TAG, Shorohou B (2016) Trend analysis for extreme rainfall at sub-daily and daily timescales in Côte d'Ivoire. Climate 4:37

Trenberth KE, Dai A, Rasmussen RM, Parsons DB (2003) The changing character of precipitation. Bull Am Meteorol Soc 84:1205–1217

Trigo RM, Pozo-Vazquez D, Osborn TJ, Castro-Diez Y, Gamiz-Fortis S, Esteban-Parra MJ (2004) North Atlantic Oscillation influence on precipitation, river flow and water resources in the Iberian Peninsula. Int J Climatol 24:925–944

Ulbrich U, Christoph M (1999) A shift of the NAO and increasing storm track activity over Europe due to anthropogenic greenhouse gas forcing. Clim Dyn 15:551–559

Ustrnul Z, Czekierda D (2001) Circulation background of the atmospheric precipitation in Central Europe (based on the Polish example). Meteorol Z (Berlin) 10:103–111

van den Besselaar EJM, Klein Tank AMG, Buishand TA (2013) Trends in European precipitation extremes over 1951–2010. Int J Climatol 33:2682–2689

Vicente-Serrano S, Beguería S, López-Moreno J, El Kenawy A, Angulo M (2009) Daily atmospheric circulation events and extreme precipitation risk in Northeast Spain: the role of the North Atlantic Oscillation, Western Mediterranean Oscillation, and Mediterranean Oscillation. J Geophys Res D114:D08106

Von Neumann J (1941) Distribution of the ratio of the mean square successive difference to the variance. Ann Math Stat 12:367–395

Wilks DS (2006) On 'field significance' and false discovery rate. J Appl Meteorol Climatol 45:1181–1189

Willems P (2013) Multidecadal oscillatory behaviour of rainfall extremes in Europe. Clim Change 120: 931–944

Zolina O, Simmer C, Kapala A, Gulev S (2005) On the robustness of the estimates of centennial-scale variability in heavy precipitation from station data over Europe. Geophys Res Lett 32:L14707

Zorita E, Kharin V, von Storch H (1992) The atmospheric circulation and sea surface temperature in the North Atlantic area in winter: their interaction and relevance for Iberian precipitation. J Clim 5:1097–1108

5

Changes in intensity of high temporal resolution precipitation extremes in Romania: implications for Clausius-Clapeyron scaling

Aristita Busuioc*, Madalina Baciu, Traian Breza, Alexandru Dumitrescu, Cerasela Stoica, Nina Baghina

National Meteorological Administration, Sos. Bucuresti-Ploiesti 97, Sect.1, Bucharest 013686, Romania

ABSTRACT: We propose a new index for quantifying the maximum rain intensity (IMAX) within a rainfall event. The goal of this paper is 2-fold: first, to analyse the characteristics of variability (trends, change points) for maximum values of this index calculated for spring (April–May), summer (June–August) and autumn (September–October) in comparison with other 2 precipitation indices (daily maximum and total amount). Secondly, to investigate the scaling of the IMAX high percentiles with temperature and compare the results to hourly precipitation extremes. The analysis was carried out at 6 Romanian stations over the period 1966–2007, extended over 1902–2007 for one station. Our results revealed a statistically significant increase of IMAX over the 20th century (intensified over the second half), in contrast with no significant trends in the other 2 precipitation indices. On the other hand, an opposite phase between IMAX variability and its corresponding duration (e.g. shorter durations correspond to higher intensities) was noted. Regarding the scaling behaviour of the IMAX percentiles, there was a difference between summer and spring/autumn. In summer, the 90th and 99th percentiles showed a Clausius-Clapeyron (CC) scaling for temperatures between 14 and 26°C and then a decrease, while the 99.9th percentile showed a super-CC scaling for temperatures lower than 18°C and then a decrease. The spring and autumn 90th and 99th percentiles showed an approximate 2CC scaling for temperatures ranging from 8–20°C. The 99.9th percentile exhibited a scaling close to the 2CC relationship for a temperatures range of 12–18°C (spring) and then a decrease, while the autumn 99.9th percentile exhibited a dependence close to 2CC scaling for temperatures <18°C and then a scaling close to CC. Comparing these results with those referring to hourly and daily precipitation extremes, we conclude that the magnitude of the CC scaling in Romania is mainly dependent on the temporal scale of the extreme precipitation event, storm intensity and season. The proposed index (IMAX) is more sensitive to temperature increases than the hourly and daily extremes, and therefore it is more appropriate to quantifying the climate signal related to intense precipitation events in a warmer climate.

KEY WORDS: Extreme precipitation index · Temperature · Scaling · Clausius-Clapeyron relation · Trends · Romania

1. INTRODUCTION

Short duration and high intensity rainfalls have a large impact on society, with one of the greatest effects being urban flooding. There is observational evidence that extreme precipitation intensity has increased in many places (Trenberth et al. 2003, Lenderink et al. 2011, Benestad 2013, O'Gorman 2015, Molnar et al. 2015) and climate simulations have revealed that this characteristic will be maintained and even intensified under future global warming (Pall et al. 2007, Allan & Soden 2008, Scoccimarro et

*Corresponding author: busuioc@meteoromania.ro

al. 2013, Ban et al. 2015, Chan et al. 2015). This behaviour is supported from a theoretical point of view by the so-called Clausius-Clapeyron relation (hereafter referred to as 'CC scaling'), which states that atmospheric humidity will increase at a rate that follows the saturation vapor pressure dependency on temperature (a rate of ~6–7% °C^{-1}), under conditions of constant relative humidity (Trenberth et al. 2003, Pall et al. 2007, Westra et al. 2014, Blenkinsop et al. 2015, Molnar et al. 2015). A recent review on this topic is presented by O'Gorman (2015).

However, recent studies have revealed that the relationship between precipitation extremes and temperature is very complex and depends on many factors, such as storm duration (Hardwick-Jones et al. 2010, Westra et al. 2014, Wasko et al. 2015), season (Ban et al. 2015, Blenkinsop et al. 2015), storm type (Berg et al. 2013, Molnar et al. 2015), microphysics (Singh & O'Gorman 2014), regional atmospheric stability (Loriaux et al. 2013) and large-scale circulation (Blenkinsop et al. 2015). In some areas, a strong difference between coastal (showing dependence close to CC scaling) and inland regions (showing a super-CC scaling) has been found (Panthou et al. 2014). The intensity of sub-daily precipitation extremes is more sensitive to temperature changes compared to the intensity of daily extremes (Westra et al. 2014). CC scaling can be also affected by local and regional factors that can limit moisture availability, especially in mountain valleys (Molnar et al. 2015).

Future projected changes in precipitation extremes under global warming can be supplied by global (GCMs) or regional (RCMs) climate models that currently operate with spatial resolutions larger than 100 or 10 km, respectively. Due to their still coarse spatial resolution, even state-of-the-art RCMs are not able to fairly reproduce short-duration precipitation extremes (convective precipitation) since some subgrid processes (e.g. deep convection) triggering such phenomena are parameterized through convection schemes (Prein et al. 2015). A solution to this problem is suggested by use of so-called 'convection-permitting models' (CPMs) (Prein et al. 2015, O'Gorman 2015).

Studies investigating the observed extreme rainfall intensity–temperature relationship have been mainly based on fixed temporal resolution precipitation data. A new challenge is an analysis carried out on storm events compared to fixed data (Wasko & Sharma 2015), but the difficulty with such a study is the lack of data regarding the internal characteristics of rainfall event progress (e.g. fractions of various intensities and durations).

Such a study is presented in this paper, where a new index quantifying the extreme precipitation intensity of rainfall events is proposed. The goal of this paper is 2-fold. First, to analyse the characteristics of variability (long-term trends, change points) for maximum values of this index calculated for spring (April–May), summer (June–August) and autumn (September–October) in comparison to 2 other precipitation indices, namely daily maximum and total amount. Second, to investigate the scaling of its high percentiles (90th to 99.9th) with temperature. The analysis was done at 6 stations in Romania. A possible connection between the scaling of this index with temperature and the CC scaling was investigated. Busuioc et al. (2016) carried out a similar analysis for the high percentiles of hourly and daily precipitation at the same stations. They found that scaling was less evident for daily extremes, while the hourly extremes exhibited a scaling close to CC or 2CC relation, depending on the storm intensity. Corroborating these results with that of the present study, we hoped to find the most sensitive temporal scale of precipitation extremes to temperature increase, at least for the Romanian area.

2. DATA AND METHODS

Measurements regarding the internal characteristics of a rainfall event progress (e.g. periods of various intensities and durations) based on graphical records (pluviograph) were used in this study to identify the period (in minutes) with the highest rainfall intensity. In this paper, a rainfall event is defined as starting with non-zero precipitation and ending when the rainfall stops, e.g. zero precipitation amount accumulated in the previous period.

As far as we know, these types of records are seldom used in climate research. Compared to the classical rainfall data set of fixed temporal resolution (sub-hourly, hourly, daily) that are usually used to investigate the observed relationship of precipitation extremes with temperature (termed 'scaling'), we propose an index quantifying the real highest rain intensity within a rainfall event, which can be derived from such data. We hypothesize that this intensity is more sensitive to temperature increases compared to the mean precipitation intensity within a rainfall event or to the maximum rainfall intensity over a fixed duration (e.g. daily or hourly). A similar idea was recently used by Wasko & Sharma (2015) through the analysis of precipitation measurements at a 6 min resolution. However, they used equal duration periods within a storm (e.g. 6 min). The division of a rainfall record into fixed duration subintervals could miss the real highest rain intensity and associated duration due to the selection procedure, as we explain in the following section. Only measurements based on a continuous rainfall record (as is the case of pluviograph records used in this study) allow identification of the highest rainfall intensity at the lowest temporal resolution of 1 min.

The types of records mentioned above at 6 weather stations across Romania representing various physical–geographical conditions (see Fig. S1 in the Supplement at www.int-res.com/articles/suppl/c072p239_supp.pdf) were used in this study. We note that these stations do not cover the entire complexity of the Romanian climate (e.g. mountain areas are not included), due to the current lack of such data. These data sets are available only for liquid precipitation, which usually occurs between April and October for Romanian climate conditions. Each rainfall event was divided into intervals of approximately constant intensity (mm min^{-1}). To explain how these intervals were defined and how the intensity for each interval was calculated, Fig. 1 presents an example of a raw graphical record regarding rainfall from 2 June 2000

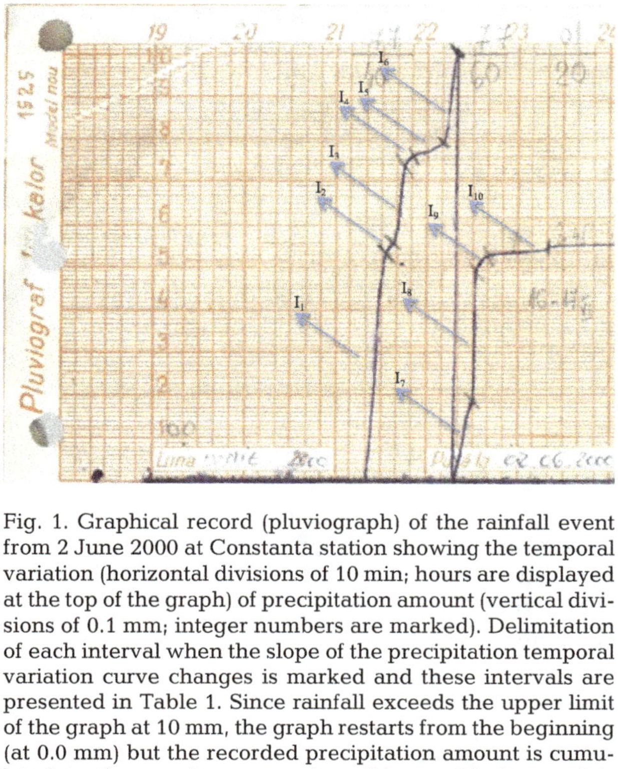

Fig. 1. Graphical record (pluviograph) of the rainfall event from 2 June 2000 at Constanta station showing the temporal variation (horizontal divisions of 10 min; hours are displayed at the top of the graph) of precipitation amount (vertical divisions of 0.1 mm; integer numbers are marked). Delimitation of each interval when the slope of the precipitation temporal variation curve changes is marked and these intervals are presented in Table 1. Since rainfall exceeds the upper limit of the graph at 10 mm, the graph restarts from the beginning (at 0.0 mm) but the recorded precipitation amount is cumulated to the previous one when this graphical record is numerically deciphered by the trained meteorological observer, so that a continuous record of the rainfall is obtained

at Constanta station, showing the curve of the precipitation temporal variation within the rainfall event. Time is displayed on the horizontal axis, while the precipitation amount is represented on the vertical. The moment when the curve slope changes (see Fig. 1) was marked by a trained meteorological observer, along with the precipitation amount recorded at that time. This way, the intervals I_j, j = 1,... ,n (where n is the number of intervals) with approximately constant intensity (mm min^{-1}) were delimited based on the assumption that an approximately constant slope of the precipitation temporal variation curve over each interval I_j leads to an approximately uniform precipitation variation over that interval. Consequently, the precipitation intensity (mm min^{-1}) is approximately constant over each I_j. To avoid repetition of the word 'approximately', it will be omitted in the following. The next step is to calculate the following information for each interval I_j: duration (D_j, in minutes; the difference between the time of the end and that of the beginning), precipitation amount (P_j) over each interval (difference between the amount recorded at the end and beginning of each interval, respectively) and rainfall intensity (IST_j) calculated as a ratio between P_j and D_j. The information presented above is summarised in Table 1, showing a complete view of the internal characteristics of the rainfall event. Only this type of record allows identifying time intervals with about the same rainfall intensity per minute, and therefore the maximum precipitation intensity on a temporal scale of 1 min can be obtained; as far as we know, this has not been presented in any other previous papers.

The interval of analysis was 1966–2007, while for the Bucuresti-Filaret station these data sets were available over a longer interval (1902–2007). The maximum intensity for each rainfall event (IMAX) is defined as follows:

$$IMAX = \max\{IST_j, j = 1, ..., n\} \quad (1)$$

where n represents the number of sub-intervals I_j with constant intensity IST_j (j = 1, ..., n) during a rainfall event. In the example presented above, IMAX = 3.0 (bold value in Table 1) and the corresponding duration (hereafter noted by DIMAX) is 1 min: here the duration of 1 min corresponding to IMAX is coincidentally equal to the temporal scale of the IST_j (mm min^{-1}). This intensity corresponds to the interval I_8 between 22:30 and 22:31 h. This procedure was applied for all rainfall events at each station. Therefore, the time series of IMAX and the corresponding DIMAX for all rainfall events at each station was retained for the analysis presented in this study.

Table 1. Temporal variation of precipitation amount during the rainfall event recorded on 2 June 2000 at Constanta station. The rainfall is divided in 10 intervals of constant intensity (mm min^{-1}). These intervals (I_j, j = 1, ...,10) are delimited by the time (t, second column) when the slope of the temporal variation curve of the precipitation amount recorded by the pluviograph changes. The precipitation amount recorded at the end of each I_j is noted (3rd column). Fig. 1 shows the graphical representation of this rainfall event and the delimitation of each interval. Duration (D_j) and total precipitation amount (P_j) recorded over each interval (I_j; calculated as the difference between precipitation amounts recorded at the end of each interval I_j and the end of previous interval I_{j-1}) are presented. Rainfall intensity (IST$_j$) is calculated as a ratio between P_j (5th column) and corresponding D_j (4th column). The row corresponding to the maximum intensity (IMAX) shown in **bold**

I_j	t	Precipitation r (mm)	D_j (min)	P_j (mm)	IST$_j$ (mm min^{-1})	Comment
	21:20 h	0.00				Rainfall start
I_1	21:33 h	5.30	13	5.30	0.41	
I_2	21:38 h	5.60	5	0.30	0.06	
I_3	21:45 h	7.40	7	1.80	0.26	
I_4	21:50 h	7.60	5	0.20	0.04	
I_5	22:10 h	7.90	20	0.30	0.02	
I_6	22:20 h	10.00	10	2.10	0.21	
I_7	22:30 h	11.90	10	11.90	1.19	Pluviograph restart
I_8	**22:31 h**	**14.90**	**1**	**3.00**	**3.00**	**IMAX**
I_9	22:40 h	15.30	9	0.40	0.04	
I_{10}	23:20 h	15.50	40	0.20	0.01	End of rainfall

Note that delimitation of the intervals I_j and the corresponding D_j and P_j were subjectively estimated by a trained meteorological observer and therefore are subject to human errors. The rainfall intensity for each interval IST$_j$ and IMAX for each rainfall event are calculated automatically through a computing programme, but they depend on D_j and P_j used as inputs into that programme. Some qualitative verification was carried out as follows: comparison between the daily precipitation amount derived from the graph records (pluviograph) as presented above and those derived from measurements made with rain gauges were performed, and large differences were analysed by verifying the data numerically deciphered from raw records on pluviograph; some outliers of IMAX and corresponding DIMAX (identified in the temporal variation of annual maximum values) have also been verified with raw data. As will be presented below, the results obtained in this paper are coherent from one station to another showing similar features. Therefore, we conclude that even if some human error may be possible in the original data, the final results are not significantly influenced.

The IMAX values and their corresponding durations (DIMAX) for all rainfall events at the 6 stations were analysed in 2 ways corresponding to the main 2 objectives of this paper. First, the linear trends and shifts in the mean for the time series of IMAX seasonal (spring, summer, autumn) maximum values and corresponding DIMAX were computed and compared to the corresponding values of another extreme precipitation index based on the daily amount (e.g. maximum daily amount, PP24) and total precipitation amount (PP). In this paper, spring months refer to April–May, while autumn refers to September–October, considering the availability of graphical rainfall records as presented above. Summer months are June–August. The statistical significance level of the linear trend and shift was estimated by the nonparametric tests Mann-Kendall (Kulkarni & von Storch 1995) and Pettitt (Pettitt 1979), respectively. Secondly, an investigation of the relationship between the highest IMAX percentiles (90th, 99th and 99.9th) and daily air temperature was carried out to find a possible connection with the CC scaling.

To reach the second objective, a technique similar to Lenderink & van Meijgaard (2008) was used. This technique was also used by Busuioc et al. (2016) to investigate the validity of CC scaling for hourly and daily precipitation extremes at 9 Romanian stations (including the 6 considered in this study). The IMAX data for all rainfall events at each station were divided into bins of 2°C width in terms of daily mean temperature. By using daily mean temperatures, the high temporal variability during a day is avoided.

The analysis was performed by combining all data over the entire country. This technique allows avoiding as much as possible the influence of local factors on extreme precipitation behaviour while retaining enough events to assess the percentiles (especially the 99.9th). Considering the fact that the CC scaling depends, among other factors, on the storm type (stratiform or convective), the analysis was carried out for each season. Rainfall in Romania is mainly of a convective type in summer, while in spring and autumn, stratiform rainfall prevails. Considering the small number of events in spring and autumn, the results referring to the combined data over these seasons are also presented. The 90th, 99th and 99.9th percentiles for each bin were computed and their dependence on temperature was compared to the CC (~7 % increase per temperature degree rise) and 2CC (~14 % increase per temperature degree rise) scaling.

3. RESULTS

3.1. Observed changes in extreme rainfall intensity in Romania

Table 2 shows the linear trend of the IMAX and DIMAX indices compared to PP24 and PP over the common period 1966–2007 (6 stations) and the long period 1902–2007 (Bucuresti-Filaret). All time series were standardized, following the procedure presented by Busuioc et al. (2015), to be compared with each other. Over the shorter period, IMAX generally exhibits an increasing trend compared to DIMAX (decreasing) but the trends are statistically significant at the 5% level only at a few stations. Since the results could be affected by the short duration of the time series, a similar analysis over more than 100 yr (1902–2007) at Bucuresti-Filaret station was carried out and the same conclusion was found. In this case, a statistically significant increase for IMAX was found in contrast to PP24 and PP (showing no significant trend) (Table 2). This result suggests different long-term variability behaviour of IMAX in contrast to PP and PP24. This conclusion is supported by the analysis of the correlation coefficient between IMAX and PP/PP24, showing much lower values (or even no statistical significance) in contrast to the higher and statistically significant correlations between PP and PP24 (Table 3). Fig. 2 shows, as an example, the temporal variability of IMAX, PP and PP24 at Bucuresti-Filaret station for spring, summer and autumn.

It can be seen that, while PP and PP24 exhibit a coherent variability with each other, IMAX shows different characteristics — especially during the most recent decades in summer, when the other indices show a decrease while IMAX exhibits a significant

Table 2. Trend in 10 years of standardized precipitation indices over the short (1966–2007) and long period (1902–2007) for 3 seasons: spring (April–May), summer (June–August) and autumn (September–October). **Bold** values show statistically significant trends and change points (at the 5% level). The change points and associated arrows (indicating upward or downward shifts) for Bucuresti-Filaret are also included. IMAX: maximum intensity of rainfall event; DIMAX: duration of IMAX; PP24: maximum daily precipitation; PP: total precipitation

Station	Spring				Summer				Autumn			
	IMAX	DIMAX	PP24	PP	IMAX	DIMAX	PP24	PP	IMAX	DIMAX	PP24	PP
1966–2007												
Bucuresti Filaret	0.15	−0.24	0.02	−0.09	**0.36**	−0.32	−0.17	−0.11	−0.08	−0.23	**0.29**	0.18
Constanta	**0.24**	−0.27	−0.03	0.02	0.18	−0.08	0.16	0.09	**0.24**	−0.27	**0.15**	0.25
Tecuci	0.11	−0.25	0.00	−0.12	0.21	−0.09	−0.09	−0.02	**0.26**	**−0.31**	**0.21**	0.14
Drobeta Tr Severin	−0.11	−0.18	−0.13	−0.15	−0.17	0.0	0.01	−0.02	0.07	−0.02	0.11	0.01
Oradea	0.12	0.09	−0.03	0.06	−0.02	0.08	−0.02	−0.05	0.06	**0.14**	0.11	0.11
Sibiu	0.07	−0.18	−0.09	−0.10	−0.08	−0.14	0.12	−0.05	0.06	0.0	0.08	0.08
1902–2007												
Bucuresti Filaret	**0.10**	−0.08	0.02	0.01	**0.12**	−0.06	−0.04	0.0	**0.08**	**−0.07**	0.06	0.04
	1961↑	**1961↓**			**1961↑**	**1987↓**			**1960↑**	**1964↓**		

Table 3. Correlation between various standardized precipitation indices over short period (1966–2007; 6 stations) and long period (1902–2007; Bucuresti-Filaret). **Bold** values are statistically significant at the 5% level. See Table 2 for abbreviation definitions

Station	Spring				Summer				Autumn			
	IMAX–PP	IMAX–PP24	PP–PP24	IMAX–DIMAX	IMAX–PP	IMAX–PP24	PP–PP24	IMAX–DIMAX	IMAX–PP	IMAX–PP24	PP–PP24	IMAX–DIMAX
1966–2007												
Bucuresti Filaret	**0.39**	0.29	**0.86**	**−0.47**	**0.39**	0.34	**0.70**	**−49**	0.20	0.16	**0.91**	**−0.44**
Constanta	−0.08	0.04	**0.79**	**−0.46**	**0.62**	**0.69**	**0.84**	**−48**	0.17	0.30	**0.77**	**−0.41**
Tecuci	**0.51**	0.31	**0.57**	**−0.44**	**0.43**	0.25	**0.70**	**−39**	**0.51**	**0.58**	**0.91**	**−0.44**
Drobeta Tr Severin	0.17	0.26	**0.66**	**−0.53**	0.17	0.06	**0.71**	**−50**	0.36	0.24	**0.87**	**−0.55**
Oradea	0.13	0.18	**0.67**	**−0.41**	0.24	0.24	**0.78**	−34	**0.37**	**0.46**	**0.81**	**−0.42**
Sibiu	0.10	0.21	**0.64**	−0.34	0.02	−0.10	**0.54**	−21	**0.41**	**0.51**	**0.75**	**−0.38**
1902–2007												
Bucuresti-Filaret	**0.23**	**0.20**	**0.74**	**−0.52**	**0.24**	0.09	**0.58**	**−0.47**	**0.31**	**0.24**	**0.87**	**−0.34**

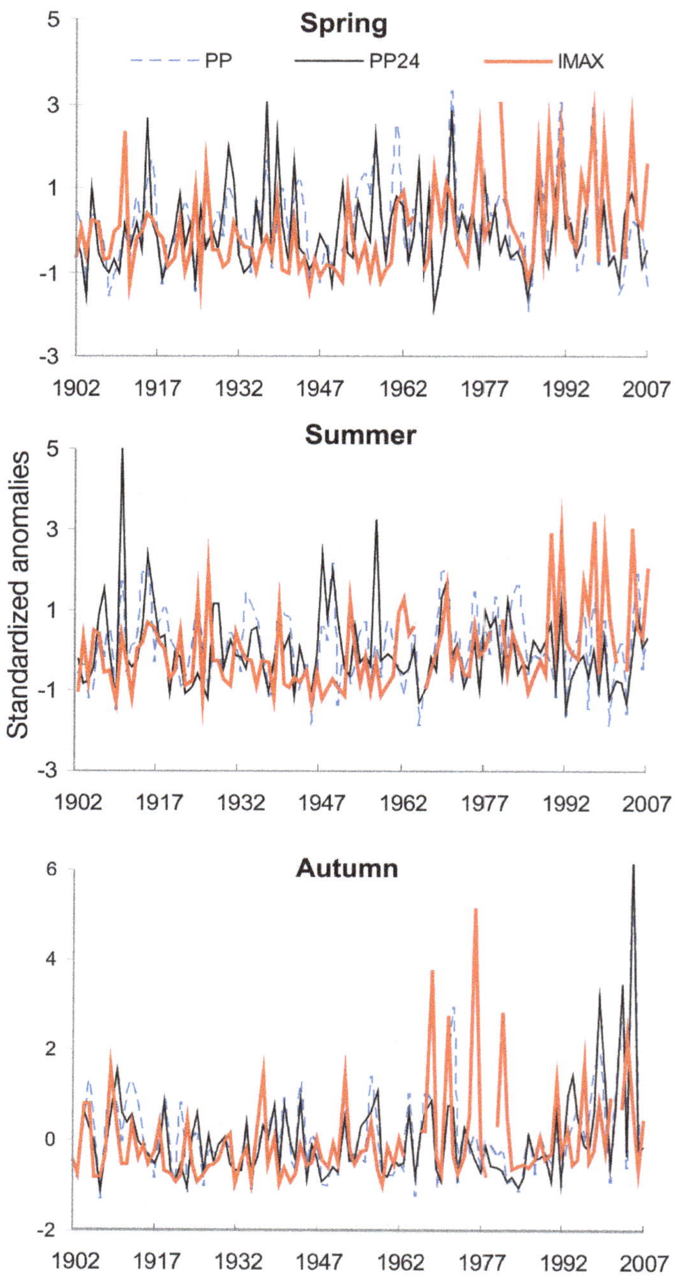

Fig. 2. Comparison between standardized extreme rainfall intensity (IMAX), extreme daily intensity (PP24) and total precipitation amount (PP) at Bucuresti-Filaret station for spring (April–May), summer (June–August) and autumn (September–October)

increase. These findings suggest that the PP and PP24 variability are governed by the same large-scale mechanisms (mainly given by the large-scale circulation in spring and autumn and by convection in summer; Busuioc et al. 2010), while for IMAX other mechanisms could be responsible for this behaviour.

One of these mechanisms could be related to CC scaling, as discussed in the next section. On the other hand, Busuioc et al. (2016) highlighted a shift in the nature of precipitation in Romania towards more showers (high intensity precipitation in a short time), in contrast with no significant increase in the frequency of rainfall days; the mechanism responsible for this behaviour is the combined effect of an increase in atmospheric instability accompanied by more precipitable water in summer and more frequent cyclonic structures crossing Romania in spring. This mechanism could be also responsible for the significant increase of IMAX found in this paper.

On the other hand, statistically significant negative correlation coefficients between IMAX and DIMAX at all stations and the 3 seasons are noted (see Table 3) on shorter (all 6 stations) as well as on longer time series (Bucuresti-Filaret). This result shows an opposite phase between the variability of IMAX and its corresponding DIMAX; namely, shorter durations correspond to a higher intensity of precipitation extremes. Fig. 3 shows, as an example, the temporal variability of IMAX and DIMAX at Bucuresti-Filaret station for spring, summer and autumn. For spring and summer, the long-term trend is not monotonous, showing an upward change point around 1960 (statistically significant at the 5% level) with IMAX increasing and DIMAX decreasing after this year (in summer a downward shift for DIMAX is recorded around 1987), and an opposite behaviour before but with a weaker negative slope for IMAX.

In conclusion, in comparison with classical indices of fixed duration, the proposed index includes, by definition, the duration associated with the fraction of the highest intensity of a rainfall event. This allows investigation of the characteristics of simultaneous variability for the intensity–duration pair associated with precipitation extremes, namely the connection between these components and possible changes under global warming. Therefore, we could assert that IMAX could be naturally associated with the intensity of extreme precipitation within a rainfall event, shortly named in the following as extreme rainfall intensity, in comparison with precipitation extremes on a fixed temporal scale (daily or hourly).

3.2. Scaling of extreme rainfall intensity

Based on the technique presented in Section 2, we investigated the scaling of IMAX highest percentiles (90th, 99th and 99.9th) with daily temperatures at the 6 selected stations in Romania, stratified over the 3

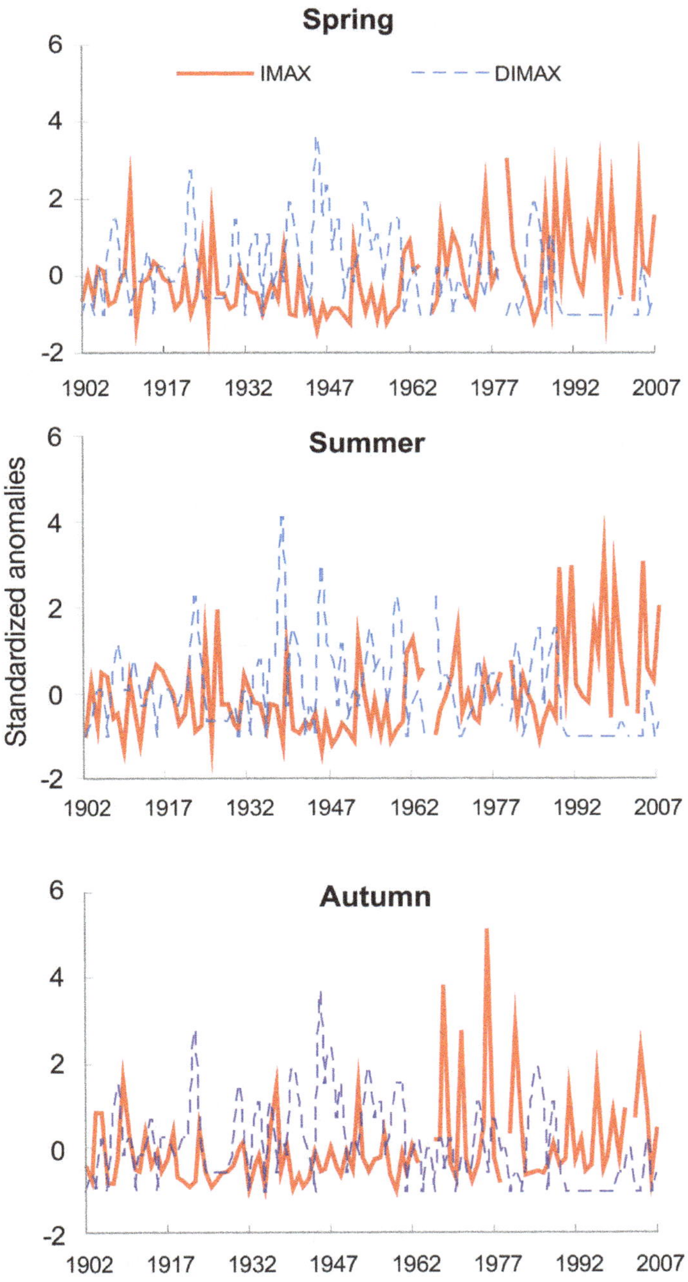

Fig. 3. Comparison between standardized extreme rainfall intensity (IMAX) and corresponding duration (DIMAX) at Bucuresti-Filaret station

seasons (spring, summer and autumn). The results are presented by combining all data over the entire country; the graphic representation of the findings are shown in Fig. 4 (left column), which have been numerically proven by computing the slope of the linear dependence between the natural logarithm of IMAX percentiles and temperature variation (see Annex 1 in the Supplement at www.int-res.com/articles/suppl/c072p239_supp.pdf) similar to Blenkinsop et al. (2015). The number (N) of events in each temperature bin is summarised in Table S1 in the Supplement. Due to the fact that summer is the wettest season in Romania and autumn is the driest, the highest number of rainfall events used to calculate the IMAX percentiles was recorded in summer (exceeding 3000 events for daily temperatures between 16 and 22°C) and lowest in autumn (the highest number accounting for around 1300 events between 12 and 16°C). In spring and autumn, for temperatures >20°C and <6°C the N values are small (<500) leading to less precise percentiles. In summer, this situation happens for temperatures >24°C and <12°C. However, to assess high quality 99.9th percentiles, N should be 1000 or more, which is the case for the temperature range of 8–18°C in spring–autumn and 14–24°C in summer (see Table S1).

Our results show that the IMAX scaling behaviour depends on season and percentiles, and it is not uniform over the temperature range. In summer, the 90th and 99th percentiles show approximately a CC scaling for the temperature range 14–26°C, and then decrease. The 99.9th percentile does not exhibit clear scaling, mainly showing a super-CC scaling for temperatures <18°C and then a decrease. The spring and autumn seasons exhibit an almost similar behaviour: the 90th and 99th percentiles show an approximate 2CC (12–14%) scaling for temperatures range of 8–20°C (see Annex 1 in the Supplement) with some inflexion points (slope change): for temperatures >18°C the spring 90th percentile follows a CC scaling, while the autumn 90th percentile for temperatures <12°C follows a CC scaling and then a 2CC scaling; the autumn 99th percentile fits better with 2CC scaling over the temperature range of 18–24°C. These details can be very well identified using the numerical computation of the percentile scaling as presented in Annex 1. The dependence of the spring 90th percentile close to CC scaling for temperatures >18°C has not been revealed in observational studies so far, but it is in agreement with CPM simulations (Singh & O'Gorman 2014), suggesting that a physical mechanism could be behind this result. However, in this case N < 1000. The spring 99.9th percentile exhibits a scaling close to the 2CC relationship (13%) for a temperature range of 12–18°C and then a decrease (but N < 1000), while the autumn 99.9th percentile exhibits a dependence close to 2CC scaling for temperatures <18°C and then a scaling close to CC. Since the number of events in spring and autumn is much smaller than that recorded in summer (see Table S1) due to the shortness of these seasons considered in this study

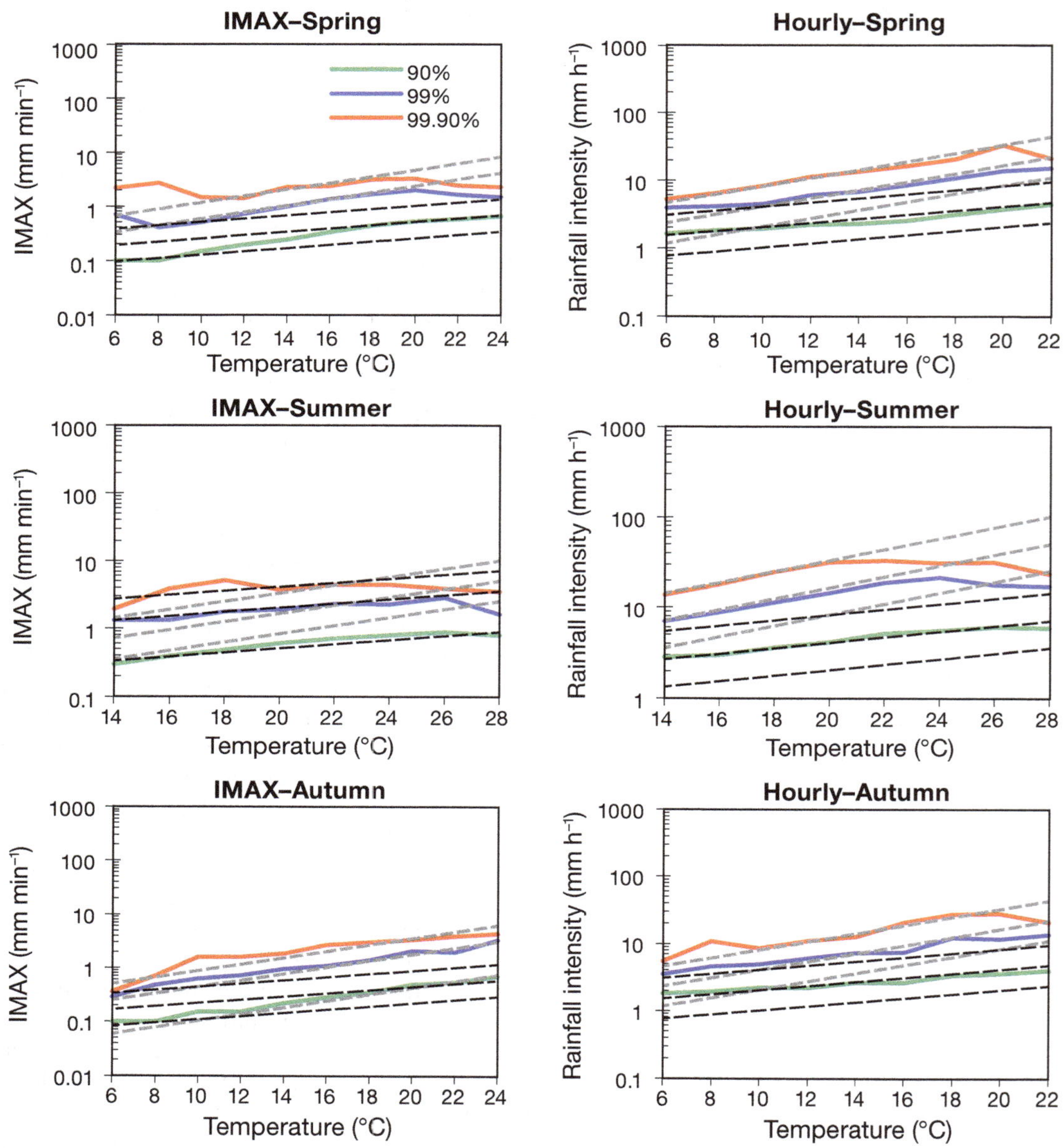

Fig. 4. Scaling of observed extreme rainfall intensity (IMAX; left) and hourly extremes (right) with temperature for 6 stations mixed up together. Shown are different percentiles (90th to 99.9th) of IMAX and hourly extreme distribution for each temperature bin. The Clausius-Clapeyron (CC) and 2CC scalings are presented through the black and grey dashed lines, respectively. Due to the logarithmic *y*-axis these exponential relations are shown as straight lines

(only 2 months as presented in Section 2) on the one hand, and the dry precipitation regime in autumn in Romania on the other hand, we could suspect that the results found for these seasons are less certain. To remove this suspicion, the IMAX data for the 2 seasons were mixed up and the scaling results presented in Fig. S2 in the Supplement. The results show similar behaviour to those presented above separately for spring and autumn, which means results are robust for those 2 seasons. To determine if possible human errors in the delimitation of intervals with constant intensity could affect the final results on CC-scaling, an analysis of the scaling behaviour on a station scale but with all data pooled together (from April–October) was carried out. For the Bucuresti-Filaret station, the short (1950–2007) and long periods (1902–2007) were considered. The results (Fig. S3 in the Supplement) show similar behaviour at all stations for each percentile, and are not overly dependent on the time series length. Therefore, we conclude that our results are robust and are not significantly influenced by possible human errors.

In conclusion, our results show a clear difference between the IMAX scaling behaviour in the summer season (representative of convective rainfalls) and spring/autumn (representative of stratiform rainfalls). Such a difference has also been reported by other studies but with different conclusions. For example, according to Berg et al. (2013), stratiform precipitation extremes increase with temperature at an approximately CC rate without a characteristic scale, while the convective precipitation extremes exceed the CC rate and show characteristic spatial and temporal scales; however, they did not use a seasonal stratification as other studies have done (Ban et al. 2015, Blenkinsop et al. 2015). Other recent studies (Westra et al. 2014, Molnar et al. 2015, O'Gorman 2015) highlighted that the relationship between precipitation extremes and temperature is much more complex than suggested by the CC relationship and is mainly dependent on factors such as temporal resolution, region, local and regional effects (e.g. orography) as well as changes in large-scale forcing. Such conclusions could also be valid in our case: namely, the CC scaling behaviour for predominant convective precipitation extremes in summer and 2CC scaling behaviour for predominant stratiform precipitation extremes in spring and autumn could be the result of the combined effect of other mechanisms specific to the Romanian area, as Busuioc et al. (2016) revealed. The combined effect of increased atmospheric instability covering Romania accompanied by more precipitable water in summer is the main mechanism responsible for the increase of summer rain shower frequency, while increased atmospheric instability in spring accompanied by changes in atmospheric circulations is responsible for the increase of spring rain shower frequency. On the other hand, the 2CC scaling rate found in spring and autumn could also be explained by a mixture of stratiform and convective precipitation (specific to the Romanian climate in these seasons) that could exaggerate the extreme precipitation–temperature relationship (Berg & Haerter 2013, Molnar et al. 2015). We must note that the rainfall events in our case are not strictly stratified in terms of storm type as Berg et al. (2013) carried out, and stress that the convective character of Romanian precipitation in summer and the stratiform one in spring and autumn is only a dominant feature; a mixture between the 2 types is possible, with the prevalence of one or the other depending on season. Future research based on storm type stratification of IMAX data and more stations are needed, among other factors, to find a more consistent explanation from a physical point of view of the results found in this study.

To see the influence of the duration of precipitation extremes on their relation with temperature, the results of this study are compared to those presented by Busuioc et al. (2016) referring to fixed durations (hourly and daily) of precipitation extremes recorded at the same stations. In this respect, considering the entire annual period with liquid precipitation (for which the hourly precipitation amounts are derived from pluviograph records) with an analysis conducted on a station scale, they found that the 90th percentile shows an approximate CC scaling for all temperatures, while the hourly 99th and 99.9th percentiles follow an almost 2CC scaling for the temperature range of 10–22°C and then a decrease. The daily precipitation extremes do not exhibit a clear scaling behaviour. In order to carry out a correct comparison with our results, the hourly precipitation amounts have been stratified in similar seasons as presented in this study and the results are presented in Fig. 4 (right column) for comparison. The number (N) of events in each temperature bin is summarised in Table S1 in the Supplement. As we can see, N values are much higher than those referring to IMAX events. The results for hourly precipitation extremes, seasonally stratified, are quite similar to those obtained for the extended period (April–October) as presented above, with some seasonal differences: the 90th percentile exhibits a CC scaling behaviour for all temperatures in all analysed seasons; the 99th percentile exceed the CC (close to 2CC) scaling for spring (10–20°C), summer (14–22°C) and autumn (10–18°C) and follows the CC scaling for temperatures below 10°C (spring); the 99.9th percentile exhibits a 2CC scaling for spring (8–20°C), summer (14–20°C) and autumn (10–18°C).

Difference in scaling behaviour of the extreme rainfall intensity conducted on rainfall events (1 min temporal scale) as proposed in this study (represented by the IMAX percentiles) and corresponding extreme precipitation intensity on hourly scale can be revealed. The most important differences are identified in the behaviour of the 90th percentile (moderate extremes) for the spring and autumn seasons: IMAX exhibits an almost 2CC scaling, in contrast to hourly extremes that follow a CC scaling. This result suggests that the moderate extremes (90th percentile) on the temporal scale proposed in this study (mm min^{-1}) are more sensitive to temperature increase than hourly extremes. The scaling behaviour of the 99th percentile is quite similar for the 2 temporal scales (Fig. 4) in the spring and autumn seasons (an almost 2CC scaling) and different in summer: CC scaling for IMAX over the temperature range 14–26°C then a decrease; 2CC scaling for hourly precipitation over the temperature range 14–22°C then a decrease. This result shows that for temperatures between 22 and 26°C, IMAX is more sensitive to temperature increase than the hourly extremes (CC scaling for IMAX and a decrease for hourly precipitation); the 2CC scaling below 22°C for hourly precipitation could be explained by the fact that at this temporal scale the convective precipitation could be mixed with the stratiform ones that may exaggerate the extreme precipitation–temperature relationship (Berg & Haerter 2013, Molnar et al. 2015).

Therefore, it can be concluded that the magnitude of the CC scaling in Romania is mainly dependent on storm duration, storm intensity and season, in agreement with other studies (Hardwick-Jones et al. 2010, Panthou et al. 2014, Wasko et al. 2015). However, some details in the CC scaling mentioned above are specific to the Romanian area, that could be explained by an additional influence of other regional factors such as an increase in atmospheric instability covering Romania simultaneously with more frequent cyclonic southern circulations crossing this area in spring and more precipitable water in summer (Busuioc et al. 2016). Future research is needed to analyse in more detail the direct dependence between the scaling of extreme precipitation intensity with temperature in connection with other factors.

4. CONCLUSIONS

A new index (IMAX) quantifying the maximum rain intensity within a rainfall event was proposed in this study. IMAX values and their corresponding durations (DIMAX) at 6 stations in Romania, seasonally stratified (spring, summer and autumn), were analysed in 2 ways corresponding to the main 2 objectives of this paper. The main conclusions are summarised in the following.

Our results revealed a robust climate signal over the entire country referring to an opposite phase between the variability of IMAX and its corresponding DIMAX (e.g. shorter durations correspond to higher intensities).

A statistically significant increase of IMAX over the 20th century (intensified over the second half) was found, in contrast with no significant trend for PP24 and PP. Different long-term variability behaviour of IMAX in contrast to PP and PP24 was revealed. This conclusion is supported by analysis of the correlation coefficient between IMAX and PP/PP24. These findings suggest that the variability in PP and PP24 is governed by the same large-scale mechanisms, while for IMAX variability, other mechanisms could be responsible. One mechanism could be related to a shift in the nature of precipitation in Romania towards more showers (high intensity precipitation in a short time), justified by the combined effect of increase in atmospheric instability over Romania accompanied by more precipitable water in summer and more frequent cyclonic structures in spring (Busuioc et al. 2016). An additional factor could be a higher IMAX sensitivity to temperature increase in contrast to PP24, as was proven in this study by the analysis of IMAX scaling.

The IMAX scaling behaviour depends on the season and storm intensity and varies over the temperature range. In summer, the 90th and 99th percentiles followed an approximate CC scaling for the temperature range 14–26°C. The 99.9th percentile (rare events) did not exhibit clear scaling, mainly showing a super-CC scaling for temperatures lower than 18°C and then a decrease. The spring and autumn 90th and 99th percentiles showed an approximate 2CC scaling for a temperatures range of 8–20°C. The 99.9th percentile exhibited a scaling close to the 2CC relationship for a temperatures range of 12–18°C (spring) and then a decrease, while the autumn 99.9th percentile exhibited a dependence close to 2CC scaling for temperatures <18°C and then a scaling close to CC.

Comparing these results with those referring to hourly and daily precipitation extremes (Busuioc et al. 2016), we can conclude that the magnitude of the CC scaling in Romania is mainly dependent on the temporal scale of the extreme precipitation event, storm intensity and season, as previous studies have revealed (Hardwick-Jones et al. 2010, Westra et al. 2014, O'Gorman, 2015). The proposed IMAX index is more sensitive to temperature increase than the hourly and daily extremes, and therefore it is more appropriate to quantify the climate signal related to intense precipitation events in a warmer climate.

All these results are in agreement with the main conclusion presented in recent review papers (Westra et al. 2014, O'Gorman 2015); that the relationship between precipitation extremes and temperature is much more complex than suggested by the CC relationship, and that it is dependent on many factors, some of them being considered in this study for Romania. Future research is needed for Romania to analyse in more detail the direct influence of other factors such as storm type, complex Romanian orography, etc.

Acknowledgements. This study was funded by the Executive Agency for Higher Education, Research, Development and Innovation Funding (UEFISCDI) through the research project CLIMHYDEX, code PNII-ID-2011-2-0073 (http://climhydex.meteoromania.ro). Three anonymous reviewers are acknowledged for their very useful comments, which considerably improved the original manuscript.

LITERATURE CITED

Allan RP, Soden BJ (2008) Atmospheric warming and the amplification of precipitation extremes. Science 321: 1481–1484

Ban N, Schmidli J, Schär C (2015) Heavy precipitation in a changing climate: Does short-term summer precipitation increase faster? Geophys Res Lett 42:1165–1172

Benestad R (2013) Association between trends in daily rainfall percentiles and the global mean temperature. J Geophys Res Atmos 118:10, 802–10, 810

Berg P, Haerter JO (2013) Unexpected increase in precipitation intensity with temperature — a result of mixing precipitation types? Atmos Res 119:56–61

Berg P, Moseley C, Haerter JO (2013) Strong increase in convective precipitation in response to higher temperatures. Nat Geosci 6:181–185

Blenkinsop S, Chan SC, Kendon EJ, Roberts NM, Fowler HJ (2015) Temperature influences on intense UK hourly precipitation and dependency on large-scale circulation. Environ Res Lett 10:054021

Busuioc A, Caian M, Cheval S, Bojariu R, Boroneant C, Baciu M, Dumitrescu A (2010) Climate variability and change in Romania. Pro Universitaria, Bucure ti (in Romanian)

Busuioc A, Dobrinescu A, Birsan MV, Dumitrescu A, Orzan A (2015) Spatial and temporal variability of climate extremes in Romania and associated large-scale mechanisms. Int J Climatol 35:1278–1300

Busuioc A, Birsan MV, Carbunaru D, Baciu M, Orzan A (2016) Changes in the large-scale thermodynamic instability and connection with rain shower frequency over Romania: verification of the Clausius-Clapeyron scaling. Int J Climatol 36:2015–2034

Chan SC, Kendom EJ, Roberts NM, Fowler HJ, Blenkinsop S (2015) Downturn in scaling of UK extreme rainfall with temperature for future hottest days. Nat Geosci 9:24–28

Hardwick-Jones R, Westra S, Sharma A (2010) Observed relationships between extreme sub-daily precipitation, surface temperature, and relative humidity. Geophys Res Lett 37:L22805

Kulkarni A, von Storch H (1995) Monte Carlo experiments on the effect of serial correlation on the Mann-Kendall test of trend. Meteorol Z NF 4:82–85

Lenderink G, van Meijgaard E (2008) Increase in hourly precipitation extremes beyond expectations from temperature changes. Nat Geosci 1:511–514

Lenderink G, Mok HY, Lee TC, Van Oldenborgh GJ (2011) Scaling and trends of hourly precipitation extremes in two different climate zones — Hong Kong and the Netherlands. Hydrol Earth Syst Sci 8:4701–4719

Loriaux JM, Lenderink G, De Roode SR, Siebesma AP (2013) Understanding convective extreme precipitation scaling using observations and an entraining plume mode. J Atmos Sci 70:3641–3655

Molnar P, Fatichi S, Gaál L, Szolgay J, Burlando P (2015) Storm type effects on super Clausius-Clapeyron scaling of intense rainstorm properties with air temperature. Hydrol Earth Syst Sci 19:1753–1766

O'Gorman PA (2015) Precipitation extremes under climate change. Curr Clim Change Rep 1:49–59

Pall P, Allen MR, Stone DA (2007) Testing the Clausius-Clapeyron constraint on changes in extreme precipitation under CO_2 warming. Clim Dyn 28:351–363

Panthou G, Mailhot A, Laurence E, Talbot G (2014) Relationship between surface temperature and extreme rainfalls: a multi-timescale and event-based analysis. J Hydrometeorol 15:1999–2011

Pettitt AN (1979) A non-parametric approach to the change-point problem. Appl Stat 28:126–135

Prein AF, Langhans W, Fosser G, Ferrone A and others (2015) A review on regional convection-permitting climate modeling: demonstrations, prospects, and challenges. Rev Geophys 53:323–361

Scoccimarro E, Gualdi S, Bellucci A, Zampieri M, Navarra A (2013) Heavy precipitation events in a warmer climate: results from CMIP5 models. J Clim 26:7902–7911

Singh MS, O'Gorman PA (2014) Influence of microphysics on the scaling of precipitation extremes with temperature. Geophys Res Lett 41:6037–6044

Trenberth KE, Dai A, Rasmussen RM, Parsons DB (2003) The changing character of precipitation. Bull Am Meteorol Soc 84:1205–1217

Wasko C, Sharma A (2015) Steeper temporal distribution of rain intensity at higher temperatures within Australian storms. Nat Geosci 8:527–529

Wasko C, Sharma A, Johnson F (2015) Does storm duration modulate the extreme precipitation-temperature scaling relationship? Geophys Res Lett 42:8783–8790

Westra S, Fowler HJ, Evans JP, Alexander LV and others (2014) Future changes to the intensity and frequency of short-duration extreme rainfall. Rev Geophys 52:522–555

6

Enhancing the resilience capacity of sensitive mountain forest ecosystems under environmental change (SENSFOR)

F. E. Wielgolaski[1,*], K. Laine[2], J. Inkeröinen[2], O. Skre[3]

[1]Department of Bioscience, University of Oslo, PO Box 1066 Blindern, 0316 Oslo, Norway
[2]Thule Institute, University of Oulu, PO Box 7300, 90014 University of Oulu, Finland
[3]Skre Nature and Environment, Fanaflaten 4, 5244 Fana, Norway

ABSTRACT: The treeline ecotone and all treeline ecosystems (i.e. mountain forests) are important indicators of environmental change. They are heavily affected by environmental drivers, in particular by changes in climate and in land use (the latter change often being land abandonment, resulting in natural reforestation). The Europe COST Action ES1203 (Enhancing the resilience capacity of SENSitive mountain FORest ecosystems under environmental change; SENSFOR) initiative focused on treeline ecosystems in relation to such changes. SENSFOR evaluated drivers and the extent of contemporary and future environmental changes in European mountain forests, and developed methods for estimating forest resilience and defining the consequences for society. The outcome of the SENSFOR initiative provides a scientific basis on which management strategies can be developed and adjusted in cooperation with regional and local stakeholders. In addition, SENSFOR provides recommendations for policy makers at European and national levels. Through application of the DPSIR (Driver, Pressure, State, Impact, Response) framework, the findings of the SENSFOR network (consisting of 24 countries) contribute to strategy development for ecosystem service preservation and biodiversity conservation in sensitive European treeline areas.

KEY WORDS: Treeline · Climate change · Land use · Resilience · DPSIR framework

Current challenges for mountain forest ecosystems

Mountain forests (i.e. treeline ecosystems) range from the area close to the upper timber line, through the treeline ecotone with a more sparse tree cover and generally lower trees, to the tree species line and low alpine areas above (Wielgolaski et al. 2017, this Special). Climatic treelines are highly responsive to changes in climate, and can be used as indicators of ecological and socio-ecological changes in all mountain forests (Zhiyanski et al. 2017). Indicators of ecological change include trees, their growth form and seedling production, biodiversity of plants and animals, and soil indicators such as carbon stock and soil biodiversity (Broll et al. 2016). The variability of stakeholder incomes—due to, for example, changes in tourism caused by land use and climate change—can be a valuable indicator of economic change (Sarkki et al. 2017c, this Special). An important indicator of socio-ecological change is the conflict level (cf. Sarkki et al. 2015, 2017a) between different user groups (stakeholders) over the use of ecosystem services (ESs).

Warmer climatic conditions and longer growing seasons through time are believed to increase the limits of the mountain forests and treelines to higher elevations and latitudes (Holtmeier 2009, Körner

*Corresponding author: f.e.wielgolaski@ibv.uio.no

2012). However, warmer climate increases the occurrence of tree insects and diseases (e.g. Ayres & Lombardero 2000, Wielgolaski et al. 2017) and—in middle and lower latitudes—also summer droughts, which have a negative impact on the wildfire resilience of the vegetation (e.g. Skre et al. 2002). At the same time, treeline ecosystems are also used as a reference for the efficacy of climate change mitigation policies (Skre et al. 2017, this Special). The altitudinal shift and/or structural change of treeline ecosystems reflect anthropogenic pressure on the environment, both through human-induced climate and land use changes. Consequently, anthropogenic treelines may be a result of changes in land use such as land abandonment with natural reforestation of strongly grazed area (Wielgolaski et al. 2017). In parts of Europe, sustainability of treeline ecosystems is threatened by erosion and soil degradation, due to overgrazing and other environmental drivers (e.g. Theurillat & Guisan 2001). Land-use-induced challenges to treeline ecosystems include marginalization of agricultural and timber production, water conflicts, changes in the protection status of forests and landscape degradation due to intensification of land use or land abandonment, frequently leading to changes in biodiversity (cf. Sarkki et al. 2015, 2017a, Wielgolaski et al. 2017).

Treeline ecosystems are important both for provisioning, regulating and cultural ESs; cf. Sarkki et al. (2017b, this Special) for definitions. The multifunctional character of mountain forest calls for an assessment of synergies and trade-offs between different ecosystems and their users. The development of policy guidelines for stakeholders of treeline ecosystems should result in a more sustainable and socially fair use of mountain forest ESs. The portfolio of ESs in treeline ecosystems may, however, vary considerably, depending on factors such as biogeographical region, elevation, relief, site conditions, population density and land use. While environmental policies are often clearly expressed from a broad-scale perspective, explicit management or policy recommendations for mountain forests are often missing (Sarkki et al. 2017a,c). Decision makers need fundamental knowledge about the ESs of these ecosystems to shape their policies and programs. Previous research has identified key ESs in European mountain areas, but there is often a lack of awareness of the importance of ESs (EEA 2010), and the policy implications that result from this. Thus there is a clear need to map ESs—along with the challenges related to these ESs—in different European treeline ecosystems, in order to enhance local governance.

The SENSFOR initiative

This CR Special is based on scientific case studies and syntheses of results from the 4 yr EU COST Action ES1203 project 'Enhancing the resilience capacity of sensitive mountain forest ecosystems under environmental change' (SENSFOR) (Zhiyanski et al. 2017) in which representatives from 24 European countries participated (Fig. 1). Several studies of European mountain treeline ecotone forests (Fig. 2) are presented in this CR Special. The studies included investigations of ecological processes in the highest elevated or northernmost patches of trees in undisturbed landscapes, the similarities and differences between treeline areas in Europe, and impacts of land use and climate change pressures on ecosystem processes.

The authors contributing to this CR Special are both social and natural scientists. Different disciplines are bridged through innovative approaches to the ES concept. Examples of research methodologies used in the Special include data processing, vegetation analyses, soil sampling, case studies and scenarios, GIS based assessments of treeline changes, and modelling of climate-change and land-use-change impacts. Investigations of multifunctional land use, synergies and trade-offs pertaining to the use of ESs and of ecosystem-based management and governance are also included. This CR Special consists of 12 contributions, including both review and original case studies, investigating the 4 SENSFOR objectives:

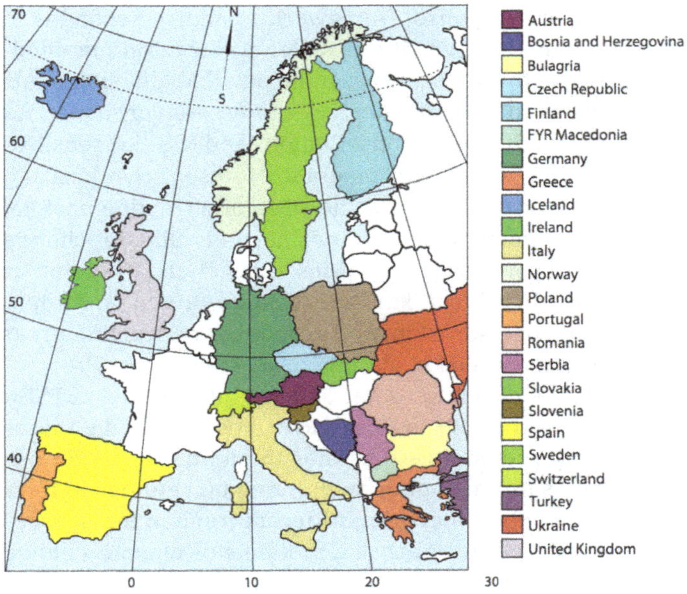

Figure 1. Countries participating in the SENSFOR project

Fig. 2. Examples of treeline ecotones. Left southern Europe (Photo: O. Skre), right northern Europe (Photo: K. Laine)

(1) Identifying main drivers of environmental change at different scales. Considering consequences for biodiversity, ecosystem functions and services, and sustainable resource use. Developing a scientific basis for best management practice in sensitive treeline ecosystems (i.e. mountain forests). Selected case studies on European high-altitude (Dinca et al. 2017, Fleischer et al. 2017, Holtmeier & Broll 2017) and on sub-polar ecosystems (Holtmeier & Broll 2017, Karlsen et al. 2017, Skre et al. 2017) were performed as part of this objective. In addition, Cudlín et al. (2017) and Wielgolaski et al. (2017) evaluated functions and processes of the treeline ecotones across Europe.

(2) Using the DPSIR (i.e. driver, pressure, state, impact, response) framework to analyse changes in treeline biodiversity, to develop monitoring methodology, to build scenarios of possible developments and land use changes, and to assess consequences for biodiversity conservation and ES in sensitive treeline areas (i.e. mountain forests). Synthesizing existing data on how the state of a treeline ecotone may function as an indicator of climate change effects on ecosystems (Moscatelli et al. 2017), and as an indicator of land-use-change-driven anthropogenic pressures on, and responses of, treeline areas (Kyriazopoulos et al. 2017, Sarkki et al. 2017b).

(3) Collecting data and re-interpreting LTSER (long-term socio-ecological research) databases using regional case studies to identify research priorities for managing environmental changes in treeline areas. Comparing different treeline forest areas in Europe in terms of ES effects of climate change and land use development, and examining which features of ecosystems will work as a basis for wide ranging services (Cudlín et al. 2017, Kyriazopoulos et al. 2017).

(4) Synthesizing knowledge of established networks of treeline specialists and local stakeholders throughout Europe. Disseminating the results in meetings, workshops and conferences. Working out policy recommendations for best management practises in these areas, including eco-sociological examinations of the synergies and trade-offs between different ES beneficiaries. Proposing ways to enhance governance at various institutional levels from the EU level to decision making at a local level (Sarkki et al. 2017c).

It is important to note that envisioning the future and working to transform potential scenarios in science and policy may be seriously limited by the pragmatic constraints of pre-existing science and policy (Nijnik et al. 2017).

In order to fulfil the SENSFOR objectives, the following working groups (WGs) were established:

WG1. Analysing the state of and changes in ecosystem structures and functions, focusing on identifying DPSIR factors in case study regions, in order to demonstrate the current status of treeline forests.

WG2. Creating a holistic set of indicators for vulnerability and resilience of coupled socio-ecological systems, based on DPSIR framework analyses.

WG3. Organizing workshops between researchers and user groups, to develop scenarios of integrated knowledge for management practices.

WG4. Disseminating knowledge from the SENSFOR COST Action to users at various levels (e.g. tourist associations, nature conservation bodies, local communities and climate change researchers).

The main results from the SENSFOR COST Action are published in a summary report (Zhiyanski et al. 2017) from the COST office, as well as in various deliverables and in this CR Special.

Acknowledgements. This article is based on work from COST Action ES 1203 (SENSFOR) supported by COST (European Cooperation in Science and Technology; www.cost.eu).

LITERATURE CITED

Ayres MP, Lombardero MJ (2000) Assessing the consequences of global change for forest disturbance from herbivores and pathogenes. Sci Total Environ 262:263–286

Broll G, Jokinen M, Aradottir AL, Cudlín P and others (2016) Working group 2: indicators of changes in the treeline ecotone. SENSFOR Deliverable 5. COST Action ES1203, www.sensforcost.eu/images/Deliverable_5%202016_FINAL.pdf

Cudlín P, Klopčič M, Tognetti R, Malis F and others (2017) Drivers of treeline shift in different European mountains. Clim Res 73:135–150

Dinca L, Nita MD, Hofgaard A, Alados CL and others (2017) Forests dynamics in the montane–alpine boundary: a comparative study using satellite imagery and climate data. Clim Res 73:97–110

European Environment Agegency (EEA) (2010) Europe's ecological backbone: recognising the true value of our mountains. European Environment Agency, Copenhagen

Fleischer P, Pichler V, Fleischer P Jr, Holko L and others (2017) Forest ecosystem services affected by natural disturbances, climate and land-use changes in the Tatra Mountains. Clim Res 73:57–71

Holtmeier FK (2009) Mountain timberlines: ecology, patchiness, and dynamics. Advances in Global Change Research 36. Springer, Dordrecht

Holtmeier FK, Broll G (2017) Feedback effects of clonal groups and tree clusters on site conditions at the treeline: implications for treeline dynamics. Clim Res 73:85–96

Karlsen SR, Tømmervik H, Johansen B, Riseth JÅ (2017) Future forest distribution on Finnmarksvidda, North Norway. Clim Res 73:125–133

Körner C (2012) Alpine treelines. Springer, Basel

Kyriazopoulos AP, Skre O, Sarkki S, Wielgolaski FE, Abraham EM, Ficko A (2017) Human–environment dynamics in European treeline ecosystems: a synthesis based on the DPSIR framework. Clim Res 73:17–29

Moscatelli MC, Bonifacio E, Chiti T, Cudlín P and others (2017) Soil properties as indicators of treeline dynamics in relation to anthropogenic pressure and climate change. Clim Res 73:73–84

Nijnik A, Nijnik M, Kopiy S, Zahvoyska L, Sarkki S, Kopiy L, Miller D (2017) Identifying and understanding attitudinal diversity on multi-functional changes in woodlands of the Ukrainian Carpathians. Clim Res 73:45–56

Sarkki S, Grunewald K, Nijnik M, Zahvoyska L and others (2015) Problems and proposals for good environmental management: empirical assessment of European treeline areas. SENSFOR Deliverable 4. COST Action ES1203, www.sensforcost.eu/images/Deliverable%204.pdf

Sarkki S, Ficko A, Grunewald K, Kyriazopoulos AP, Nijnik M (2017a) How pragmatism in environmental science and policy can undermine sustainability transformations: the case of marginalized mountain areas under climate and land use change. Sustain Sci 12:549

Sarkki S, Ficko A, Wielgolaski FE, Abraham EM and others (2017b) Assessing the resilient provision of ecosystem services by social-ecological systems: introduction and theory. Clim Res 73:7–15

Sarkki S, Jokinen M, Nijnik M, Zahvoyska L and others (2017c) Social equity in governance of ecosystem services: synthesis from European treeline areas. Clim Res 73:31–44

Skre O, Baxter R, Crawford RMM, Callaghan TV, Fedorkov A (2002) How will the tundra-taiga interface respond to climate change? Ambio (Special Report 12):37–46

Skre O, Wertz B, Wielgolaski FE, Szydlowska P, Karlsen SR (2017) Bioclimatic effects on different mountain birch populations in Fennoscandia. Clim Res 73:111–124

Theurillat JP, Guisan A (2001) Potential impact of climate on vegetation in the European Alps: a review. Clim Change 50:77–109

Wielgolaski FE, Hofgaard A, Holtmeier FK (2017) Sensitivity to environmental change of the treeline ecotone and its associated biodiversity in European mountains. Clim Res 73:151–166

Zhiyanski M, Bratanova-Doncheva S, Kyriazopoulos A, Broll G and others (2017) Final leaflet: enhancing the resilience capacity of sensitive mountain forest ecosystems under environmental change (SENSFOR). COST Action ES1203, www.sensforcost.eu/images/Leaflet_6.pdf

7

Human–environment dynamics in European treeline ecosystems: a synthesis based on the DPSIR framework

A. P. Kyriazopoulos[1], O. Skre[2,*], S. Sarkki[3], F. E. Wielgolaski[4], E. M. Abraham[5], A. Ficko[6]

[1]Department of Forestry and Management of the Environment and Natural Resources, Democritus University of Thrace, 193 Pantazidou str., 68200 Orestiada, Greece

[2]Skre Nature and Environment, Fanaflaten 4, 5244 Fana, Norway

[3]Cultural Anthropology, Faculty of Humanities, PO Box 1000, University of Oulu, 90014 Oulu, Finland

[4]Department of Bioscience, University of Oslo, PO Box 1066 Blindern, 0316 Oslo, Norway

[5]Laboratory of Range Science, Department of Forestry and Natural Environment, Aristotle University of Thessaloniki, 54124 Thessaloniki, Greece

[6]Biotechnical Faculty, Department of Forestry and Renewable Forest Resources, University of Ljubljana, Vecna pot 83, 1000 Ljubljana, Slovenia

ABSTRACT: The state of, and changes to, altitudinal and polar treeline ecosystems and their services in selected mountain regions in Europe were analyzed using the drivers-pressures-state-impacts-responses (DPSIR) framework. The analysis was based on 45 responses of experts from 19 countries to 2 semi-structured questionnaires on treeline ecosystem services (ESs), stakeholders and the DPSIR factors, and 11 case study descriptions of best management practices. The experts recognized climate and land-use changes as the main drivers, resulting in various pressures that contrasted among the regions. The impacts of the pressures were mainly considered as negative (e.g. loss of biodiversity, root rot diseases, moth and bark beetle outbreaks, wild fires, decrease of (sub)alpine grasslands, browsing), but also as positive (e.g. increase in forested area). The influence of climate warming, altered precipitation regimes, a longer growing season, annual variation in winter climate and increased ground-level ozone concentrations were considered less critical for recent treeline dynamics than land abandonment, increased tourism and livestock pressure. Current policy responses to emerging pressures and stakeholder demands were considered insufficient and incoherent. Mitigation, adaptation and restoration actions were rare and with no evident long-term impact. We conclude that (1) locally-specific human–environment interactions have greater influence on treeline dynamics than global warming; (2) ecological and social sustainability of the treeline areas can be enhanced by simultaneously promoting traditional land use and regulating tourism development; (3) ES users should look for new opportunities arising from environmental change rather than trying to sustain current levels of ESs indefinitely; and (4) to safeguard the unique ecological and social values of treeline areas, more coherent and proactive policies are needed.

KEY WORDS: Mountain forests · DPSIR framework · Climate change · Land-use change · Ecosystem services · Disturbances · Biodiversity

1. INTRODUCTION

Mountain forest ecosystems are among the most endangered ecosystems in the world (Broll & Keplin 2005). In Europe and elsewhere, climate change (Theurillat & Guisan 2001, Grace et al. 2002, Skre et al. 2002, Kullman & Öberg 2009, Smith et al. 2009), and land-use change (Bryn & Daugstad 2001,

*Corresponding author: oddvar@nmvskre.no

§Advance View was available online May 18, 2017

Gehrig-Fasel et al. 2007, Hofgaard et al. 2013, Ameztegui et al. 2016) are 2 important drivers of environmental change. The treeline ecotone (Holtmeier 2009) can be particularly useful as an early indicator of environmental change (Kupfer & Cairns 1996). Altitudinal and polar treeline ecotones are primarily controlled by climate, as seasonal mean temperature decreases with increasing elevation and latitude. The isotherm theory for natural treeline formation has been confirmed in several empirical studies and modelling exercises (Paulsen & Körner 2014). However, centuries of human disturbances have altered the climatic position of treelines. To capture the complexity of factors influencing the dynamics of treeline ecotones, the ecological definition of a treeline as an ecotone should be extended to include 'treeline-related administrative areas, and associated landscapes and ecosystems' (Sarkki et al. 2016a, p. 2020).

The contributions that ecosystems make to human well-being arising from the interaction of biotic and abiotic processes (Grunewald & Bastian 2015) have been described as ecosystem services (ESs). To systematically arrange the benefits that humans receive from nature, several ES classifications have been proposed (e.g. Boyd & Banzhaf 2007, Luck et al. 2009) and used in studying the treelines (e.g. Sarkki et al. 2016a), such as classification into provisioning, regulation and maintenance, and cultural services.

The subalpine and subarctic forests close to the treeline provide a number of resources for local communities with traditionally low land-use intensity. They provide regulating ESs, such as erosion and flood control and reduction of environmental risks such as avalanches and forest fires. Access to exclusive and non-exclusive ESs may potentially cause conflicts among stakeholders due to overuse of some ESs by some stakeholders. Treeline ESs have been analyzed in small areas at or near the treeline and for certain ESs in mountainous regions (e.g. Grabherr 2009, Hastik et al. 2015). However, there is little information regarding Europe-wide mapping of ESs combined with stakeholder and governance analysis. Sarkki et al. (2016a), for instance, identified key ESs in European treeline areas, along with treeline-relevant stakeholders and the threats they pose to, and benefits they receive from, ESs. They found context- and treeline-specific behavior of stakeholders and governance structures that insufficiently addressed the sustainability of these areas. A similar finding was obtained by Sarvašová et al. (2014). However, to be able to understand the pressures on treeline ESs and adaptive responses, a holistic framework is required for studying cause–effect relationships in treeline areas (see Sarkki et al. 2017a, this Special).

In this context, the drivers-pressures-state-impacts-responses (DPSIR) framework (see Section 2.1 below) is useful for describing the interactions between society and the environment. It has been adopted by the European Environment Agency (EEA 2016) to show how social–ecological systems function in a dynamic way and what the interactions are among different factors. The 'drivers' (e.g. climate and land-use change) create 'pressures' that are causing social–ecological changes in treeline areas. The resulting change in the 'state' of the social–ecological system has 'impacts' on the functioning of the system, but societal 'responses' may restore the desired state or reduce pressure, trying to make the social–ecological system resilient to change (cf. Fig. A1 in the Appendix)

This study examines factors influencing human–environment dynamics at or near European treelines by synthesizing various results of the SENSFOR project ('Enhancing the resilience capacity of SENSitive mountain FORest ecosystems under environmental change'). Empirical results summarized in this paper have been previously published as technical reports, but with no overall synthesis. In this study, pressures to treeline ESs (Kyriazopoulos et al. 2014), the state of ESs in treeline areas (Sarkki et al. 2013) and potential policy responses (Sarkki et al. 2015, 2016a,b, 2017b, this Special) are synthesized under the DPSIR framework for the first time. We summarize the most important DPSIR factors and their threats to, and benefits for, the coupled human–environmental system and ESs. This synthesis allows us to draw conclusions about the relationships between ecological and social sustainability in European treeline areas and their relationships to governance responses.

2. METHODOLOGY

2.1. The DPSIR framework

The DPSIR framework is a well-established causal framework for describing the interplay between the environment and socio-economic activities (EEA 2016) that has its roots in the stress-response environmental statistical system (S-RESS) proposed by Friend (1979). The basic logic behind the DPSIR framework is that any status or change in a social–ecological system is related to its driving forces and pressures through a number of feedbacks. Humans play a determinant role in steering the social–ecological system and structuring its state.

The DPSIR framework belongs to the family of systems analysis frameworks. It shares many similarities with the Framework for the Development of Environment Statistics (FDES), the Pressure-State-Response (PSR) framework, the Pressure-State-Response/Effect (PSR/E) framework, the Pressure-State-Impact-Responses (PSIR) framework and the Driver-Pressure-State-Welfare-Response (DPSWR) framework (see Cooper 2013 for a review).

Applications of the DPSIR framework in practice differ with regard to the interpretation of DPSIR factors, and may include a number of modifications. We used an interpretation that closely follows the original definition by the EEA (2016). For each of the categories in the DPSIR framework, we present the most important factors and issues as reported by the respondents and discuss how these categories link to each other. Responses were grouped into restoration practices, adaptation practices and mitigation measures.

2.2. Surveys

To identify the DPSIR factors, we used the multiple case study analysis, which is a qualitative technique of data gathering when representative sampling is difficult to conduct. The multiple case study analysis enabled us to generalize causalities between the drivers, pressures, state, impact and responses based on similarities in patterns and processes.

We sent 2 questionnaires and a call for detailed case study descriptions to more than 100 experts from 20 European countries to obtain the widest possible assessment on the state of altitudinal and polar treelines in Europe. The aims of these surveys were to (1) analyze the present state of, and potential changes in, ecosystems with a particular focus on identifying DPSIR factors in each case study region; (2) map the drivers and pressures of ecosystem change; (3) analyze the history and present state of ecosystems and their services for different land uses and management practices, including the identification of trends in ES delivery capacity and socio-political adaptation; (4) identify current societal responses to ES-related challenges; and (5) generate a holistic understanding of treeline ecotone ecosystems and their services, based on the DPSIR framework.

The first questionnaire consisted of mainly closed-ended questions on the relevance of ESs, stakeholder benefits from, and threats to, ESs, and governance and science–society relationships. The questionnaire was sent in 2013 to approximately 100 SENSFOR consortium members (Sarkki et al. 2013, 2016a, Table 1). The assessment of ESs was based on the Common International Classification of Ecosystem Services using 5-point Likert scale (Sarkki et al. 2016a). Altogether, 22 treeline case study descriptions from 15 countries were received, describing 20 different treeline areas in Europe (see Sarkki et al. 2013, 2016a for detailed description of the methodology and the questionnaire). Stakeholder benefits from, and threats to, treeline ESs were then cross-tabulated with pressures identified in the second questionnaire.

The second questionnaire was distributed to more than 100 experts in late 2013 using the SENSFOR and Mountain Research Initiative (MRI) network. This mostly open-ended questionnaire (see Kyriazopoulos et al. 2014 for the questions) used the DPSIR framework, and also included a request for a detailed description of the size, altitude, latitude, climate, geology, dominant tree species and the major wild, semi-domestic and domestic herbivores in the area. The questionnaire included several questions regarding land-use change and climate change trends to estimate possible effects of these 2 drivers (see Table A1 in the Appendix). A total of 26 responses from different European treeline case study regions were received. A follow-up questionnaire was distributed in mid-2014 to all who responded to this questionnaire, focusing on the state and responses in the treeline areas. Eight responses from different treeline areas were received in the second round (Kyriazopoulos et al. 2014).

A call for detailed case study descriptions was sent in 2015 to SENSFOR consortium members asking them to describe best management practices in European treeline areas and to suggest proposals for enhanced governance of treeline areas. Eleven descriptions of successful and failed practices were received, identifying 75 proposals for enhanced governance. These proposals were inductively clustered into several groups (Sarkki et al. 2015, 2017b).

3. RESULTS AND DISCUSSION

3.1. Drivers

Climate change and land-use change were recognized as the main drivers of treeline ecotone dynamics (cf. Kyriazopoulos et al. 2014). Although high-resolution spatio-temporal data indicating directional impact of climate or land-use change were not provided, both drivers were unanimously suggested as being of major importance for all case study regions.

Table 1. Case study areas used for the assessment of human–environment dynamics in European treeline areas (adapted from Kyriazopoulos et al. 2014, Sarkki et al. 2015, 2016a, 2017b). ES: ecosystem services; DPSIR: drivers-pressures-state-impacts-responses framework, NP: National Park

Country	Case study area	Type	Questionnaire on:		
			ES, stakeholders and governance	The DPSIR	Good environmental management
Iceland	Almenningar, Southern Iceland	Polar	✓	✓	
Ireland	Burrishoole	Non-climatic	✓		
Spain	Cantabrian range, Iberian peninsula	Altitudinal		✓	
Slovakia	Carpathian Mts.	Altitudinal		✓	
Russia	Caucasus Mts.	Altitudinal		✓	
Italy	Central Alps	Altitudinal		✓	
Spain	Central Pyrenees	Altitudinal	✓		✓
Slovenia	Dinaric Mts, Sneznik Mt.	Altitudinal	✓		
Serbia	Dinaric Mts., Tara NP	Altitudinal		✓	
Norway	Dovre Mt., Central Norway	Altitudinal		✓	
Czech Republic	Hercynian Mts.	Altitudinal		✓	
Ukraine	Khust and Rachiv regions, villages Nyzhniy Bystry and Bohdan, Carpathians	Altitudinal			✓
Russia	Khybini Mts., Kola Peninsula	Polar		✓	
Finland	Kilpisjärvi-Käsivarsi region, NW Finnish Lapland	Altitudinal	✓		✓
Sweden	Lake Torneträsk Catchment (Abisko)	Altitudinal	✓		
UK	Lochaber Forest District, North Western Scotland	Non-climatic			✓
Ukraine	Lviv, Chernivtsi and Ivano-Frankivsk regions, Carpathians	Altitudinal			✓
Italy	Majella NP, central Apennines	Altitudinal	✓	✓	
Spain	Montseny, NE Spain	Altitudinal		✓	
UK	Moray, Aberdeenshire and the Dee river catchment, Cairngorms NP	Non-climatic	✓	✓	✓
Greece	Mount Parnon, South Greece	Altitudinal		✓	
Finland	Muonio, North western Lapland	Altitudinal			✓
Finland / Norway	Northern Fennoscandia	Polar		✓	
Norway	Northern Norway	Polar		✓	
Spain	Northern Pyrenees	Altitudinal		✓	
Sweden / Norway	Northern Scandinavia	Polar	✓		
Germany	Berchtesgaden-Alps NP	Altitudinal	✓		
Bulgaria	Central Balkan NP	Altitudinal	✓		
Romania	Piatra Craiului NP	Altitudinal	✓		
Italy	Paneveggio-Pale di San Martino NP, Eastern Alps	Altitudinal	✓	✓	
Portugal	Peneda do Geres NP	Altitudinal		✓	✓
Greece	Pindos Mts.	Altitudinal	✓	✓	✓
Greece	Psiloritis, Crete	Altitudinal		✓	
Bulgaria	Rila and Pirin Mts.	Altitudinal	✓	✓	✓
Spain	Sierra Nevada	Altitudinal		✓	
Slovenia	South-Eastern Alps	Altitudinal		✓	
Norway	Southern Norway (Hardangervidda)	Altitudinal	✓		
Spain	Southern Pyrenees	Altitudinal		✓	
Bulgaria	Stara Planina	Altitudinal		✓	
Czech Republic	Sudeten Mts.	Altitudinal	✓		
Slovakia	Tatra NP, Dolina Parichvost	Altitudinal			✓
Slovakia	West Carpathians	Altitudinal	✓	✓	
Slovakia	Západné Tatry Mts. (Western Tatras)	Altitudinal	✓		

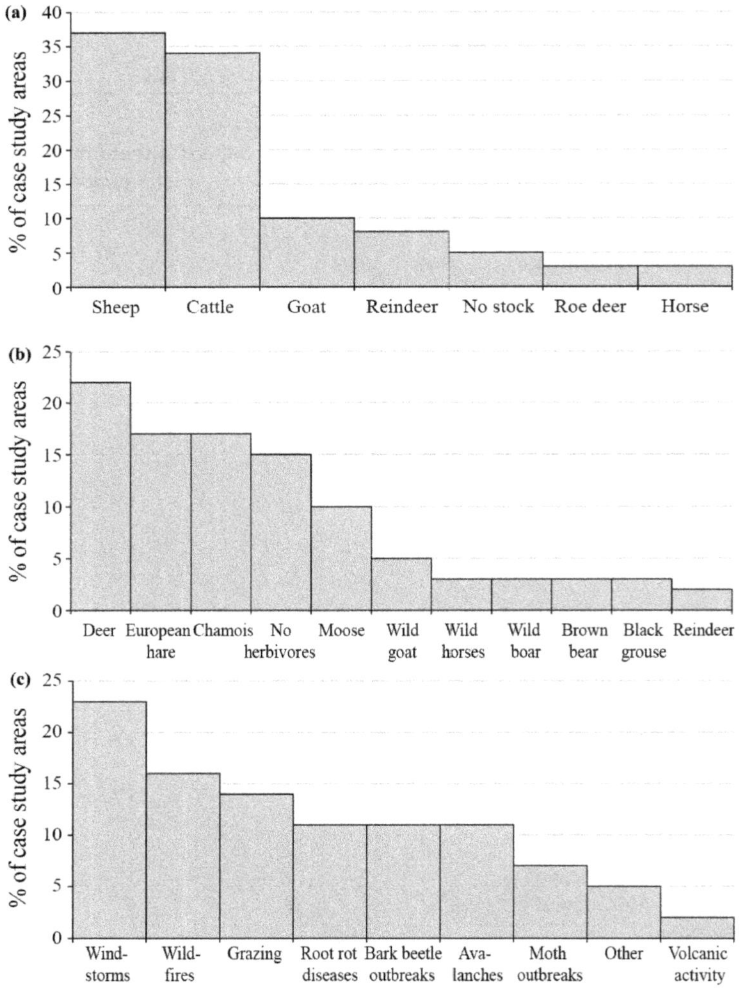

Fig. 1. (a) Main grazing stock, (b) wild herbivores and (c) main disturbances at and near treeline ecotones in Europe (based on Kyriazopoulos et al. 2014)

It is well-documented that treeline ecotones are sensitive to climatic fluctuations and, consequently, are useful indicators of climate change (Theurillat & Guisan 2001, Grace et al. 2002, Smith et al. 2009, Kullman & Öberg 2009). However, Kupfer & Cairns (1996), among others, warned against the limitations of using treelines as climate change indicators in highly altered anthropogenic landscapes. Firstly, the response of an ecotone to changing climate is at a timescale of several decades, in which these impacts may be difficult to disentangle from impacts of more short-term duration. Moreover, although tree growth in general becomes more and more constrained by the harsher environmental conditions associated with increasing altitude or latitude (Körner 2012), the altitudinal or latitudinal forest- and tree-limits may show different responses to climate change (Hofgaard et al. 2013), mostly due to differences in upslope edaphic limitations and competitive interactions with established canopy and subcanopy individuals (Kupfer & Cairns 1996).

The main changes that can counteract climate-driven forest expansion, especially at the local level (Debussche et al. 1999), are land-use changes. Land-use changes in treeline areas have been strongly linked to the dynamics of the socio-economic systems in Europe over the last century (e.g. reduced summer farming in Scandinavia; Bryn & Daugstad 2001). Locally-specific impacts of land-use changes are described in detail in Section 3.2. In Iceland, volcanic activity was identified as a driver causing specific natural disturbances. Other disturbances, such as insect attacks, wildfires and windthrows were classified as pressures related mostly to climate change.

3.2. Pressures

The pressures acting on the treeline ecotones were divided into 3 categories according to the forces driving them: (1) climate change, (2) land-use changes and (3) volcanic activity. Our findings suggest that in the past decades, land-use changes have been more critical for treeline dynamics than climate change.

According to the respondents, the main pressure related to climate change was climate warming, which was sometimes accompanied with increased precipitation, resulting in a longer growing season, summer drought, and inter-annual variation in winter climate as well as increased ground-level ozone concentrations (cf. Kyriazopoulos et al. 2014). A longer growing season was reported as a positive result of the warmer climate in most of the northern and the central European study areas. The predicted warmer climate could potentially move the treeline ecotone toward more northern latitudes. Kullman (2004) reported that seedlings of white birch *Betula pubescens*, Norway spruce *Picea abies* and Scots pine *Pinus sylvestris* have recently become established 500–700 m above their recent tree limits. However, occurrence of tree seedlings above the tree limit may also be interpreted as part of the natural dynamics of the zone, with largely ephemeral seedling populations (Hofgaard et al. 2013). An advance of the treeline ecotone will probably cause an increase in forest cover and create the potential for development of local forestry-dependent economies, but this process could take decades or centuries (Skre et al. 2002, Van Bogaert et al. 2011, De Wit et al. 2014). In contrast, inter-annual variation in winter climate and summer drought was reported as a negative result of climate change, particularly in the southern European regions (cf. Kyriazopoulos et al. 2014).

Land abandonment was identified as the major pressure arising from land-use change. This pressure manifests through abandonment of traditional pastoral activities mirroring the socio-economic changes in mountainous areas. The most common pressure was abandonment of pastures due to a decrease in livestock (54 % of cases). The impact of this change was considered positive in all cases except in the southern part of the Balkan Peninsula and some parts of the Carpathian Mountains. Some respondents reported land abandonment in general (13 %) and an increase in grazing (11 %), the latter reported only from northern Europe and in Crete, Greece. Increased livestock farming was related to an increasing demand for high-quality reindeer and goat products, and local livestock farming traditions (Kyriazopoulos et al. 2014). Our results confirmed the conclusions of several studies, that the current treeline position mostly occurs well below its potential location due to the long history of anthropogenic disturbances such as clear-cutting, burning and grazing (e.g. Gehrig-Fasel et al. 2007, Palombo et al. 2013, Ameztegui et al. 2016).

Populations of grazing livestock were found to be decreasing in 67 % of the study regions, increasing in 12 % and stable in 17 %; there was no data for the rest of the cases. The most important grazing stocks in European treeline areas (Fig. 1) include cattle and sheep (34 and 37 %, respectively), goats (10 %) and semi-domestic reindeer (8 %). Horses and roe deer were considered less important pressures (3 % each).

Livestock grazing was recognized as a positive pressure, since high grazing intensity reduces or prevents tree establishment (Potthoff 2009). In some cases in Spain and Greece, the abandonment of grazing in treeline ecosystems was recognized as unfavorable due to shrub encroachment, which could pose a serious threat to forage productivity and biodiversity of the subalpine grasslands. In these regions, grazing was recognized as an ecological tool to restore or conserve these habitats (Papanastasis 2009). Using similar justification, some respondents described increased livestock grazing as negative because of disturbances to vegetation dynamics and an increase in soil erosion. However, a negative effect of livestock grazing was highlighted in only a few case study areas in northern Europe and Crete, Greece.

Wild herbivores were described as a pressure in 90 % of the case studies (Fig. 1b); Cervidae (deer family), European hare *Lepus europaeus* and chamois *Rupicapra* spp. were among the main species in 22, 17 and 17 % of the case studies, respectively. Additionally, elk *Alces alces* was listed as a pressure in 10 % of the cases, but in certain areas (e.g. Scandinavia) it occurs along with wild reindeer *Rangifer tarandus*. Wild goat *Capra aegagrus*, brown bear *Ursus arctos*, wild boar *Sus scrofa* and black grouse *Tetrao tetrix* were reported at lower percentages (3–5 % of the cases).

For some areas in Europe, i.e. Scotland and northern Fennoscandia (Danell et al. 1991, Oksanen et al. 1995, Hester et al. 2004) the impact of wild and semi-domestic herbivorous species on treeline vegetation is well-documented. In addition to the negative effect of grazing on treeline vegetation, large ungulates may also contribute to local livelihood through income obtained from hunting and tourism (e.g. Hester et al. 1996, Tolvanen et al. 2005, Sarkki et al. 2013). Furthermore, environmental policies may multiply pressures on treeline ES and stakeholders by applying strict conservation guidelines for predators (*U. arctos*, grey wolf *Canis lupus*, Eurasian lynx *Lynx lynx*, eagles) and support for tourism development.

Intensive tourist activity was considered as a negative pressure in 19 % of the cases. Although increased numbers of tourists may financially benefit local communities, it can also negatively impact sensitive ecosystems and species, mainly due to the construction of infrastructure. Furthermore, in some cases industrial development (including mining, and the construction of hydropower plants and windmills) was recognized as a pressure. A majority of the respondents identified limited or no logging activities in the treeline ecotones. The only exception was reported from Caucasus, and was considered a negative effect. Pressures related to volcanic activity were identified only in Iceland. They were reported as negative, as lava and ash can destroy the treeline vegetation.

3.3. State

The state of treeline ecosystems varies greatly by country and region. Here, we describe exemplary states of treelines in 8 case studies (Kyriazopoulos et

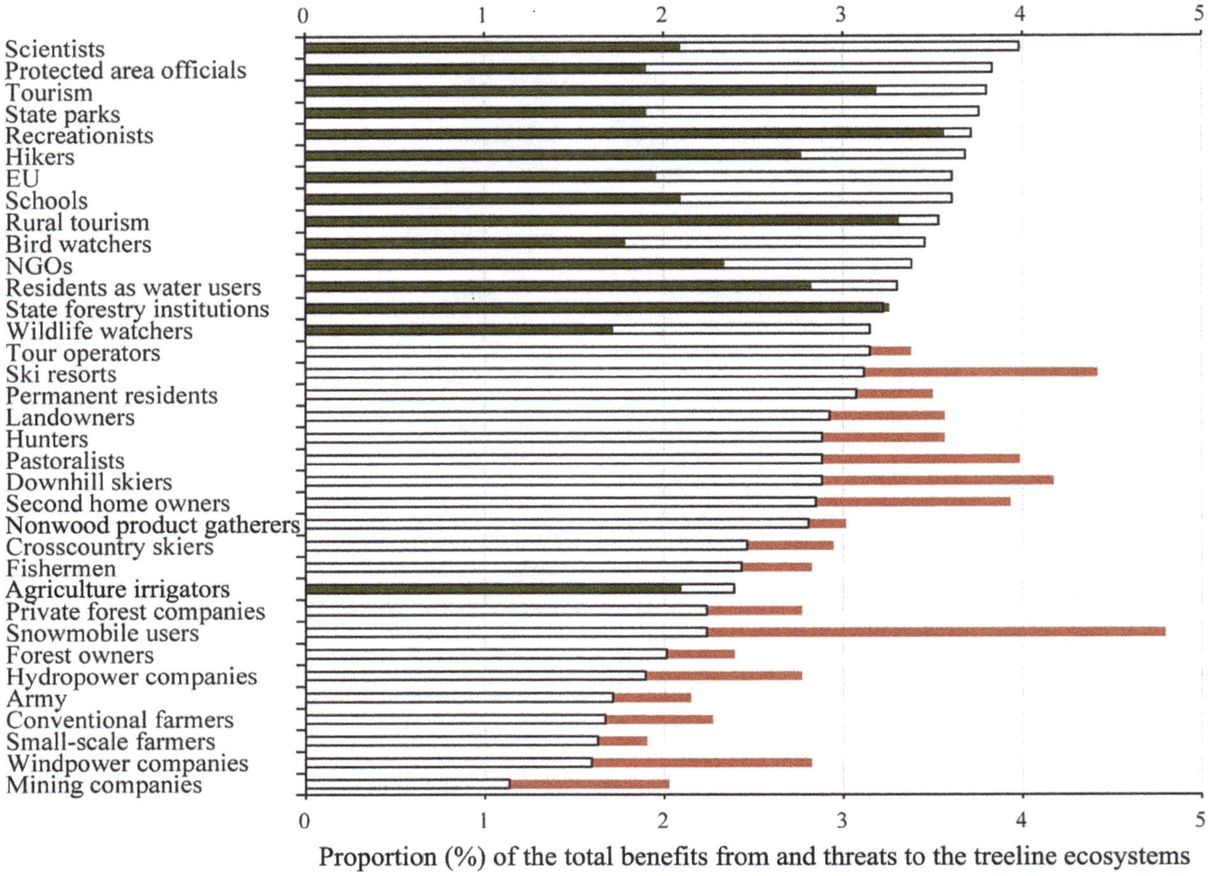

Fig. 2. Main stakeholders in European treeline areas sorted top-down on a scale from 0 to 5 by the benefits they obtain from treeline ecosystems (white bars: the degree a stakeholder gets of the total benefits produced by treeline ecosystems). Colored bars (green or red): degree of the total threats to the treeline ecosystems that is attributable to a certain stakeholder. If this percentage is lower than the percentage the stakeholder gets of the total benefits produced by treeline ecosystems, the bar is colored green (net receivers), else red (net-givers). 'Ski resort business', 'Downhill skiers' and 'Tour operators' are different forms of the tourism industry. 'Recreationists', 'Hikers', 'Bird watchers' and 'Crosscountry skiers' are different forms of nature-based tourism

al. 2014) to illustrate the diversity of the pressures in different parts of Europe.

Most of the Pyrenean treeline ecotones (Spain) were severely disturbed by humans in the past (e.g. grazing and logging), but in the period from the 1930s until 1980, sheep densities decreased in concert with a 61% decrease in the human population (Alados et al. 2014). As a result, 24% of the subalpine grasslands under 2100 m a.s.l. in the Central Pyrenees reverted to forest between the 1980s and 2000s (Gartzia et al. 2014). Since 1960, industrial development, the intensification of agriculture and development of the skiing industry have been the prevailing human activities. Several areas show destruction of the vegetation due to overgrazing and extreme erosion, which is usually conditioned by local topography and disturbances such as snow avalanches (Cudlín et al. 2017, this Special). Elsewhere, encroaching scrublands (horrible broom *Echinospartum horridum*, snow rose *Rhododendron ferrugineum*) dominate former grassland above the current treeline (Camarero et al. 2015). The expansion of woody plants has affected biodiversity and the goods and services provided by grasslands. It has also altered the spatial distribution of wild animals. *S. scrofa*, for instance, is now found to be feeding on grasslands, which rarely happened before.

In the Central Stara Planina Mountains (Bulgaria), burning and grazing in areas near the treeline has resulted in the replacement of conifers by European beech *Fagus sylvatica* and reduced soil carbon storage capacity (Grunewald & Scheithauer 2011). Forest conversion has also decreased soil organic matter. However, climate warming is expected to contribute to upslope advance of conifers as well as *F. sylvatica* (Dakov et al. 1980). Land-use intensity also influenced the quality and composition of soil humus in mountain grasslands. The alteration of natural vegetation was mostly due to fire and grazing (Cudlín et al. 2017). Today, only few cattle graze on Bulgarian high-altitude pastures. In the last few decades, changes to the timberline have mainly been caused by winter sport facilities (Grunewald & Scheithauer 2011).

The current state of Pindos Mountain treeline ecotones (northern Greece) indicates intense human activities in the past, which have since been strongly reduced. As a result, soil erosion risk is being minimized. As a result of land abandonment, shrub encroachment by junipers *Juniperus* spp. in formerly grazed grasslands is common. Further decreases in livestock grazing may negatively affect both floristic and faunistic diversity, especially populations of raptors (Bakaloudis 2016). Conversely, in Psiloritis (Crete), decades of overgrazing have contributed to severe soil erosion. Desertification has become a serious problem, particularly due to fires set by the shepherds to combat unpalatable species, followed by continuous, heavy grazing. However, pastoral activities have positively influenced biodiversity. There is still a relatively high population of raptors in Crete (Xirouchakis 2004). Reduced tree regeneration due to overgrazing has also been reported in North Caucasus (Ukraine). The threat of soil erosion is huge as a result of the overexploitation of forests, fire and intense grazing (Kyriazopoulos et al. 2014).

In some areas of the Carpathians (Czech Republic) the treeline ecotone was subjected to several disturbances, particularly originating from the construction or enlargement of ski resorts, recreation facilities and infrastructure (Cudlín et al. 2013). On the other hand, large tracts of the treeline ecotone are regenerating well via secondary succession due to a reduction in sheep and cattle grazing. (cf. Cudlín et al. 2017). Large-scale windfalls and bark beetle outbreaks in forests below the treeline represent constant pressure on treeline ecosystems resulting in a decrease in the aesthetic qualities of the cultural landscapes and possibly distracting tourists and hikers. These areas are at risk of fire due to forest operations such as road construction and burning wood residuals from salvage logging.

In alpine space, conflicts between stakeholders about ES are common (von der Dunk et al. 2011). In the treelines of the Central Alps (Switzerland) there has been a strong impact from grazing by sheep and cattle, and to some extent, goats (Bebi 1999), followed by a treeline advance as a result of warmer climate and abandonment of cottage farming. In the south-eastern Alps, livestock grazing was historically high at the beginning of the 19th century but decreased after World War II, while pressures from large wild ungulates (mostly red deer *Cervus elaphus*) remained noticeable in most of the central European mountain forests (e.g. Ficko et al. 2016). One of the most important ES provided by the treeline forests in the Alps is avalanche protection (Bebi et al. 2009), often combined with erosion control (Brang et al. 2006, Huber et al. 2013).

In the northern Scandes on the subarctic mountain plateau Finnmarksvidda and adjoining areas in Finnish Lapland (Kevo), birch forest covers a 300 km wide zone north of and above the conifer treeline. In contrast to most treeline areas in Europe, these birch forests have been partly overgrazed by domestic reindeer over the last 50 yr (Tømmervik et al. 2005, Cudlín et al. 2017). Since *Cladonia* lichen is the preferred winter food for reindeer, this overgrazing has dramatically reduced the lichen cover throughout northern Fennoscandia (Helle 2001, Lempa et al. 2005). Because of the warming climate, birch forests are increasingly subjected to attacks by various moth species, particularly *Epirrita autumnata* (Neuvonen et al. 2005, Skre et al. 2017, this Special), because more insect eggs can survive in milder winters (Tenow et al. 2005).

3.4. Impacts

Climate warming has led to increased occurrences of tree diseases (root rot diseases) and insect outbreaks (e.g. moths, bark beetles), which are considered particularly problematic for the northern and central European regions. Several projections have indicated that climate warming is likely to affect the frequency, geographical extent and intensity of insect population outbreaks, with potentially severe consequences for the affected ecosystems (Neuvonen et al. 1999, Ayres & Lombardero 2000, Volney & Fleming 2000, Logan et al. 2003, Battisti et al. 2006). However, some disturbances (e.g. insect outbreaks, forest fires on cold soils) may also be beneficial because they could increase recycling of nutrients and production in the long-term.

Increasing risk of wildfires was interpreted mainly as a consequence of summer drought, which was considered particularly critical for the treeline ecosystems during drought years in southern European regions (Hoff 2013) and in central Europe. Such fires strongly impact tourism in the affected areas.

An upward shift of the treeline ecotone and consequently a potential increase in forested area were associated with both warmer climate and abandonment of traditional agricultural and pastoral activities in mountainous areas. The majority of the respondents interpreted the upward shift of the ecotone positively, while in a few study cases (e.g. Norway, Spain and Greece) an increase in forest cover was recognized as negative as it would result in a decrease in grassland area, loss of biodiversity and possibly negative influences on tourism.

Loss of biodiversity was interpreted as a negative impact in the treeline ecosystems as a result of intense human activities such as overgrazing and massive tourism. Moreover, climate change might also have negative impacts on biodiversity, as it affects ecosystems and the ability of species to adapt (Gitay et al. 2002).

Among the impacts of climate and land-use changes, wind (23 % of the cases), wildfires (16 %) and grazing (14 %) were mentioned as significant (Fig. 1a–c). Avalanches, root rot diseases and bark beetle outbreaks were considered important in 11 % of case studies. Moth outbreaks were identified as a major disturbance, especially in northern Fennoscandinavia and southern Norway. In Iceland, volcanic activity was identified as causing changes to vegetation assemblages. In all cases, except for the Carpathian Mountains, the impacts of natural disturbances were identified as a negative. Direct human activity by logging was nearly absent in most of the study regions (84 %) but was recognized in the Caucasus Mountains (4 %).

Pressures in treeline areas may also be induced by stakeholders and ES users. While using the ES, stakeholders pose threats to ecosystems and could potentially change their state (Fig. 2). The strongest beneficiaries of ESs are scientists, who were considered to be the least threatening stakeholders. Much of the benefit also goes to protected area officials, tourism, state parks, hikers and recreationists. Several stakeholders were considered as net beneficiaries, meaning that the benefits they receive from treeline ecosystems are higher relative to the threats. Stakeholders posing relatively high threats were snowmobile users, ski resorts and windmill companies. However, the negative impact of most of the stakeholders was rather low, and the vast majority of stakeholders in the treeline area operate in a sustainable way (Sarkki et al. 2016a).

3.5. Responses

Linkages between human responses to the outlined drivers, pressures, and their impacts on the current state of treeline ecosystems and their services (Table 2) are based on the practices used in the same regions as outlined in Section 3.3.

Mitigation measures were rarely practiced. Among the mitigation measures, adaptations of silvicultural practices as well as re-zoning and re-assessment of the Natura 2000 sites were mentioned. These measures mostly targeted specific objects, and were operating at the local level, such as prescribed fires near National Parks in Spain. The locally or regionally focused actions were not really mitigation measures based on a precautionary principle, but rather were risk-reduction activities taking place after the occurrence of a large windfall event or bark beetle outbreak (e.g. Slovakia, Slovenia, Spain). Moreover, the effectiveness of mitigation measures was considered to be very limited. Most of the mitigation measures resulted from international or national commitments on nature protection or national legislation on reserves and national park territories. Current mitigation measures are focused on improvement and maintenance of the regulating and supporting ESs, but without a systematic approach. No specific mitigation measures at any level were reported from Russia, Norway, Greece and the Kola Peninsula.

Adaptation practices were extremely rare and of limited effectiveness, and they took place mainly at the local level with no evident long-term impact. The general problem of adaptation was that the adaptation measures tried to alleviate the maintenance of all ESs, which is in contradiction to ecosystem functioning. The trade-offs between different ESs may not always be resolved at the same place and same time, which should be accounted for in the projections of future ES portfolios (Sarkki et al. 2016a). No adaptation practices at any level were reported from Russia, Norway, Slovenia, Greece and the Kola Peninsula.

Restoration was also seldom practiced. Soil erosion was recognized as the main problem that required restoration. However, information on the successfulness of the restoration was missing. This is partly due to the limited data about the changes near the treeline ecotone. GIS and remote sensing methods will be helpful to address this issue. Future monitoring of ecosystem changes will answer the question of whether restoration actions are feasible in a highly dynamic social–ecological system such as the treeline.

Governance and political instruments specifically designed for treeline ecosystem restoration or adaptation to changes were not common either. Detailed case study descriptions on best management practices highlighted various methods for land-use planning (Sarkki et al. 2015, 2017b). Many of the

Table 2. Adaptive measures applied in treeline areas for maintaining ecosystem services (ES). LSU: livestock unit

Pressures	State	Impact	Responses
Local climate fluctuations	Reduced number of days with snow cover	Fewer positive impacts (benefits) for skiing resorts due to shorter ski season	Use of artificial snow
Large predator conservation policies	Increased number of predators in the treeline areas	Predation on herds has negative impacts on pastoralism	Intensive herding practices (e.g. shepherd dogs, electric fences), state compensation to cover losses
Building ski slopes	Erosion	Negative impact on aesthetic value	Land-use zoning, priority areas
Use of motorized vehicles	Trampling effects; experience of nature disturbed	Ecotourists dislike this, attractiveness of the area decreases	Rules on how protected areas can be accessed
Land abandonment	Encroachment, nitrification in the areas with excess LSUs	Decreasing proportion of pastures	Agri-environmental schemes, e.g. management of low-intensity pasture systems, min.–max. LSU load control, *Rumex alpinus* control in nitrate-rich environments
Warming climate	Insect attacks	Negative impact on economic and aesthetic value, higher CO_2 emissions	Adapted silviculture, climate-smart forestry
		More rapid succession and faster recycling of nutrients (positive impact)	
Industrial forestry	Treeline shift downwards	Negative impact on aesthetic value; impacts on pastures	Collaborative land-use planning, promoting the principles of continuous-cover forestry
Less precipitation	Summer droughts	Increased risk of wildfires; deteriorating pastures and deteriorating aesthetic value	Hazardous fuels reduction programs, vulnerability zoning, simulation models for wildfire risk assessment

European treeline areas include protected areas. The management planning in these areas is not sufficiently linked to other land-use planning and often suffers from inadequate participation of local actors and government sectors, e.g. in municipal zoning to control infrastructure building, forest planning, environmental impact assessment processes linked to industrial developments.

4. CONCLUSIONS

Based on 45 case studies, we have come to several conclusions about the social and ecological sustainability of European treeline areas. Firstly, climate change and land-use change represent the 2 most significant drivers in treeline areas. However, unlike most previous analyses of treeline dynamics, our synthesis suggests that land-use change is a more immediate threat to the social and ecological sustainability of European treeline areas than climate change. This calls for paying more attention to human and local-specific pressures instead of fine-tuning climate change scenarios. Secondly, examples from different parts of Europe have evidenced that maintaining traditional pastoral practices in treeline areas in combination with sustainable tourism could enhance social and ecological sustainability. Thirdly, in analyzing the social–ecological systems with the DPSIR framework, it is crucial to be aware that the state dimension should be regarded as a state of a social–ecological system, not as a state of an ecological system (cf. Sarkki et al. 2017a). This distinction is not entirely academic, and has implications for the sustainability of ESs, which are not solely the benefits that ecosystems provide to human well-being, but benefits co-evolved through centuries of human–environment coexistence. Consequently, society must adapt to the changing nature of treeline ESs and look also for new opportunities arising from environmental change, rather than trying to sustain the current level of ESs indefinitely. Finally, policy and governance systems that have so far been incoherent and not specifically focused on treeline areas should be proactive without being bureaucratic (Heikkinen et al. 2010). More emphasis should be given to local stakeholder participation in decision making, enhancing local acceptability of decisions. The results from this study indicate that without proactive response to changes, some ESs may be lost in the following decades, changing historical landscapes, reducing financial and other benefits, decreasing biodiversity and consequently, reducing the opportunity to maintain ESs in the future.

Acknowledgements. This study is based upon work from the COST Action ES1203 'Enhancing the resilience capacity of SENSitive mountain FORest ecosystems under environmental change' (SENSFOR), and supported by European Cooperation in Science and Technology (COST) (www.cost.eu). Svetla Bratanova-Doncheva, Pavel Cudlín, Concepcion Alados, Karsten Grunewald, Johann Thorsson and Annika Hofgaard are particularly acknowledged for valuable input to the study.

LITERATURE CITED

Alados CL, Errea P, Gartzia M, Saiz H, Escós J (2014) Positive and negative feedbacks and free-scale pattern distribution in rural-population dynamics. PLOS ONE 9: e114561

Ameztegui A, Coll L, Brotons L, Ninot JM (2016) Land-use legacies rather than climate change are driving the recent upward shift of the mountain treeline in the Pyrenees. Glob Ecol Biogeogr 25:263–273

Ayres MP, Lombardero MJ (2000) Assessing the consequences of global change for forest disturbance from herbivores and pathogens. Sci Total Environ 262:263–286

Bakaloudis DE (2016) Livestock grazing, openings and raptors conservation in the Dadia-Lefkimi-Soufli forest national park. Options Mediterraneennes Ser A 114: 437–440

Battisti A, Stastny M, Buffo E, Larsson S (2006) A rapid altitudinal range expansion in the pine processionary moth produced by the 2003 climatic anomaly. Glob Change Biol 12:662–671

Bebi P (1999) Erfassung von Strukturen im Gebirgswald als Beurteilungsgrundlage ausgewählter Waldwirkungen. PhD thesis, ETHZ, Zürich

Bebi P, Kulakowski D, Rixen C (2009) Snow avalanche disturbances in forest ecosystems—state of research and implications for management. For Ecol Manage 257: 1883–1892

Boyd J, Banzhaf S (2007) What are ecosystem services? The need for standardized environmental accounting units. Ecol Econ 63:616–626

Brang P, Schönenberger W, Frehner M, Schwitter R, Thormann JJ, Wasser B (2006) Management of protection forests in the European Alps: an overview. For Snow Landsc Res 80:23–44

Broll G, Keplin B (eds) (2005) Mountain ecosystems: studies in treeline ecology. Springer, Berlin Heidelberg

Bryn A, Daugstad K (2001) Summer farming in the subalpine birch forest. In: Wielgolaski FE (ed) Nordic mountain birch ecosystems. UNESCO, Paris, p 307–315

Camarero JJ, Garcia-Ruiz JM, Sangüesa-Barreda G, Galvàn JD and others (2015) Recent and intense dynamics in a formerly static Pyrenean treeline. Arct Antarct Alp Res 47:773–783

Cooper P (2013) Socio-ecological accounting: DPSWR, a modified DPSIR framework, and its application to marine ecosystems. Ecol Econ 94:106–115

Cudlín P, Seják J, Pokorný J, Albrechtová J, Bastian O, Marek M (2013) Forest ecosystem services under climate change and air pollution. In: Matyssek R, Clarke N, Cudlín P, Mikkelsen TN, Tuovinen JP, Wiesner G, Paoletti E (eds) Climate change, air pollution and global challenges: understanding and perspectives from forest research. Developments in Environmental Science, Vol 13. Elsevier, Oxford, p 521–546

Cudlín P, Klopčič M, Tognetti R, Malis F and others (2017) Drivers of treeline shift in different European mountains. Clim Res in press73:135–150

Dakov MI, Iliev DA, Donov V, Dimitrov S (1980) Increase of upper forest line. Zemizdat, Sofia (in Bulgarian)

Danell LS, Niemelä P, Varvikko T, Vuorisalo T (1991) Moose browsing on Scots pine along a gradient of plant productivity. Ecology 72:1624–1633

De Wit HA, Bryn A, Hofgaard A, Karstensen J, Kvalevåg M, Peters G (2014) Climate warming feedback from mountain birch forest expansion: reduced albedo dominates carbon uptake. Glob Change Biol 20:2344–2355

Debussche M, Lepart J, Dervieux A (1999) Mediterranean landscape changes: evidence from old postcards. Glob Ecol Biogeogr 8:3–15

EEA (European Environmental Agency) (2016) Environmental terminology and discovery service (ETDS). http://glossary.eea.europa.eu/terminology/concept_html?term=dpsir

Ficko A, Roessiger J, Bončina A (2016) Can the use of continuous cover forestry alone maintain silver fir (*Abies alba* Mill.) in central European mountain forests? Forestry 89:412–421

Gartzia M, Alados CL, Pérez-Cabello F (2014) Assessment of the effects of biophysical and anthropogenic factors on woody plant encroachment in dense and sparse mountain grasslands based on remote sensing data. Prog Phys Geogr 38:201–217

Gehrig-Fasel J, Guisan A, Zimmermann NE (2007) Tree line shifts in the Swiss Alps: climate change or land abandonment? J Veg Sci 18:571–582

Gitay H, Suarez A, Watson R, Dokken D (eds) (2002) Climate change and biodiversity. IPPC Technical Paper V. Intergovernmental Panel on Climate Change, Geneva

Grabherr G (2009) Biodiversity in the high ranges of the Alps: ethnobotanical and climate change perspectives. Glob Environ Change 19:167–172

Grace J, Berninger F, Nagy L (2002) Impacts of climate change on the tree line. Ann Bot 90:537–544

Grunewald K, Bastian O (eds) (2015) Ecosystem services—concept, methods and case studies. Springer-Verlag, Heidelberg

Grunewald K, Scheithauer J (2011) Landscape development and climate change in Southwest Bulgaria (Pirin Mountains). Springer, Heidelberg

Hastik R, Basso S, Geitner C, Haida C and others (2015) Renewable energies and ecosystem service impacts. Renew Sustain Energy Rev 48:608–623

Heikkinen HI, Sarkki S, Jokinen M, Fornander DE (2010) Global area conservation ideals versus the local realities of reindeer herding in northernmost Finland. Int J Bus Glob 4:110–130

Helle T (2001) Mountain birch forests and reindeer husbandry. In Wielglaski FE (ed) Nordic mountain birch ecosystems. UNESCO, Paris, p 279–291

Hester AJ, Miller DR, Towers W (1996) Landscape scale vegetation change in the Cairngoms, Scotland 1946-1988: implication for land management. Biol Conserv 77:41–52

Hester AJ, Millard P, Baillie GJ, Wendler R (2004) How does timing of browsing affect above- and belowground growth of *Betula pendula*, *Pinus sylvestris* and *Sorbus aucuparia*? Oikos 105:536–550

Hoff H (2013) Vulnerability of ecosystem services in the Mediterranean region to climate changes in combination with other pressures. In: Navarra A, Tubiana L (eds) Regional assessment of climate change in the Mediterranean. Advances in Global Change Research, Vol 51. Springer, Dordrecht, p 9–29

Hofgaard A, Tømmervik H, Rees G, Hanssen F (2013) Latitudinal forest advance in northermost Norway since the early 20th century. J Biogeogr 40:938–949

Holtmeier FK (2009) Mountain timberlines: ecology, patchiness and dynamics. In: Beniston M (ed) Advances in Global Change Research, Vol 36. Springer, Dordrecht, p 5–167

Huber R, Rigling A, Bebi P, Brand FS and others (2013) Sustainable land use in mountain regions under global change: synthesis across scales and disciplines. Ecol Soc 18:36

Körner C (2012) Alpine treelines: functional ecology of the high elevation tree limits. Springer, Basel

Kullman L (2004) The changing face of the alpine world. Global Change Newsl 57:12–14

Kullman L, Öberg L (2009) Post-Little Ice Age tree line rise and climate warming in the Swedish Scandes: a landscape ecological perspective. J Ecol 97:415–429

Kupfer JA, Cairns DM (1996) The suitability of montane ecotones as indicators of global climatic change. Prog Phys Geogr 20:253–272

Kyriazopoulos AP, Abraham E, Hofgaard A, Sarkki S (2014) DPSIR for treeline ecosystem and their services. Progress Report COST Action ES1203, COST/EU, Brussels

Lempa K, Neuvonen S, Tømmervik H (2005) Effects of reindeer grazing on pastures—a necessary basis for sustainable reindeer herding. In: Wielgolaski FE (ed) Plant ecology, herbivory and human impact in Nordic mountain birch forests. Ecological Studies, Vol 180. Springer, Berlin Heidelberg p 157–164

Logan JA, Regniere J, Powell JA (2003) Assessing the impacts of global warming on forest pest dynamics. Front Ecol Environ 1:130–137

Luck GW, Harrington R, Harrington PA (2009) Quantifying the contribution of organisms to the provision of ecosystem services. Bioscience 59:223–235

Neuvonen S, Niemelä P, Virtanen T (1999) Climatic change and insect outbreaks in boreal forests: the role of winter temperatures. Ecol Bull 47:63–67

Neuvonen S, Bylund H, Tømmervik H (2005) Forest defoliation risks in birch forest by insects under different climate and land use scenarios in northern Europe. In: Wielgolaski FE (ed) Plant ecology, herbivory and human impact in Nordic mountain birch forests. Ecological Studies, Vol 180. Springer, Berlin Heidelberg, p 125–138

Oksanen L, Moen J, Helle T (1995) Timberline patterns in northernmost Fennoscandia: relative importance of climate and grazing. Acta Bot Fenn 153:93–105

Palombo C, Chirici G, Marchetti M, Tognetti R (2013) Is land abandonment affecting forest dynamics at high elevation in Mediterranean mountains more than climate change? Plant Biosyst 147:1–11

Papanastasis VP (2009) Restoration of degraded grazing lands through grazing management: Can it work? Restor Ecol 17:441–445

Paulsen J, Körner C (2014) A climate-based model to predict potential treeline position around the globe. Alp Bot 124: 1–12

Potthoff K (2009) Grazing history affects the tree-line ecotone: a case study from Hardanger, western Norway. Fennia 187:81–98

Rapport D, Friend A (1979) Toward a comprehensive framework for environmental statistics: a stress-response approach. Statistics Canada, Ottawa

Sarkki S, Grunewald K, Alados CL, Nijnik M and others (2013) Questionnaire-based investigation of stakeholder needs as linked to ecosystem services, governance and science in European treeline areas, Progress Report, COST Action ES1203, COST/EU, Brussels

Sarkki S, Grunewald K, Nijnik M, Zahvoyska L and others (2015) Problems and proposals for good environmental management: empirical assessment of European treeline areas. Progress Report, COST Action ES1203, COST/EU, Brussels

Sarkki S, Ficko A, Grunewald K, Nijnik M (2016a) Benefits from and threats to European treeline ecosystem services: an exploratory study of stakeholders and governance. Reg Environ Change 16:2019–2032

Sarkki S, Ficko A, Grunewald K, Kyriazopoulos AP, Nijnik M (2016b) How pragmatism in environmental science and policy can undermine sustainability transformations: the case of marginalized mountain areas under climate and land-use change. Sustain Sci 4, doi:10.1007/s11625-016-0411-3

Sarkki S, Ficko A, Wielgolaski FE, Abraham EM and others (2017a) Assessing the resilient provision of ecosystem services by social-ecological systems: introduction and theory. Clim Res 73:7–15

Sarkki S, Jokinen M, Nijnik M, Zahvoyska L and others (2017b) Social equity in governance of ecosystem services: synthesis from European treeline areas. Clim Res 73:31–44

Sarvašová Z, Cienciala E, Beranová J, Vančo M, Ficko A, Pardos M (2014) Analysis of governance systems applied in multifunctional forest management in selected European mountain regions. For J 60:159–167

Skre O, Baxter R, Crawford RMM, Callaghan TV, Fedorkov A (2002) How will the tundra-taiga interface respond to climate change? Ambio (Spec Rep) 12:37–46

Skre O, Wertz B, Wielgolaski FE, Szydlowska P, Karlsen SR (2017) Bioclimatic effects on different mountain birch populations in Fennoscandia. Clim Res 73:111–124

Smith WK, Germino MJ, Hancock TE, Johnson DM (2009) The altitude of alpine treeline: a bellwether of climate change effects. Bot Rev 75:163–190

Tenow O, Bylund H, Nilsen AC, Karlsson PS (2005) Long-term influences of herbivores on northern birch forests. In: Wielgolaski, FE (ed) Plant ecology, herbivory and human impact in Nordic mountain birch forests. Ecological Studies, Vol 180. Springer, Berlin, p 165–182

Theurillat JP, Guisan A (2001) Potential impact of climate change on vegetation in the European Alps: a review. Clim Change 50:77–109

Tolvanen A, Forbes B, Wall S, Norokorpi Y (2005) Recreation at tree-line and interactions with other land-use activities. In Wielgolaski FE (ed) Plant ecology, herbivory and human impact in Nordic mountain birch forests. Ecological Studies, Vol 180. Springer, Berlin, p 203–217

Tømmervik H, Wielgolaski FE, Neuvonen S, Solberg B, Høgda KA (2005) Biomass and production at a landscape level in the northern mountain birch forests. In: Wielgolaski FE (ed) Plant ecology, herbivory and human impact in Nordic mountain birch forests. Ecological Studies, Vol 180. Springer, Berlin, p 53–70

Van Bogaert R, Haneca K, Hoogesteger J, Jonasson C, De Dapper M, Callaghan T (2011) A century of treeline changes in subarctic Sweden shows local and regional variability and only a minor influence of 20th century climate warming. J Biogeogr 38:907–921

Volney WJA, Fleming RA (2000) Climate change and impacts of boreal forest insects. Agric Ecosyst Environ 82: 283–294

von der Dunk A, Gret-Regamey A, Dalang T, Hersperger AM (2011) Defining a typology of peri-urban land-use conflicts—a case study from Switzerland. Landsc Urban Plan 101:149–156

Xirouchakis S (2004) Causes of raptor mortality in Crete. In: Meyburg BU, Chancellor R (eds) Raptors worldwide. WWGBP/MME (Birdlife Hungary), Berlin, p 849–860

Appendix

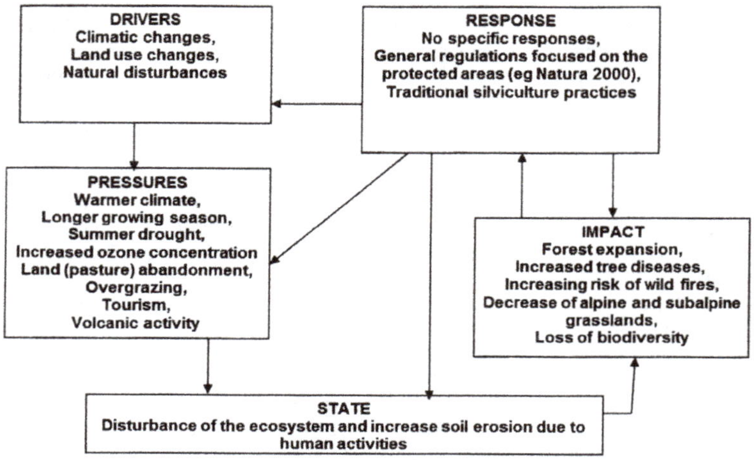

Fig. A1. Application of the drivers-pressures-state-impacts-responses (DPSIR) framework for treeline ecotones in Europe (Kyriazopoulos et al. 2014)

Table A1. Example of the description protocol for drivers, pressures, state, impacts and responses in the second survey (Kyriazopoulos et al. 2014). Colour shading—grey: location, climate and species; green: land use types; blue: climate change types; red: DPSIR factors

Location	Western Carpathian Mts N 48° 48′ to 49° 35′, E 19° 18′ to 21° 00′ Elevation >1200 to 1850 m
Climate	Mountainous
Dominant tree species	*Pinus mugo, Pinus cembra, Picea abies, Fagus sylvatica*
Wild herbivores	*Cervus elaphus, Rupicapra rupicapra, Ursus arctos*
Land use changes	Pasture abandonment
Grazing / livestock	Decreased / sheep and cattle
Logging pressure	No
Climatic changes	Climate warming
Temperature trend	Increase
Precipitation trend	Slightly increase
Driving forces	Climatic changes, land-use changes
Pressure	Climate warming, decreased grazing
State	Treeline is being increasingly disturbed, particularly by construction/enlargement of ski resorts and other recreation facilities. Regeneration via secondary succession due to abandonment of grazing
Impact	Longer growing season, bark beetle outbreaks
Responses	Creation of nature protection zones, or specific parts of Rural Development Programme dedicated to treeline ecosystems. Silviculture measures

8

Identifying and understanding attitudinal diversity on multi-functional changes in woodlands of the Ukrainian Carpathians

Albert Nijnik[1], Maria Nijnik[2,*], Serhiy Kopiy[3], Lyudmyla Zahvoyska[3], Simo Sarkki[4], Leonid Kopiy[3], David Miller[2]

[1]Environmental Network Limited, The Hillocks, Tarland, Aboyne AB34 4TJ, Scotland, UK
[2]The James Hutton Institute, Craigiebuckler, Aberdeen AB15 8QH, Scotland, UK
[3]Ukrainian National Forestry University, Gen. Chuprynky 103, Lviv 79057, Ukraine
[4]Thule Institute, PO Box 7300, 90014 University of Oulu, Finland

ABSTRACT: This paper advances existing knowledge of stakeholder attitudes towards ecosystem-based forest strategies and management practices in marginalized mountain areas of the Ukrainian Carpathians. The principal research question was to analyse the state of affairs regarding multi-functional changes in upland woodlands, as viewed by relevant stakeholders. An ultimate goal was to assist decision-makers in evaluation (e.g. through stakeholder evaluation using the suggested method) and implementation (through stakeholder engagement) of sustainable policy and management decisions. The Q-method, with the sequential application of its correlation and factor analytical tools, was applied to elucidate a range of existing attitudes (e.g. towards the expansion of woodlands under climate change and other drivers) and the spectrum of sustainability (its ecological, economic and social components) to which they relate. Dominant attitudes of representatives of relevant stakeholder groups towards multi-functional changes in forestry in the Carpathian Mountains were revealed and analysed. Key factors influencing the attitudinal diversity were explained. The results improve our understanding of stakeholder priorities and of commonalities and differences in existing attitudes/perceptions, providing some indication of how the diversity of attitudes towards forestry changes (e.g. integration of woodlands in mountain landscapes) could potentially influence sustainable forestry decisions. At times, entirely opposite attitudes (e.g. of the so-called Conservationists and Productivists attitudinal groups) towards forestry practices and key objectives of forestry in the Carpathian uplands were revealed. However, people put strong emphasis on multi-functional forestry offering a range of benefits to people, the environment and the economy.

KEY WORDS: Public attitudes and preferences · Ecosystem-based adaptation practices · Carpathian Mountains · Q-method

INTRODUCTION

International, regional and national environmental policies are drivers of changes in European mountains. Globalization processes, technological and economic advances, social innovations, and environmental and climatic changes also influence the development of mountain regions. Specifically, the Paris Agreement—negotiated at the 21st Conference of the parties (COP) of the UNFCCC and adopted on 12 December 2015 (European Commission 2015)—recognized the role of non-party stakeholders in addressing climate change through the up-scaling of efforts and support actions to reduce carbon emis-

*Corresponding author: maria.nijnik@hutton.ac.uk

§Advance View was available online May 18, 2017

sions, build resilience and decrease vulnerability to the adverse effects of climate change. At the Third Meeting of the Conference of the Parties to the Framework Convention on the Protection and Sustainable Development of the Carpathian Mountains (Bratislava, 25–27 May 2011), environment ministers and experts from Central and Eastern Europe discussed the drivers and responses to manifold changes affecting the region, and adopted a protocol on sustainable forest management. Identification of areas with high conservation value, and the establishment and better management of an ecological network under Natura 2000 predefined a key policy issue in the Carpathian Mountains, which are considered to be '*a unique natural treasure of great beauty and ecological value, an important reservoir of biodiversity, the headwaters of major rivers, an essential habitat and refuge for many endangered species...*' (The Carpathian Convention 2003).

Ukraine signed the Carpathian Convention accepting the strategy of sustainable development of this region. It foresees that the level of production in forestry and its specialization are to match the natural resource regeneration potential. Additional measures that stipulate the environmental role of forests were added to The Forest Code of Ukraine (2006). However, adopting the sustainable development considerations in a country in transition to a market economy, such as Ukraine, requires adjustment of the complexity of institutions to the requirements of a market economy (Krott 2008). It also requires the promotion of involvement of local communities in decision-making and policy implementation (Nijnik et al. 2009a).

However, 'implementation deficits' (common to young democracies) demonstrate a gap between the formulation of policy goals and their implementation (Krott 2008, Keeton & Crow 2009). These include the following: lack of will of policy actors; a shortage of skilled experts (e.g. trained for tackling climate change challenges); misperceptions of the policy targets, measures and instruments by local stakeholders (Zahvoyska et al. 2014); a deficit of resources (especially financial); weak market infrastructure (tenure rights on land and forest resources are not properly ensured and enforced; market failures are common; Nijnik & Oskam 2004; Soloviy et al. 2012); and, particularly observed in remote mountain areas, deficient democratic institutions and therefore an incomplete participation process (Bizikova et al. 2012, Sarkki et al. 2016).

In this context, social innovation (www.simra-h2020.eu) is considered to be pivotal. While aiming at improving human wellbeing, it responds to fresh changes in society and to pressing social demands that are traditionally not addressed by existing institutions. Social innovation manifests itself in new institutions, related relationships, networks, interactions and collaborations between people. It promotes the development and uptake of new services and new fields of activity, such as social entrepreneurship and social enterprises (i.e. businesses trading for social purposes within the social economy, which also includes foundations, charities and cooperatives that improve the quality of life of individuals and communities). Examples are community-owned renewable energy inititatives, and social enterprises which help disadvantaged groups gain access to work as well as to healthcare. Social innovation in the Ukrainian Carpathians may also respond to the new structural disadvantages that marginalised rural areas have seen in the frame of rising agglomeration economies resulting from globalisation.

However, boosting social innovations and linked governance mechanisms is difficult in marginalized rural areas. This is particularly challenging in the Ukrainian Carpathians, where the observed deficiency of stakeholder knowledge and capacity in the climate-changed and carbon-constrained world (Elbakidze & Angelstam 2013, Krynytskyy & Chernyavskyy 2014) is multiplied by 'path dependency' (on the former, i.e. command-and-control economy), and the legacy of 'politonomy' (the nexus between economic agents and politicians) in this country (Nijnik & Oskam 2004).

Forestry in the Ukrainian Carpathians remains regimented. Practically all forests are publicly owned and managed by the state based on centralized planning at the ministerial level (Soloviy et al. 2012). However, the presence of 'black' and 'grey' timber markets (i.e. different degrees of tax avoidence in a 'shadow' economy) implies the absence of governance frameworks necessary to warrant sustainable forestry development (Krott 2008). The detected mismatches between national, regional and local policies, and existing gaps between multi-functional forest policy goals and the state of affairs on the ground, contribute to the challenges (Soloviy 2010).

Stakeholders in multi-functional forestry are communities of place (e.g. forest owners, users and local people) and wider communities of interest (e.g. governments and authorities, non-governmental organizations). Priorities of these stakeholder groups may substantially differ (Nijnik & Miller 2014). Further, notions of sustainable provision of ecosystem services become even more subjective when viewed from the perspectives and goals of an individual. Therefore, complexity, multiplicity and at times the uncertainty related to simultaneous production of multiple forest ecosystem services to be delivered to their multiple, and at times very diverse, beneficiaries may result in conflicts of stakeholder interests, and it is important to reduce and/or resolve possible conflicts (Vogel & Lowham 2007).

A way forward could be to use new methods to engage with stakeholders and develop a better understanding of their perspectives of land use changes and their requirements, expectations and aspirations, and the role of woodlands in such perspectives (Prell et al. 2009). A deeper understanding of stakeholder perceptions is essential (Zahvoyska et al. 2015). It is needed, for example, for linking forest management practices (including carbon forestry) to biodiversity conservation requirements influencing policy design in forests that are managed for multiple purposes. Stakeholder engagement is becoming an instrument for targeting sustainability in multi-functional forestry where the trade-offs between forest ecosystem services are a matter of concern.

Current literature on stakeholder attitudes to multi-functional mountain forest policy and management is limited (Sarkki et al. 2016). However, stakeholder engagement can be considered part of adaptive innovative governance, where decision-making is based on learning from past experiences and where stakeholder evaluation is a basis for capacity building and participatory decision-making (Muñoz-Rojas et al. 2015). These ideas come close to those of adaptive co-management, which implies learning from past experiences, being flexible to changes, using science to reflect governance, and co-constructing knowledge with stakeholders (Berkes 2009). Capacity and trust building, knowledge sharing and learning-by-doing, strengthening of science–society–policy relationships, and boosting social innovation are considered essential in this regard (Funtowicz & Ravetz 1993, Nijnik & Oskam 2004). An improved understanding of stakeholder attitudes and perceptions could be seen as a first step in getting manifold actors engaged in the decision-making.

Thus, our main purpose in this paper is to examine the heterogeneity of stakeholder attitudes towards multi-functional changes in upland woodlands of the Ukrainian Carpathians. We examine the attitudinal diversity concerning the services that forest ecosystems can provide and the trade-offs between these services. We map similarities and divergences in stakeholder attitudes towards ecosystem-based forest policy and management decisions in upland woodlands of the Carpathian region in the belief that this research could be particularly useful in identifying potential conflicts (e.g. caused by differing stakeholder interests) so as to avoid and/or manage them, or for assisting relevant decision-makers with the incorporation of public perceptions into policy design. Considerations around the forest policy and management, decision-making and governance in the region form the key focus of this paper, while we seek to advance the knowledge of ecosystem-based adaptation practices in the Carpathian uplands, as seen through the eyes of people on the ground.

The Methods section briefly presents the Q-method applied. In the Results section, key outputs of quantitative analyses are interpreted as social discourses in order to construct the typologies of experts' perspectives on adaptation policy and practices in the Ukrainian Carpathians. Further, we explain the attitudinal heterogeneity from contextual and personal factors, and proceed with a brief Discussion section. Findings show a broad discourse of stakeholder perspectives. This diversity was identified and examined to help respond to existing challenges, including climate change, with an ultimate aim of devising better-informed and more sustainable ecosystem-based forest policy and management decisions. Finally, conclusions are drawn to capitalize on the methods used and our key findings.

METHODS

The Q-method is a participatory, systematic and rigorous scientific tool (Stephenson 1963) designed to reveal and examine existing attitudes and perspectives, provide in-depth insights into attitudinal diversity and human preferences, identify the key criteria concerning the issues in question that are important to people, and to explain major factors influencing the heterogeneity of stakeholder attitudes and perspectives (Watts & Stenner 2012).

The Q-method (Schmolck 2012) combines qualitative and quantitative tools (Brown 1996). It incorporates elements of behavioural studies into action research (Miller et al. 2009, Nijnik et al. 2014) and can be viewed as a quantitative method for doing discourse analysis (van Exel & De Graaf 2005). Unlike the better-known R-method used in surveys, the Q-method enables researchers to discover similarities and differences across the attitudes and perceptions of individuals (rather than among traits) by sorting the statements.

Statements on the issue in question are derived through a concourse analysis, i.e. from interviews or written narratives, and can comprise discourses and media, encompassing considerations spoken and written (Davies et al. 2005). Commonly, statements emerge from interviews and communication with stakeholders. However, they can be derived from secondary sources (McKeown & Thomas 2013, Stevenson 2015). Then, in order to examine individual opinions on specific issues presented in statements and assess these statements in relation to each other, respondents are asked to sort statements according to the level of agreement/disagreement in a fixed, normally distributed scale, which is commonly used in this type of research (Watts & Stenner 2012).

The sorting exercise results in Q-sorts. Each Q-sort, which is the formal model of each respondent's attitude to the statements, is then correlated with every other Q-sort, and their inter-correlation matrix is factor-analysed (e.g. using principal component analysis [PCA] and Varimax rotation). This leads to attitudinal clusters (Brown 1996) and to linking these clusters to individual characteristics of respondents (Nijnik et al. 2014). However, the most important outcome is that the identified factors permit the capture of the variety of individual attitudes and permit the structuring of them in such a way that different 'common discourses' emerge (Stevenson 2015). The resulting factor arrays, i.e. typical Q-sorts, represent the attitudes people have, enabling researchers to describe and improve their understanding of the attitudinal heterogeneity (Nijnik et al. 2010, McKeown & Thomas 2013, Nijnik et al. 2014).

The Q-method allows for a rather simple data set because most of the data derive from how much information is implicit in each participant's Q-sort, i.e. the formal model of each individual's understanding of the points of view at issue (Barry & Proops 1999). The Q-method is useful for examining the areas of consensus and conflict between people, and for specifying, selecting and evaluating policy options (Nijnik et al. 2016). However, it does not allow for the extrapolation of results (McKeown & Thomas 2013).

To better match the research objectives and be time- and resource-efficient, for this study we decided to draw on existing material (i.e. secondary sources). Rigorous and extensive data collection was conducted.

Table 1. Distinguishing statements for Group 1 (Conservationists). In Tables 1 to 5, the rank is derived from the weighted composites, where $^{*}p < 0.01$

No.	Statement	Rank	Score
1	Woodlands are an important part of our national heritage and should be preserved whatever the cost	4	2.17*
12	The rights of people to enjoy the beauty of treeline areas are more important than making profits from the land	2	1.07*
6	I support woodland expansion in treeline areas for multiple purposes	1	0.34*
11	Woodlands should be planted if they improve visual beauty of the landscape	0	−0.17*
25	Shortage of investment is the main challenge in treeline areas	0	−0.22*
9	Forest where trees are planted very close together is unacceptable for wildlife and should be cut down	−2	−0.89*
23	Woodland expansion in treeline areas to create jobs is more important than protecting nature	−3	−1.18*
14	It should be possible to cut the trees down, if the land is needed for other purposes	−3	−1.20*
20	The currently wooded areas could be made wealthy through clear-cutting	−4	−1.79*

Table 2. Distinguishing statements for Group 2 (Productivists)

No.	Statement	Rank	Score
25	Shortage of investment is the main challenge in treeline areas	4	2.19*
16	Planting forests to produce timber should be a policy priority	3	2.03*
22	Creation of forest is to be promoted as this provides employment opportunities for remote rural communities	3	0.98*
18	Letting wildlife take care of itself should be a principle of forest management	2	0.94*
24	Forest management in the uplands should be profitable	2	0.69*
8	We should only harvest forests in treeline areas within ecological limits	1	0.44*
1	Woodlands are an important part of our natural heritage and should be preserved whatever the cost	0	0.29*
3	We should expand woodlands today to make sure that future generations have plenty of timber	−1	−0.29*
10	I support the creation of new woodlands, but any new planting must be in tune with the character of the landscape	−1	−0.63*
12	The rights of people to enjoy the beauty of treeline areas are more important than making profits from the land	−2	−1.23*
6	I support woodland expansion in treeline areas for multiple purposes	−3	−1.37*
7	Public access should be given to all forests	−4	−1.89*

Table 3. Distinguishing statements for Group 3 (Non-Green Technocrats)

No.	Statement	Rank	Score
25	Shortage of investment is the main challenge in treeline areas	2	1.09*
16	Planting forests to produce timber should be a policy priority	2	1.20*
18	Letting wildlife take care of itself should be principle of forest management	−1	0.80*
13	I support government policies of woodland expansion because woodlands improve the landscape and attract tourists	−2	−1.11*
12	The rights of people to enjoy the beauty of treeline areas are more important than making profits from the land	−4	−2.05*

Table 4. Distinguishing statements for Group 4 (Challengers)

No.	Statement	Rank	Score
16	Planting forests to produce timber should be a policy priority	4	2.58*
12	The rights of people to enjoy the beauty of treeline areas are more important than making profits from the land	3	1.08*
23	Woodland expansion in treeline areas to create jobs is more important than protecting nature	0	−0.14*
14	It should be possible to cut the trees down, if the land is needed for other purposes	−1	−0.64*
6	I support woodland expansion in treeline areas for multiple purposes	−2	−0.69*
25	Shortage of investment is the main challenge in treeline areas	−2	−1.25*
18	Letting wildlife take care of itself should be a principle of forest management	−3	−1.38*

Table 5. Distinguishing statements for Group 5 (Green Protectionists)

No.	Statement	Rank	Score
9	Forest where trees are planted very close together is unacceptable for wildlife and should be cut down	4	1.71*
6	I support woodland expansion in treeline areas for multiple purposes	2	1.33*
25	Shortage of investment is the main challenge in treeline areas	2	1.18*
12	The rights of people to enjoy the beauty of treeline areas are more important than making profits from the land	1	0.49*
24	Forest management in the uplands should be profitable	0	−0.13*
23	Woodland expansion in treeline areas to create jobs is more important than protecting nature	−2	−0.99*
14	It should be possible to cut the trees down, if the land is needed for other purposes	−4	−2.17*

For this purpose, we used available research results (e.g. surveys in the form of reports, blogs, etc.) and relevant literature sources, including Ukrainian ones that addressed forest policy and governance, multi-functional forestry, decision-making, and science–society relationships in particular (Nijnik et al. 2009b, Sarkki & Karjalainen 2012, Soloviy et al. 2012, Sarkki et al. 2016). A set of Q-statements reflecting different sub-elements of the issue in question was designed. The statements (somewhat shortened, with examples in Tables 1 to 5) reflect stakeholder needs in the Carpathians, as these are linked to the ecosystems and their services (e.g. '*Woodlands are an important part of our national heritage*', or '*Woodlands should be planted, if they improve visual beauty of the landscape*'); forest policy (e.g. '*Planting more forests to produce timber should be a policy priority*') and adaptation strategy (e.g. '*I support woodland expansion for multiple purposes*', or '*We should only harvest forests within ecological limits*'); forest management (e.g. '*Forest management should be profitable*'); decision-making processes in forestry (e.g. '*It should be possible to cut the trees down, if the land is needed...*'); and policy measures and governance (e.g. '*Shortage of investment is the main challenge in treeline areas*'). We came up with 25 statements and, correspondingly, 25 boxes in the Q-sort chart (Fig. 1).

After pre-testing and improving the statements with colleagues from the Earth System Science and Environmental Management (ESSEM) European Cooperation in Science and Technology (COST) Action ES1203, Enhancing the Resilience Capacity of Sensitive Mountain Forest Ecosystem under Environmental Change (SENSFOR), the set of statements was finalized and presented to respondents, i.e. people living and working in the Carpathian Mountains. Respondents were identified, following van Asselt et al. (2001), who distinguished 5 types of participant considering public decisions: (1) government/decision-makers; (2) citizens; (3) interest groups; (4) business-people; and (5) scientific experts. Forty respondents, selected to be as diverse as possible in their social, economic, political and educational backgrounds, performed the Q-sorting. They were approximately equally distributed among the 5 types

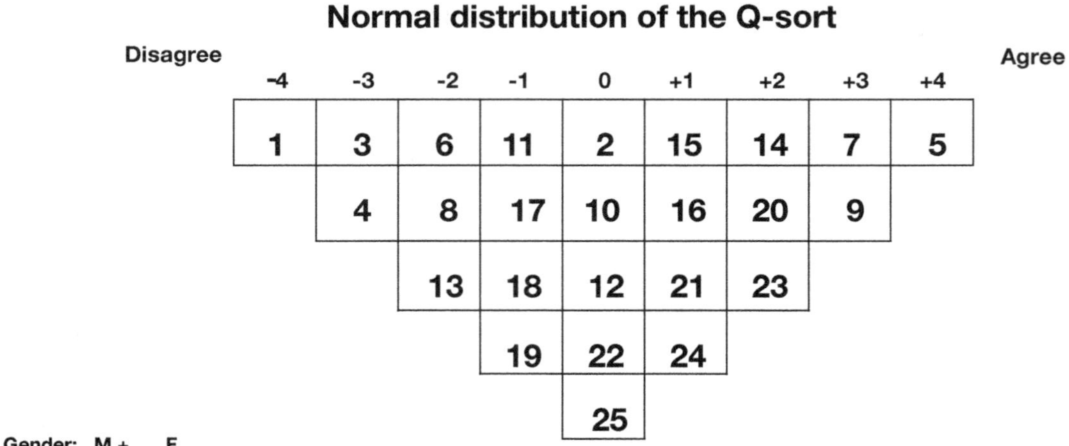

Fig. 1. Example of a Q-sort completed by a participant. The number in each box represents the statement number. The position of each statement in the Q-sort chart indicates the participant's agreement/disagreement with each of the 25 statements and his/her judgement on the statements in relation to each other

identified by van Asselt et al. (2001), with local people representing 'citizens', but only 6 respondents representing 'business people'.

We translated the questionnaire into Ukrainian and approached the selected respondents in person, initially explaining the purpose of this research, briefly presenting its method and providing information about this study. Each respondent was asked to distribute the statements across boxes in the score sheet according to his/her agreement/disagreement to each one and his/her assessment of the statements in relation to each other. Respondents therefore ranked statements on the scale ranging from +4 to –4, where +4 indicated full agreement and –4 full disagreement with the statement, and 0 indicated a neutral attitude to it (but could also mean 'ambivalent', or 'don't know', depending on the respondent) (Fig. 1).

The output data were assessed using the sequential application of correlation and factor analyses (i.e. PCA and Varimax rotation). Correlation analysis was used to compare the views among respondents, while PCA was used to categorize correlated expert views under different 'factors' by creating new uncorrelated choice variables that captured the common essence of the individual Q-sorts (Brown 1980, Brown 1996, Nijnik et al. 2013). This approach ultimately enabled us to identify the prevailing attitudinal typologies and compare the key differences and similarities between the identified groups of attitudes.

Additionally, and fully complying with the requirements stipulated by scientific ethics (i.e. without disclosing any personal information), we asked respondents about gender, age, work experience, education, occupation, income level, partnership status and number of children. Finally, we provided an interpretation of the social discourses uncovered by the quantitative analysis, contrasted the value outputs with the above-named socio-economic characteristics of respondents, verified our results with respondents, and communicated the final results to them and a wider audience, i.e. at the SENSFOR meeting of October 2015, through a press release (in Ukrainian) and through this paper.

RESULTS

The modelling outputs received through factor extraction (PCA), Varimax rotation and interpretation demonstrated that 4 factors (with correlations ranging from –0.20 to 0.37) provide the best representation of distinctive types of existing attitude, or discourse. Approximately 50% of respondents are uniquely associated with one of the factors (for which they have their only significant factor loading). However, almost 50% of respondents are mixtures of 2 or more types. The number of factors is, therefore, not totally deterministic—not every individual is a member of only one factor (i.e. attitudinal group). It also means that we have purely loading Q-sorts and confounding Q-sorts. However, our 5 factors cumulatively captured 69% of the total variance (i.e. the full meaning and variability within the data). The values of variance (% expl.Var) in the total dataset of individual Q-sorts, as explained by the factors, are 42, 9, 7, 6 and 5, respectively.

Characterization of the identified attitudinal groups

Based on a series of output tables, the distinguishing statements (playing the principal role in identifying the attitudinal groups) were analysed for each factor to better understand the substance of the prevailing attitudinal groups (Tables 1 to 5). All statements concern the Carpathian upland areas, but this context is omitted to save space.

Group 1 respondents (Conservationists) believe that '*woodlands are an important part of the heritage and should be preserved whatever the cost*' (derived Q-method value: +4) with the assurance that '*the rights of people to enjoy the beauty of treeline areas are more important than making profits from the land*' (+2). These experts support '*woodland expansion for multiple purposes*' (+1) (Table 1). They reject '*tree-felling in the Carpathian uplands*' (–3), especially '*clear-cutting*' (–4). These respondents are not supportive of the '*creation of jobs at the expense of endangering wildlife*' (–3) or the '*increase in timber production*' (–2). They voted for the '*preservation of woodlands*' and seem not to be interested in an increase in '*investment*'.

Group 2 respondents (Productivists) believe that '*shortage of investment is the main challenge in treeline areas*' (+4). They strongly '*support planting trees for timber production*' (+3) rather than '*woodland expansion for multiple purposes*' (–3). They support woodlands expansion '*to provide employment opportunities*' (+3) and agree that '*forest management should be profitable*' (+2). They rated '*beauty of landscapes*' (–2) and '*planting of more trees for the sake of future generations*' (–1) fairly low (Table 2).

Group 3 respondents (Non-Green Technocrats) experts believe that '*shortage of investment is the main challenge*' (+2) and that woodlands expansion '*to create jobs is more important than protecting wildlife and nature*' (+2). These respondents, labelled as Non-Green Technocrats, support woodland expansion '*for timber production*' (+2) but not '*for multiple purposes*' (–1) or '*to improve the landscape and attract tourists*' to remote areas (–2) (Table 3).

Group 4 respondents (Challengers) express thought-provoking attitudes to forest management in upland areas. They consider that '*planting forests to produce timber should be a policy priority*' (+4). However, at the same time, they support '*the rights of people to enjoy the beauty of treeline areas, which they consider to be more important than making profits from the land*' (+3). They are against poorly justified felling '*of trees*' (–1), but also against expansion of woodlands '*for multiple purposes*' (–2). Through their attitudes, Challengers seem to strive for reaching a balance between managing and using woodlands for timber, in upland woodlands, with their promotion of wider ecosystem services considerations (Table 4).

Group 5 respondents (Green Protectionists) consider '*that forests where trees are planted very close together are unacceptable for wildlife*' (+4). They support expansion of woodlands '*for multiple purposes*' (+2). They also think that shortage of '*investment is the challenge in treeline areas*' (+2) and suggest '*that rights of people to enjoy the beauty of treeline areas are more important than making profits from the land*' (+1). These respondents reject any felling of trees in the Carpathian uplands (–4), as well as the planting of forest either for timber production (–3), or to create jobs (–2). They have no clear opinion of whether profitability in forestry is desirable, or whether or not the wildlife should take care of itself (Table 5).

Research also identified social-economic background characteristics of respondents across the atti-

Table 6. Social-economic characteristics of respondents across the identified attitudinal groups

Group 1: Conservationists	Junior and post-doc researchers, male/female, under 30 yr old, with university education, 1 to 5 yr of experience in forestry, relatively low salaries and no children
Group 2: Productivists	Stakeholders, representatives of forest industry, senior researchers, members of local communities, male/female, 30 to 50 yr old, with various types of education, 5 to 20 yr working in forestry, with middle/high salaries and several children
Group 3: Non-Green Technocrats	Stakeholders, representatives of forest industry, local communities, junior researchers, male/female, under 40 yr old, with university education, no more than 10 yr of working/studying in forestry, average salaries and 0 to 1 child
Group 4: Challengers	Representatives of forest industry, local businesses, junior researchers and local communities, only male, under 30/40 yr old, with university education, 1 to 10 yr working/studying in forestry, with middle/high salaries and 0 to 1 child
Group 5: Green Protectionists	Primarily junior researchers and local businessmen, male/female, under 30 yr old, with university education, with 1 to 5 yr of working/studying experience in forestry, low salaries and no children

tudinal groups (Table 6). The discourse appeared to be somewhat influenced by occupation, education, age and gender of respondents. For example, the observation that Group 4 (Challengers) consists exclusively of men, or that Green Protectionists are represented by people under 30 yr, may be of interest to scientists from other disciplines (e.g. psychologists) to explore, but it should be kept in mind that our observations cannot be scaled up or applied to other cases.

Analysing group attitudes

The findings indicate commonalities in the attitudes of Conservationists (Group 1) and Green Protectionists (Group 5), with a factor correlation of 0.37. Both groups '*support woodland expansion in treeline areas for multiple purposes*' (+1, +2, respectively) and both reject the '*felling of trees in treeline areas, if the land is needed for other purposes*' (–3, –4). Alongside a number of commonalities (largely illustrated in Fig. 2), there are attitudinal differences between the 2 groups, a principal one being the statement that '*forest where trees are planted very close together is unacceptable for wildlife and should be cut down*' (–2, +4) (Tables 1 & 5).

There are also certain commonalities between the perceptions of Productivists (Group 2) and Non-Green Technocrats (Group 3), with examples seen in Fig. 2. These commonalities, among others, concern the statements on '*planting of forests to produce timber as a policy priority*' (+3, +2) and that woodland expansion '*to create jobs is more important than protecting nature*' (+3, +2). Examples of differing perceptions include the statements that '*letting wildlife take care of itself should be a principle of forest management*' (+2, –1) (Fig. 2, Table 3) and '*forest management should be profitable and financially viable*' (+2, 0) (Table 2).

Finally Group 4, labelled as Challengers, contests the opinions of the 4 other groups and strives for an ideal win–win position, where '*planting forests to produce timber is a policy priority*' (+4), but '*the rights of people to enjoy the beauty of treeline areas are recognized to be more important than making profits from the land*' (+3) (Fig. 3, Table 4). These respondents try to find the middle ground and to somehow balance different priorities of the multipurpose forestry.

On the whole, the results provide an indication that key areas of potential conflict concerning ecosystem-based adaptation of forest management practices in the upland woodlands of the Ukrainian Carpathians (as seen through the eyes of our respondents) encompass (1) expansion of woodlands for multiple purposes, (2) tree planting to produce timber as a policy priority, (3) rights of people to enjoy the beauty of landscapes as a more important objective than making profits from the land, (4) letting wildlife take care of itself and (5) shortage of investments as the main challenge in upland areas.

DISCUSSION AND CONCLUSIONS

Numerous studies on public attitudes towards forestry have been carried out worldwide (Jacobsen & Koch 1995, Jensen & Koch 2000, Karjalainen & Tyrvainen 2002, Forestry Commission 2003). These studies either examined or generalized existing attitudes. Sometimes, they categorized them. However, rarely (Nijnik et al. 2013)—and never for the Ukrainian Carpathians—has scientific research managed to explain attitudinal diversity from the standpoints of people being observed. The questions of why stakeholders think the way they do, and what stakeholders perceive in their understanding of the interactions between manifold socio-economic and nature protection activities in upland woodlands that are managed for multiple purposes, have largely remained out of focus (Nijnik et al. 2010). Our research seeks to fill the existing knowledge gap. Stakeholder engagement in consultation processes and the uncovering of perceptions concerning multi-functional forestry development in the Ukrainian Carpathians provides an indication of social choices in support of the decision-making.

Co-constructing knowledge with various stakeholders, including local communities, and respecting their conventional culture and traditional ways of life, are essential requisites for the success of decision-making processes in uplands (Sarkki et al. 2016). Owing to the 'path-dependency' legacy (Nijnik & Oskam 2004), this observation particularly applies to the Carpathian Mountains, where the development of tailor-made solutions and more active participation of stakeholders in decision-making processes is crucial (Bizikova et al. 2012).

Therefore, our methodological approach went beyond a traditional analysis of public attitudes, as it aimed to identify and explain a variety of factors influencing the decision-making processes. Through this study, our consultation and communication with stakeholders belonging to the Carpathian Mountains can be seen as a first step in promoting their engage-

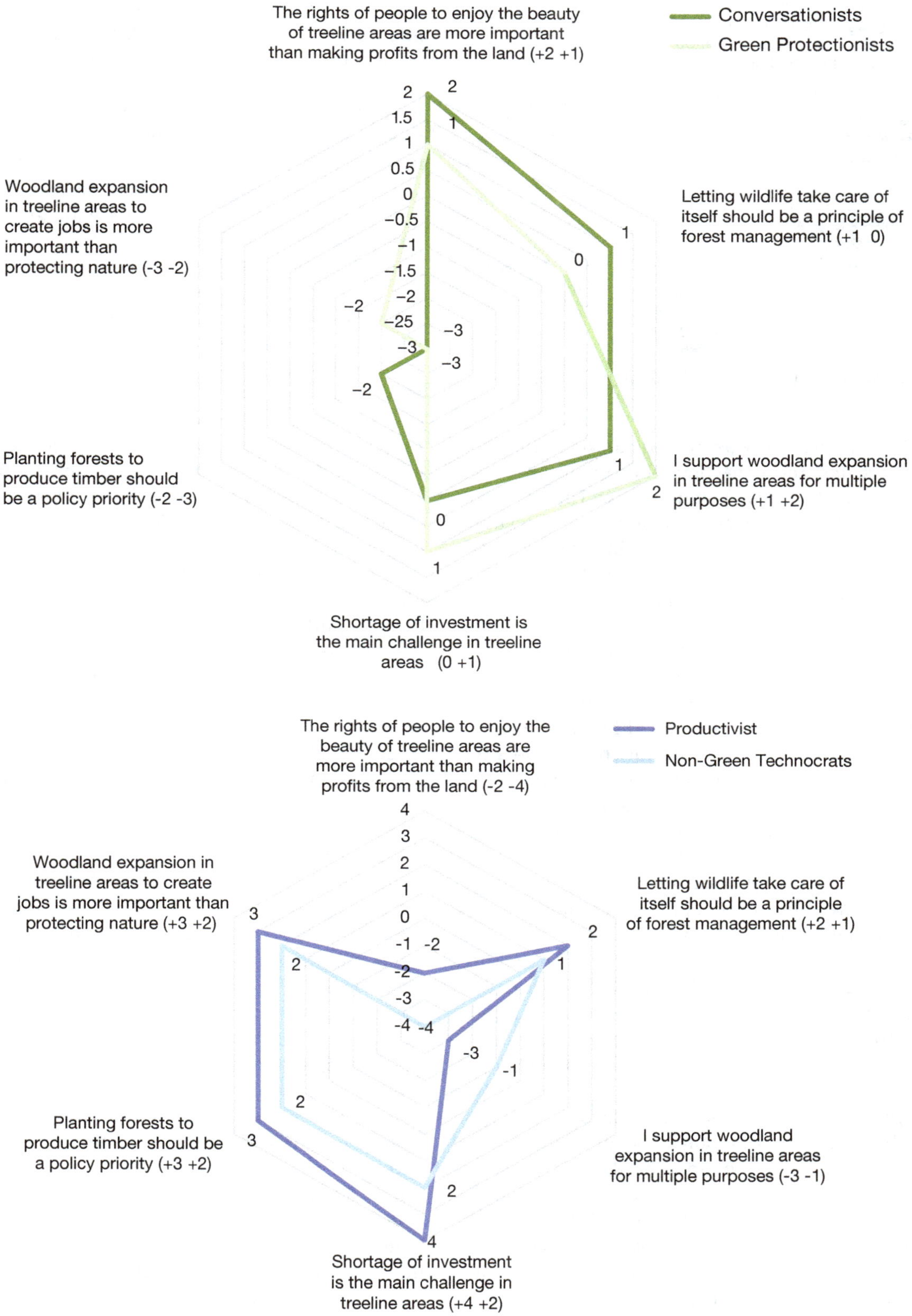

Fig. 2. Attitudinal commonalities and differences between Conservationists (Group 1) and Green Protectionists (Group 5); and between Productivists (Group 2) and Non-Green Technocrats (Group 3)

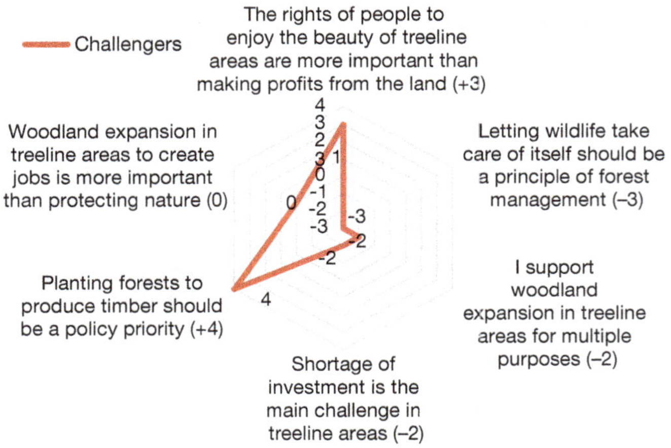

Fig. 3. Attitudinal diversity of Challengers (Group 4)

ment. The approach was beneficial both for the researchers and those participating in discussions, and indirectly for local communities. The method explained in this paper could potentially be useful elsewhere, allowing for mutual learning and capacity building, while key results from such studies could help to provide relevant policy actors and practitioners with some guidance for strategic planning and a more sustainable development of mountain areas.

Our approach, however, focused on perception analyses. It provided a discursive evaluation of major issues that respondents expressed. Although the identified and explained typologies likely reflect the attitudinal diversity of people beyond the respondent groups, the current research is not a general public. Other methods can be used (and be combined with the Q-method) for finding out how many people (within the 'stakeholder population' or for the public in general) have one opinion or another about multifunctional changes in upland woodlands.

Research findings suggest that people in the Ukrainian Carpathians feel attachment to woodlands and pay attention to biodiversity conservation and nature preservation. This is likely because woodlands are embedded in the local culture, and because people want to maintain their national heritage (Zahvoyska et al. 2014). Moreover, local stakeholders clearly benefit from woodlands beyond just using their cultural, supporting and regulatory services. The provisioning services of trees are considered to be important, e.g. the use of non-wood forest products and fire wood. This is likely because practically all forests in the region remain publicly owned and state managed (Soloviy 2010).

The results provided an understanding of the differences and similarities in the dominant attitudes of upland area stakeholders (those who are residing and working in the Ukrainian Carpathians). Acknowledging the different stakeholder reasonings was a stride towards making science that is more relevant to society. It was also a step towards gaining access to a 'real world' understanding of problems pertaining to forestry in the Carpathian uplands (through the eyes of people on the ground) and finding potential solutions to these problems (e.g. to avoid and manage possible conflicts).

In this regard, our findings concerning the differing (and at times conflicting) stakeholder attitudes towards woodland expansion, timber production, the importance for people to enjoy landscapes, wildlife conversation and investment policy, identified across the 5 attitudinal factors, may offer useful insights to assist decision-makers in addressing the opinions of each attitudinal group on a case-by-case basis, and applying the most suitable solutions to problems where consensus can hardly be reached. However, the question of whose preferences are most important remains challenging.

Also, as research findings indicate that some key differences in perceptions are caused by value conflicts, a participatory decision-making process may help to raise awareness of such conflicts and indicate the way towards reaching consensus, where possible. Similarities in the attitudes identified across some of the attitudinal groups (as seen in Fig. 2) could be helpful for reaching consensus and, eventually, enabling science to assist decision-makers in making forest policy and management in the Capathian Mountains more socially acceptable, inclusive and robust.

The findings from this paper specifically identified the following: similarities in attitudes with regard to native woodland regeneration; heterogeneity of attitudes and, therefore, key areas of potential conflict between stakeholders concerning forest policy and management practices in the uplands; improved participation in decision-making as a key challenge in the region; and a requirement for sustainable provision of forest ecosystem services. The identified attitudinal similarities could be of help in designing policy measures and better targeting of forest policy and management decisions.

Key findings from an analysis of socio-economic characteristics of respondents across the identified attitudinal groups indicated, in this particular research, that Conservationists are primarily young people. This is likely because this category of stakeholders is more idealistic and not too concerned with the current economic situation in Ukraine, e.g. the shortage of investment in the forestry sector, which is negatively affecting the Carpathian region. Also, the observation that Challengers are primarily men leaves us with speculations that men, more often than women, are willing to balance various dimensions of sustainability in the multi-functional forestry development in this region.

To conclude, it can be argued that the applied approach offers a credible means of performing stakeholder evaluations relevant to forest policy and management. We should be careful, however, regarding any attempt to transfer benefits across case studies or to extrapolate them. The results of stakeholder evaluation derived from the Q-method application are very much case- and context-dependent. Therefore, we suggest complementing this research with analyses of other complex institutional matters in the region, such as social innovation (e.g. using social network analysis), common pool resource and tenure rights problems.

Acknowledgements. The article is based on work from COST Action ES1203 SENSFOR (www.cost.eu), coordinated by Prof. Kari Laine and Vice Chair, Prof. Oddvar Skre. We also thank the Scottish Government, who supported this research through their Rural Affairs and the Environment Strategic Research Programme and the European Commission for support provided to the project on Social Innovation in Marginalised Rural Areas (SIMRA). This project has received funding from the European Union's Horizon 2020 Research and Innovation Programme under Grant Agreement No. 677622. It is coordinated by the James Hutton Institute and includes the University of Oulu. We are also grateful to respondents who contributed to the survey, to suggestions by the Guest-Editor, Prof. Frans Wielgolaski, on an earlier draft, to our colleague from the James Hutton Institute, Ms Gillian Donaldson-Selby, for the English language proofreading of the manuscript and to the anonymous reviewers for their comments.

LITERATURE CITED

Barry J, Proops J (1999) Seeking sustainability discourses with Q methodology. Ecol Econ 28:337–345

Berkes F (2009) Evolution of co-management: role of knowledge generation, bridging organisations and social learnings. J Env Manage 90:1692–1702

Bizikova L, Nijnik M, Kluvanková-Oravská T (2012) Sustaining multifunctional forestry through the developing of social capital and promoting participation: a case of multiethnic mountain communities. Small-scale For 11: 301–319

Brown SR (1980) Political subjectivity. Applications of Q methodology in political science. Yale University Press, New Haven, CT

Brown S (1996) Q methodology and qualitative research. Qual Health Res 6:561–567

Davies BB, Sherlock K, Rauschmayer F (2005) 'Recruitment', 'composition' and 'mandate' issues in deliberative processes: should we focus on arguments rather than individuals? Environ Plan 23:599–615

Elbakidze M, Angelstam P (2013) Sustainable forest management from policy to landscape and back again: a case study in the Ukrainian Carpathians Mountains. In: Kozak J, Ostapowicz K, Bytnerowicz A, Wyzga B (eds) The Carpathians: integrated nature and society towards sustainability. Springer-Verlag, Berlin

European Commission (2015) Paris Agreement. Available at http://ec.europa.eu/clima/policies/international/negotiations/paris_en (accessed on 1 December 2016)

Forestry Commission (2003) UK public opinion survey of forestry. Forestry Commission, Edinburgh

Funtowicz SO, Ravetz J (1993) Science for the postnormal age. Futures 25:739–755

Jacobsen CH, Koch NE (1995) Summary report on ongoing research on public perceptions and attitudes on forestry in Europe. Danish Forest and Landscape Research Institute, Horsholm

Jensen CH, Koch NE (2000) Measuring forest preferences of the population—a Danish approach. Schweiz Z Forstwesen 151:11–16

Karjalainen E, Tyrvainen L (2002) Visualization in forest landscape preference research: a Finnish perspective. Landsc Urban Plan 59:13–28

Keeton WS, Crow SM (2009) Sustainable forest management alternatives for the Carpathian Mountain region: providing a broad array of ecosystem service. In: Soloviy I, Keeton WS (eds) Ecological economics and sustainable forest management: developing a trans-disciplinary approach for the Carpathian Mountains. UNFU Press, Lviv, p 109–126

Krott M (2008) Forest government and forest governance within a Europe in change. In: Cesaro L, Gatto P, Pettenella D (eds) The multifunctional role of forests—policies, methods and case studies. EFI Proc. No. 55. Gummerus Printing, Saarijärvi, p 13–25

Krynytskyy GT, Chernyavskyy MV (eds) (2014) Close to nature and multifunctional forest management in the Carpathian region of Ukraine and Slovakia. PE "Circle", Uzhorod

McKeown B, Thomas B (2013) Q methodology. Sage Publications, Thousand Oaks, CA

Miller DR, Vogt N, Nijnik M, Brondizio E, Fiorini S (2009) Integrating analytical and participatory techniques for planning the sustainable use of land resources and landscapes. In: Geerteman S, Stillwell J (eds) Plan support systems: best practice and new methods. Springer, Dordrecht

Muñoz-Rojas J, Nijnik M, Puente MG, Garcia FC (2015) Synergies and conflicts in the use of policy and planning instruments for implementing forest and woodland corridors and networks in Scotland. For Policy Econ 57:47–64

Nijnik M, Miller D (2014) Targeting sustainable provision of forest ecosystem services with special focus on carbon sequestration. In: Matyssek R, Clarke N, Cudlín P, Mikkelsen T, Tuovinen JP, Wiesser G, Paoletti E (eds): Climate change, air pollution and global challenges: understanding and perspectives from forest research. Elsevier, Oxford, p 547–565

Nijnik M, Oskam A (2004) Governance in Ukrainian forestry: trends, impacts and remedies. Int J Agric Resour Gov Ecol 3:116–133

Nijnik A, Nijnik M, Soloviy I (2009a) Challenges and opportunities for sustainable development in the Ukrainian

Carpathians. MRI Newsl 2:21–24

Nijnik M, Soloviy I, Nijnik A, Deyneka A (2009b) Challengers and potential policy responses towards sustainable mountain development and nature conservation in the Ukrainian Carpathians. In: Chmielewski T (ed) Nature conservation management: from idea to practical result. AlterNet, Lublin, p 132–149

Nijnik M, Nijnik A, Lundin L, Staszewski T, Postolache C (2010) Stakeholder attitudes to multi-functional forests in Europe. For Trees Livelihoods 19:341–358

Nijnik M, Miller D, Nijnik A (2013) Linking multi-functional forestry goals with sustainable development objectives: a multi-national study. J Settl Spat Plann 2:185–190

Nijnik M, Nijnik A, Bergsma E, Matthews R (2014) Heterogeneity of experts' opinion regarding opportunities and challenges of tackling deforestation in the tropics. Mitig Adapt Strategies Glob Change 19:621–640

Nijnik M, Nijnik A, Brown I (2016) Exploring the linkages between multi-functional forestry goals and the legacy of spruce plantations in Scotland. Can J For Res 46: 1247–1254

Prell C, Hubacek K, Reed M (2009) Stakeholder analysis and social network analysis in natural resource management. Soc Nat Resour 22:501–518

Sarkki S, Karjalainen TP (2012) Science and issue advocacy in a forestry debate in northern Finland. Polar J 2: 125–138

Sarkki S, Ficko A, Grunewald K, Nijnik M (2016) Benefits from and threats to European treeline ecosystem services: an exploratory study of stakeholders and governance. Reg Environ Change 16:2019–2032

Schmolck P (2012) PQMethod version 2.33. Available at http://schmolck.userweb.mwn.de/qmethod/

Soloviy IP (2010) Policy for forest sector sustainable development: paradigm and instruments. Ukrainian National Forestry University Press, Lviv

Soloviy I, Nijnik M, Ilkiv K (2012) Mountain forestry policy and strategies in response to climate change challenges. J Ecol Econ 111–120

Stephenson W (1963) Independency and operationizm in Q-sorting. Psychol Rec 13:269–272

Stevenson H (2015) Contemporary discourses of green political economy: a Q method analysis. J Env Policy Plan. http://dx.doi.org/10.1080/1523908 X.2015.1118681

The Carpathian Convention (2003) www.carpathianconvention.org/documents-carpathian-convention.html (accessed on 7 December 2015), UNEP, Vienna

The Forest Code of Ukraine (2006) http://zakon.rada.gov.ua/cgi-bin/laws/anot.cgi?nreg=3852-12 Verkhovna Rada, Kyiv (accessed on 7 December 2015)

van Asselt MBA, Mellors J, Rijkens-Klomp N, Greeuw SCH, Molendijk KGP, Beers PJ, van Notten P (2001) Building blocks for participation in integrated assessment: a review of participatory methods. International Centre for Integrated Studies, Maastricht

van Exel J, De Graaf G (2005) Q methodology: a sneak preview. Erasmus University, Rotterdam

Vogel J, Lowham E (2007) Building consensus for constructive action: a study of perspectives on natural resource management. J For 105:20–27

Watts S, Stenner P (2012) Doing Q methodological research: theory, method and interpretation. Sage, London

Zahvoyska L, Sarkki S, Zhyla T (2014) Perceptions of treeline ecosystem services and their governance: focus on Ukrainian Carpathians. In: Kruhlov I, Prots B (eds) Local responses to global challenges. Proc. Forum Carpaticum, Sept 16–19, 2014. Ukrainian Bestseller, Lviv, p 65–69

Zahvoyska L, Nijnik M, Sarkki S, Nijnik A, Pelyuch O (2015) Insights into treeline ecosystem services of the Ukrainian Carpathians from a stakeholder's perspective: an analysis of challenges for adaptive governance. J Proc For Acad Sci Ukr: Coll Sci Pap 13:193–200

Forest ecosystem services affected by natural disturbances, climate and land-use changes in the Tatra Mountains

Peter Fleischer[1,2], Viliam Pichler[1], Peter Fleischer Jr.[1], Ladislav Holko[3], František Máliš[1,4], Erika Gömöryová[1,*], Pavel Cudlín[5], Jan Holeksa[6], Zuzana Michalová[7], Zuzana Homolová[2], Jaroslav Škvarenina[1], Katarína Střelcová[1], Pavol Hlaváč[1]

[1]Faculty of Forestry, Technical University in Zvolen, 96053 Zvolen, Slovakia
[2]Research Station of TANAP, State Forest of TANAP, 059 60 Tatranská Lomnica, Slovakia
[3]Institute of Hydrology, Slovak Academy of Sciences, 031 05 Liptovský Mikuláš, Slovakia
[4]Forest Research Institute Zvolen, National Forest Centre, 960 92 Zvolen, Slovakia
[5]Institute of Systems Biology and Ecology, Academy of Sciences, 370 05 České Budejovice, Czech Republic
[6]Faculty of Biology, Adam Mickiewicz University, 61 614 Poznań, Poland
[7]Faculty of Forestry and Wood Technology, Czech University of Life Sciences, 16 521 Prague, Czech Republic

ABSTRACT: The consequences of large-scale disturbances magnified by climate extremes and land-use changes in Norway spruce forests in the Tatra Mountains (Slovakia) are assessed in this study. The study area is part of the territory of Tatra National Park (TANAP). The driver–pressure–state–impact–response (DPSIR) framework was applied to evaluate how the ecosystem and its services are affected. The state of the ecosystem and its potential for provisioning ecosystem services before and after disturbances is expressed by a set of indicators derived mostly from long-term ecological research conducted in TANAP. The differences are classified by a standardised change index (CI). Ten years after the major windthrow disturbance in 2004, all ecosystem services were still below the pre-disturbance state. The most pronounced declines were found in cultural (average CI = 0.69) and provisioning (average CI = 0.86) ecosystem services. Regulating services are recovering faster (average CI = 0.97), with some indicators exceeding the state before the disturbances. Significant changes took place at the tree line, which is a new phenomenon not known from previous disturbances. Despite a gradual recovery of the ecosystem state and functioning, this analysis confirms that there is a serious risk of decline in forest ecosystem benefits according to regional climate change projections. It also indicates the increasing importance of sustainable forest management for safeguarding ecosystem services under changing conditions.

KEY WORDS: Forest ecosystem state · Bark beetle outbreak · Long-term research

1. INTRODUCTION

Ecosystem services (ES) are the benefits that people receive from nature, natural processes or ecosystem functions. As such, they reflect societal values, goals, desires and benefits (TEEB 2010, Kelble et al. 2013). ES are generally categorised into provisioning (PES), regulating (RES) and cultural services (CES), as well as ecosystem functions. The capacity of forest ecosystems to provide ES depends on their state (Sing et al. 2015), which can be described by various indicators. PES provide goods (e.g. timber, biomass, water, game, etc.), while RES are the benefits obtained from the regulation of ecosystem processes

*Corresponding author: gomoryova@tuzvo.sk

§Advance View was available online June 28, 2017

(e.g. water regulation and purification, carbon balance, nutrient content, etc.). CES are non-material benefits achieved through cognitive development, reflection, recreation, etc. Ecosystem deterioration by external pressures may reduce both the quantity and quality of expected services.

Forest ecosystem services are threatened by increasing natural disturbances in many parts of the world, particularly in relation to climate change (e.g. Schelhaas et al. 2003, Seidl et al. 2014). At the same time, a higher susceptibility of forests to disturbance has been interpreted as a consequence of past land use (Seidl et al. 2009). However, recent large-scale dieback in remote natural forests, even at the alpine treeline, confirms the complexity and still rather limited knowledge of the interactions among disturbances, climate and land-use changes that seriously affect ecosystem functions and services (Dale 1997, Thom & Seidl 2016). The emerging conflict between increasing demands and a compromised or limited ES capacity is one of the key challenges for forest management and protection.

Different approaches have been proposed to resolve conflicts between the demands for ES and their provisioning capacity. 'Cause–effect' models are widely used to illustrate the relation among indicators describing pressures and impacts in ecosystems (Smeets & Weterings 1999). The DPSIR (driver–pressure–state–impact–response) framework (Levin et al. 2008) is an example of such a model. It is a conceptual tool for structuring and communicating policy-relevant environmental research, which expresses how human society impacts on ecosystems (EEA 2005). Its application to certain problems helps to identify crucial points in the causal chain of DPSIR and supports decision-making processes to mitigate the negative effects that the driving forces may have. Social and economic developments or natural disturbances (driving forces) exert pressures on the environment and, as a consequence, the state of the environmental changes. This leads to impacts on ecosystems, human health and society, which may elicit a societal response to the driving forces, state or impacts via various mitigation, adaptation or curative actions (Smeets & Weterings 1999).

The objective of this study was to assess the impacts of key drivers that cause changes in mountain forests and the alpine treeline on ES provisioning in the Tatra Mountains. The study area, protected as a national park, has been heavily affected by repeated natural disturbances, of which the severity has been magnified by climate and land-use changes. We investigated how ES changed 5 and 10 yr after an extraordinary large-scale windthrow disturbance in 2004, the role of changing land use and the possible impacts of changing climate. The analysis was based on an assessment of indicators that were derived from published long-term ecological research and monitoring data (Fleischer 2008). We applied the DPSIR framework to assess the ecosystem state and its capacity to provide ES. To our knowledge, this study represents the first application of this type of analysis in the Tatra Mountains. The results could assist National Park managers in balancing societal requirements with the potential of the forest ecosystems to provide the benefits of nature to people under changing environmental conditions.

2. MATERIALS AND METHODS

2.1. Study site and disturbances

The Tatra Mountains are the highest part of the Carpathians and are situated along the border between Slovakia and Poland. Although the altitude of many peaks exceeds 2000 m above sea level (a.s.l.), the mountains cover a relatively small area (341 km^2) and the main ridge is only 26 km long. Despite century-long use by humans for wood and grazing, the forests still cover a large continuous area, covering around 400 km^2. An abrupt change in land use occurred in 1949 when the Tatra National Park was established. Various management approaches were applied afterwards with the aim to reconstruct ecosystems disturbed by earlier human overuse and to safeguard an array of ES, such as tourism, recreation, steep slope stabilisation, water regulation, habitat protection and wood production. Since the mid-1990s, a non-intervention approach has been applied on approximately 30% of TANAP forests, which has had a significant negative effect on traditional land-use (Koreň 2015).

The Tatra Mountains are frequently hit by damaging downslope winds that storm their lee side. During the past 200 yr, massive forest destruction caused by such winds occurred in 1834, 1868, 1898, 1915, 1919, 1925, 1941, 1971, 1981, 2004 and 2014 (Koreň 2015, Holeksa et al. 2016). On a regional scale, the windthrow event of 2004 was the most severe known wind disturbance in the history of the Tatra Mountains (Zielonka et al. 2010). The wind gusts exceeded 230 km h^{-1} and laid down some 2.3 million m^3 of the forests' standing volume across 120 km^2 (30% of the forest area), making it one of the 10 most devastating storms in Europe in recent decades (Gardiner et al.

2010). Along with this harmful event, the extraordinarily high summer temperatures recorded after the 2004 disturbance caused forest fires and unprecedented European spruce bark beetle (*Ips typographus*, hereafter bark beetle) outbreaks. During the last decade, air temperatures have exceeded the long-term climatological mean (1930–1960) for the growing season several times, particularly in 2007, 2012 and 2015 when the temperature was higher than the climatological mean by 1.6, 1.5 and 1.9 °C, respectively. The combination of windthrow disturbance and elevated temperature has resulted in an increase in various pressures that have severely affected the ecosystem and its services. High numbers of fallen trees combined with warm weather lead to an increased risk of bark beetle population outbreak (e.g. Schroeder & Lindelöw 2002, Wermelinger 2004). Indeed, unlike in the past, a massive bark beetle outbreak spread upslope up to the treeline (1500–1650 m a.s.l.) and damaged more than 70 km^2 of mature, mostly natural, Norway spruce *Picea abies* (L.) Karst and European larch *Larix decidua Mill*) forest (Fleischer et al. 2016). Warm weather and dry periods caused forest fires on the scale of several square kilometres.

Although we focus on natural drivers, the recent socioeconomic developments in the study area should also be mentioned. Large investments in winter sport resorts made in the last decade (€200 million) and a 50% increase in the transport capacity and area of ski slopes have dramatically increased the pressure on supramontane and subalpine forests in the Tatra Mountains, which has particularly affected the treeline.

Five research sites, each 100 ha in size, were established in TANAP in 2006–2008 to assess the response of ecosystems and their functions and services to the wind, fire and bark beetle disturbances as well as to different land-use practices, with different types of human intervention. The EXT and FIR sites represent salvage-logged windthrow and windthrow burnt by incidental fire, respectively. The sites without human intervention are NEX (no timber extraction) and IPS (site with no timber extraction disturbed by bark beetle). The REF site represents intact reference conditions. For a detailed description see Fleischer (2008).

2.2. DPSIR model, ecosystem services and indicators

We applied the DPSIR framework to evaluate the impacts of disturbance (windthrow, fire, bark beetle), land use (natural unmanaged forest, semi-natural managed forest) and climate (increasing temperature) on ES. To apply the DPSIR framework, we adhered to the generally applicable criteria for discerning drivers, pressures, states and impacts as established by Oesterwind et al. (2016). Fig. 1 illus-

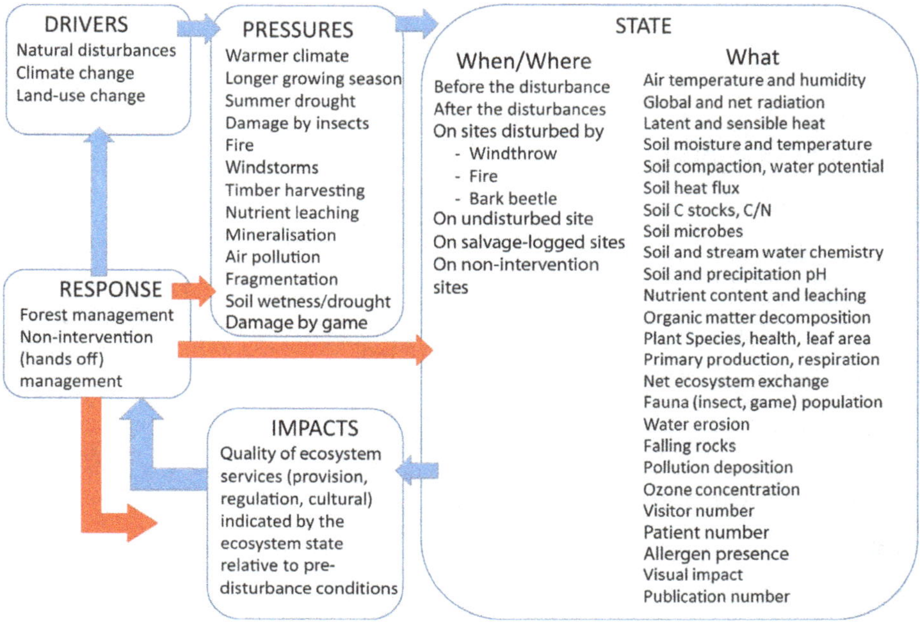

Fig. 1. Conceptual model for evaluation of forest ecosystem services in the Tatra Mountains using the driver–pressure–state–impact–response (DPSIR) approach (blue arrows) and possible societal response (red arrows)

trates the concept of progressive environmental changes induced by the driving forces, i.e. natural disturbances, climate and land-use changes, and their impacts on ES.

We understand pressures as processes (air and water warming, mineralisation, leaching, fragmentation, insect outbreaks, drought, erosion, etc.) stimulated by the drivers. The magnitude of the pressures determines the state and the functioning of an ecosystem. The state of an ecosystem is its capacity for providing ES on a certain level that we interpret as impact. The state evaluated in time intervals is cru-

Table 1. List of ecosystem state indicators categorised according to the ecosystem services (PES, provisioning; RES, regulating; CES, cultural) and their relative normalized change index (CI), when compared with the reference (REF) value (pre-disturbance state recorded at the REF site in 2004 [REF0] or alternatively, a site-specific value if recorded at the distinct research site before the disturbance) according to land-use types (EXT: salvage-logged windthrow; FIR: windthrow burnt by incidental fire; NEX: no timber extraction; IPS: no timer extraction but disturbed by bark beetle occurrence including treeline) and the time of assessment (5 and 10 yr after the disturbance). +/– indicates the interpretation of increasing ecosystem state indicator value on ecosystem functioning and ES provision based on the authors' evaluation. See Section 2.2 for further details. Key indicators are in **bold**

Ecosystem service category	Indicator (unit)	ES trend (+/–)	REF REF0 or site value	CI_{ij}				
				REF 5	REF 10	EXT 5	EXT 10	FIR 5
PES	**Gross primary production (g C m^{-2})**	+	1282	0.92	0.85	0.43	0.93	0.96
	Leaf area index (m^2 m^{-2})	+	6	0.92	0.67	0.2	0.63	0.33
	Net primary production (g C m^{-2})	+	460–505	0.81	0.18	0.63	0.62	0.83
	Natural regeneration (ind. ha^{-1})	+	2200	1.09	1.18	2.04	1.63	2.00
	Dead wood (m^3 ha^{-1})	+	77	1.09	1.17	1.43	1.43	0.39
	Fragmentation (%)	–	15	0.88	0.75	0.16	0.16	0.16
	Yield (m^3 ha^{-1})	+	280	1.04	0.25	0.00	0.04	0.00
	Soil fauna diversity (n)	+	5	1.00	0.60	2.80	1.00	3.20
	Game (n ha^{-1})	–	0.5	1.00	1.00	0.50	0.33	0.50
	Bark beetle population (% trees infested)	–	30	0.75	0.60	1.00	1.00	1.00
	Tree species (n)	+	2	1.00	1.00	2.00	2.50	2.50
	Plant species (n)	+	22	1.09	1.05	0.36	0.77	0.32
RES	**Respiration (g C m^{-2})**	–	1235	1.07	1.16	1.15	0.93	0.96
	C balance (g C m^{-2})	+	47	0.74	0.40	0.16	0.21	0.22
	Total organic C (mg l^{-1})	–	375	0.99	0.99	0.99	0.99	1.00
	C/N	–	22	0,.6	0.96	1.00	1.05	1.31
	C stock (Mg C ha^{-1})	+	92	1.09	1.12	1.02	0.94	0.94
	Leaching (index)	–	4.6	1.00	0.92	0.58	0.92	0.55
	Soil pH	+	3.9–4.8	1.00	1.00	0.98	0.98	0.98
	Microbial diversity	+	226	0.39	0.51	0.33	0.56	0.30
	Cellulose decomposition (%)	+	25	1.00	1.00	1.60	1.60	2.40
	Naturalness (% of natural stands)	+	20–90	1.00	1.00	0.25	0.40	0.25
	High flow pulses (n)	–	90	0.75	0.69	0.82	0.69	0.64
	Flow reversals (n)	–	7	1.00	1.00	1.00	1.00	0.46
	Flow minimum (index)	–	0.15–0.4	0.33	0.33	0.87	1.00	0.60
	Interception (% of rain retained by canopy)	+	42	0.90	0.67	0.24	0.24	0.12
	Ground water level (m)	–	2.5	0.40	0.58	0.40	0.58	0.40
	Surface runoff (%)	–	0.61–1.57	0.87	0.76	0.32	0.41	0.27
	Erosion (kg ha^{-1} yr^{-1})	–	0.007	0.43	0.43	0.70	1.00	0.70
	O_3 (ppb)	–	39–48	1.08	1.05	1.08	1.05	1.08
	Precipitation pH	+	4.0–4.1	1.07	1.36	1.20	1.38	1.05
	Soil temperature (°C)	–	7.4–9.0	0.98	0.78	0.72	0.82	0.80
	Air temperature (°C)	–	11.0–11.5	0.94	0.92	0.85	0.84	0.86
	Soil moisture (%)	+	33–38	1.00	1.08	0.86	0.70	1.06
	Snow accumulation (cm)	+	150–250	1.00	1.10	1.00	1.08	1.00
	Sensible heat (% in energy balance)	–	45.9	0.96	0.92	0.83	0.84	0.79
	Wind velocity (m s^{-1})	–	0.2	0.67	0.50	1.06	1.2	1.11
	Fire risk (index)	–	1	1.00	0.50	0.14	0.25	0.33
	Rockfall risk (index)	–	1	1.00	1.00	1.00	1.00	1.00
	Treefall risk (index)	–	1	0.50	0.33	1.00	1.00	1.00
CES	**Visitors (n)**	+	18 000	0.88	1.17	0.88	1.17	0.88
	Patients (n in climatic spa)	+	3000	0.63	0.40	0.63	0.40	0.63
	Attractiveness (index)	+	1	1.00	0.80	0.50	0.55	0.50
	Allergen, disease (index)	–	1	0.66	0.50	0.28	0.25	0.67

cial for the detection of changes in state, in our case before and after (5 and 10 yr after the windthrow disturbance in 2004). The ecosystem state before the disturbance was used to classify ES modified by different disturbance and land-use types. The impact on PES, RES and CES was analysed.

The identification and description of the ecosystem changes was based on the indicators published by scientists involved in 'windstorm research' (Fleischer 2008). The indicators presented in this study (Table 1) are physical, chemical and biological variables that describe the current status of drivers, pressures and impacts on the studied ecosystem. We summarised all published data, derived indicators and assigned them to the DPSIR categories. A total of 44 indicators were selected, which were grouped into 3 categories characterising PES (12 indicators), RES (28 indicators) and CES (4 indicators) (Table 1). Then, we classified the indicators by increasing or decreasing tendency (before the disturbance, 5 and 10 yr after) using the change index (CI). Indicators at all the sites were standardised with the reference values, applying the rule 'the higher the better' (Wang et al. 2015). We chose 2004 and the REF site as the reference for

CI_{ij}					Source
FIR 10	NEX 5	NEX 10	IPS 5	IPS 10	
0.92	0.95	1.11	0.63	0.39	Fleischer (2016)
0.73	0.35	0.73	0.50	0.28	Fleischer (2016)
0.81	0.76	0.90	0.13	0.28	Kyselová & Homolová (2012), Pajtík et al. (2015), Fleischer (2016)
2.51	3.63	3.77	1.33	2.13	Jonášová et al. (2010), Šebeň et al. (2011b), Michalová (2015)
0.39	3.25	3.38	3.25	3.38	Šebeň et al. (2011a)
0.16	0.13	0.11	0.30	0.18	Kopecká & Nováček (2009), Havašová et al. (2017)
0.04	0.01	0.04	0.50	0.11	State Forests of TANAP, Tutka (2009)
1.00	1.00	0.40	0.75	0.75	Hloška et al. (2016)
0.33	0.50	0.50	0.50	0.25	State Forests of TANAP
1.00	0.86	0.75	0.50	0.40	Nikolov et al. (2014)
3.50	0.50	0.50	1.00	1.00	Šebeň et al. (2011b), Fleischer (2016)
0.54	0.63	1.00	1.12	1.31	Máliš et al. (2015)
0.97	0.98	0.88	0.73	0.78	Fleischer (2016)
0.25	0.45	0.49	0.06	0.04	Fleischer (2016)
1.00	1.00	1.00	1.00	1.00	Ziegler (2007), Frič & Škvarenina (2008)
1.29	1.16	1.10	1.00	1.05	Don et al. (2012), Hanajík & Šimonovičová (2015), Gömöryová et al. (2017)
0.96	1.00	0.98	1.00	0.97	Don et al. (2012), (2015)
0.90	0.55	0.94	0.84	0.82	Bischoff et al. (2008), Fleischer (2014)
1.00	1.13	1.08	1.00	1.00	Šimkovič et al. (2009), Gömöryová et al. (2017)
0.52	0.74	0.81	0.39	0.40	Gömöryová et al. (2017)
1.60	2.60	1.60	1.25	1.40	Kyselová & Homolová (2011), Z. Kyselová (pers. comm.)
0.75	1.29	1.29	1.00	1.00	www.lesytanap.sk
0.56	0.64	0.56	0.64	0.56	Holko et al. (2012), Holko & Škoda (2016)
0.35	0.46	0.35	0.46	0.46	Holko et al. (2012), Holko & Škoda (2016)
0.25	0.70	0.25	0.70	0.40	Holko et al. (2012), Holko & Škoda (2016)
0.36	0.52	0.66	0.53	0.36	Fleischer (2014)
0.58	0.40	0.58	0.40	0.58	Bičárová (2011), S. Bičárová (pers. comm.)
0.34	0.76	0.87	0.89	0.89	Sitko et al. (2011)
1.00	0.70	1.00	0.70	0.09	Sitko et al. (2011)
1.05	1.05	1.08	1.06	1.17	Bičárová et al. (2015)
1.34	1.13	1.35	1.12	1.35	Fleischer (2014)
0.89	0.92	0.88	0.93	0.88	Fleischer (2014), Fleischer (2016)
0.84	0.85	0.83	0.85	0.79	Matejka & Fleischer (2011), Fleischer (2014)
1.09	1.02	0.95	1.11	1.05	Fleischer (2016),
1.08	1.00	1.07	1.00	1.00	Bartík et al. (2014)
1.56	0.75	3.85	0.88	0.81	Fleischer et al. (2012), Fleischer (2014)
1.25	1.06	1.13	0.28	0.15	P. Fleischer (unpubl.)
0.33	0.10	0.25	0.14	0.11	www.lesytanap.sk
1.00	1.00	1.00	0.25	0.25	www.lesytanap.sk
1.00	0.50	0.50	0.14	0.11	Kajba et al. (2013), www.lesytanap.sk
1.17	0.88	1.17	0.88	1.17	www.lesytanap.sk
0.40	0.63	0.40	0.63	0.40	Božíková (2009), J. Božíková (pers. comm.)
0.55	0.50	0.55	0.50	0.50	Vyskot et al. (2007), Čekovská (2013)
0.25	0.66	0.33	0.50	0.50	Turčeková et al. (2014), Homolová et al. (2015)

this analysis (hereafter REF0). In a limited number of cases (where available), pre-disturbance values at disturbed sites were used as a reference. Two types of formula were applied to standardise the indicator values, reflecting their different temporal dynamics (increasing, decreasing) and impact (positive, negative) on the ES:

$$\mathrm{CI}_{ij} = \frac{\mathrm{REF0}}{x_{ij}} \text{ (increasing value increases ES)}$$
$$\mathrm{CI}_{ij} = \frac{x_{ij}}{\mathrm{REF0}} \text{ (increasing value decreases ES)} \quad (1)$$

where x_{ij} is the state indicator real value, i is disturbance and land-use type (EXT, FIR, NEX, IPS, REF) and j is temporal category (5 and 10 yr after the disturbance).

CI value <1 indicated a decline and values >1 indicated an increase in ES compared with pre-disturbance conditions. The average CI value was calculated for each ES category from all the indicators. The development trajectories (differences among the study sites) were evaluated using principal component analysis (PCA). A descriptive classification of the indicator change (positive or negative) on the ES was based on a literature review (Thom & Seidl 2016) and the authors' interpretation (Table 1).

The relevance of the indicators for the ES was assessed *post priori*. The key indicators were identified in 2 steps. First, using exploratory factor analysis (EFA; R Development Core Team 2008), the groups of indicators were identified following the approach outlined by Bryce et al. (2016). Alternatively, in cases where the statistical interpretation was uncertain, we applied an expert estimation approach (Grêt-Regamey et al. 2015). Second, the most relevant indicators were selected by correlation analysis. The indicator with the highest correlation to other indicators inside the statistically or intentionally derived groups was denoted as the key indicator.

3. RESULTS

Repeated disturbances, climate and land-use changes have had a significant impact on the state of the forest ecosystem and its functioning in the study area. Large windthrow events (in 2004 and 2014) and consequent bark beetle outbreaks damaged 200 km^2 of the forest during the study period, which represents half of the TANAP forest area. In recent decades, increasing air temperature has been recorded at meteorological stations located in the vicinity of the Tatra Mountains. As an example, Fig. 2 shows data from Tatranská Lomnica (elevation 830 m a.s.l.), where measurements started in 1898. The largest windthrow disturbances in the Tatra Mountains during the last 100 yr were plotted against the backdrop of the annual average temperature and indicate that the most recent windthrow disturbances (2004 and 2014) coincided with extremely high air temperatures. This synergy of drivers has a large impact on the forest in the Tatra Mountains. As shown later, higher temperatures resulted in a larger bark beetle population, increased fire risk, etc.

Key results regarding the positive or negative values of ecosystem state indicators are presented in Table 1. A visual interpretation of the ES before and after the disturbances is given in Fig. 3. ES was calculated for each ES category (PES, RES and CES) as an average of all the indicators. Distinct groups of indicators resulted from the application of the statistical EFA and were supplemented by the expert knowledge method. In both cases, a correlation analysis identified the most representative indicators (Table 1). A more detailed description of the ES indicators is elaborated below.

3.1. PES

Timber yield after disturbances declined to almost zero and the value of timber losses reached €18 million (Tutka 2009). Gross primary production (GPP) was initially strongly reduced by the wind disturbances (50% reduction), but the recovery in recent years has been very fast. Ten years after the disturbances, the GPP for some land-use types (FIR, NEX) reached or even exceeded the pre-disturbance state (Fleischer 2016). Increasing GPP was closely related to the increase in the vegetation cover indicated by

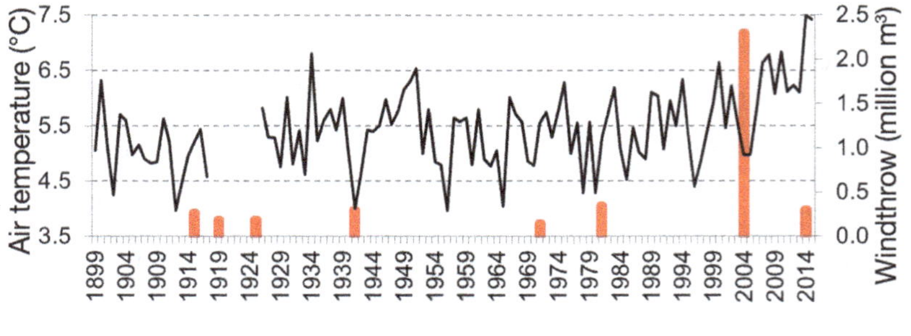

Fig. 2. Mean annual air temperature (Tatranská Lomnica, elevation 830 m a.s.l., period 1898–2015) and windstorm damage (million m^3 of fallen trees). Source: State Forests of TANAP and the Slovak Hydrometeorological Institute

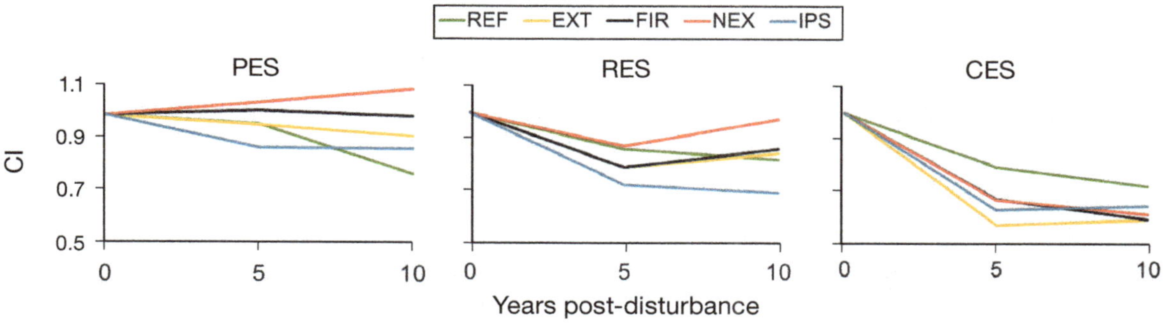

Fig. 3. Trajectories of provisioning (PES), regulating (RES) and cultural (CES) ecosystem services before the disturbance (0) and 5 and 10 yr after the disturbance at sites with different disturbance and land-use practices: REF: reference; EXT:extracted windthrow; FIR: burnt extracted windthrow; NEX: non-extracted windthrow; IPS: non-extracted site disturbed by bark beetle. Temporal values represent the average normalised change index (CI) of ecosystem service provision indicators (n = 12 for PES, n = 28 for RES and n = 4 for CES)

the leaf area index (LAI; Fleischer et al. 2015, Fleischer 2016). A different situation occurred at the treeline, which was disturbed by bark beetle occurrence. The current GPP is 39 % of the pre-disturbance value and the vegetation cover features sparse successional vegetation in formerly closed canopy stands.

The ground vegetation biomass increased after the windthrow disturbance at all sites. The aboveground biomass was almost identical on both salvage-logged and non-intervention windthrow sites (9300 kg ha^{-1}). The annual biomass stocks culminated 5–7 yr after the disturbance, with the highest value observed on the burnt windthrow (FIR) site (Kyselová & Homolová 2012). In the following years, the biomass declined, mostly in the salvage-logged sites (Fleischer 2016). The decline resulted in a slight reduction of PES. Both grass and herb species assemblages did not differ considerably but the salvage logging in the EXT site favoured colonisation rate and an increase in the abundance of early successional tree species. The differences in the natural regeneration of Norway spruce ranged from a few more up to 6-fold more trees per study site in favour of the non-intervention sites compared with the salvage-logged site during early post-disturbance years. In more recent years, the difference has begun to disappear due to factors such as competition from the surrounding vegetation (e.g. Jonášová et al. 2010, Šebeň et al. 2011b, Máliš et al. 2015, Michalová 2015). Jonášová et al. (2010) mentioned mechanical destruction during logging as responsible for a poorer regeneration of Norway spruce in EXT. At the same time, the salvage-logged sites EXT and FIR have accommodated a considerably larger number of pioneer broadleaved species, i.e. birch, willow and rowan. Due to reforestation in sites with intervention (EXT, FIR), the number of so-called climax species has been much higher than in both reference (REF) and non-intervention (NEX) sites. Nonetheless, Norway spruce remains the dominant tree species for both land-use types (Fleischer 2016).

With respect to dead wood, Šebeň et al. (2011a) found 346 ± 75 m^3 ha^{-1} of dead wood in the NEX site. In the managed EXT site, the amount was 86 ± 12 m^3 ha^{-1}. Necromass in the undisturbed forest with traditional management was 49 ± 4 m^3 ha^{-1}. The amount of dead wood increased in the IPS site where no pest control measures were applied. In this context, the European spruce bark beetle *Ips typographus* is the most important insect that affects the forests in the Tatra Mountains. During the first 2 yr after the windthrow disturbance, its population developed in fallen trees with no outward impact. The largest area was disturbed during the following 2 yr and hit even the treeline (Økland et al. 2016). It should be noted that according to Nikolov et al. (2014), unmanaged sites in nature reserves (roughly 400 ha) were a source of an enormous increase in the bark beetle population. Overall, more than 70 km^2 of mature Norway spruce stands were destroyed by the bark beetle (www.lesytanap.sk, Nikolov et al. 2014). Surprisingly, natural bark beetle enemies (e.g. *Thanasimus formicarius*, *Scoloposcelis pulchela*, *Medetera signaticornis*) were unable to exert any control function (Gubka et al. 2010).

PES were substantially reduced by forest fragmentation. The length of forest edges immediately after the windfall of 2004 remained almost the same as in the pre-disturbance state (Kopecká & Nováček 2009). However, considerable forest fragmentation emerged later due to bark beetle outbreaks (Hlásny et al. 2010), reducing PES mostly at the IPS site.

Water availability, water balance and drinking water quality were included in RES, but might be partly considered as PES. Since these indicators remained almost unchanged (Holko et al. 2009, 2012, Holko & Škoda 2016), water-related PES remained stable. Streams flowing through the windthrow areas showed a higher content of nitrogen compounds for 2–3 yr after the disturbance (Fleischer & Homolová 2011).

Average CI values in PES after 5 and 10 yr slightly decreased from 0.97 to 0.93 (Fig. 3). A larger decline was observed in salvage-logged (EXT, FIR) than in non-intervention sites (NEX, IPS). The largest difference (CI = 0.75) was documented at REF 10 yr after the disturbance, which nevertheless was caused by another windthrow event that occurred in May 2014.

3.2. RES

Generally, RES declined after the disturbances more profoundly than PES. The average CI values were 0.81 and 0.84, 5 and 10 yr after the disturbance, respectively (Fig. 3). In the first 5-yr period, the average CI declined in all the study sites. In the second period, the decline continued at the IPS and REF sites. The largest declines were observed in carbon sequestration, natural structure, interception, surface runoff and fire risk (Table 1). Less evident declines were observed in parameters describing soil water quality, such as total organic carbon and nitrogen content (Ziegler 2007, Bischoff et al. 2008, Frič & Škvarenina 2008, Fleischer & Homolová 2011). Soil physical properties (Fleischer & Koreň 2009, Hanajík & Šimonovičová 2015) remained almost unchanged. Stable soil prevented an increase in overland flow generation, soil moisture changes and erosion (Šimkovič et al. 2009, Novák & Šurda 2010). These processes are generally expected as a consequence of dramatic change in interception and transpiration after canopy destruction (Kňava et al. 2007, Fleischer 2014). Although the surface runoff and particle transport increased several-fold on salvage-logged sites (EXT, REF), the magnitude of erosion was still classified as negligible (<0.7 kg ha^{-1} yr^{-1}) for all the disturbance and land-use types (Sitko et al. 2011). Rojan (2015) found pronounced erosion on forest roads used for timber transport, but the area affected was only a few hectares and the recovery was very fast. Water balance, runoff and retention were primarily controlled by precipitation regime and soil-geology properties that were not affected by the disturbance type and subsequent land-use practices (Kňava et al. 2007, Holko et al. 2009, 2012, Novák & Kňava 2009, Novák & Šurda 2010, Bičárová & Holko 2013, Bartík et al. 2014).

Some RES indicators were significantly affected by air temperature. An increase in air temperature caused by the windthrow disturbance was recorded at all the sites. The average air temperature during the growing season increased by 2 °C in all disturbed sites compared with the intact forest in the first 5 yr (Fleischer et al. 2012). The increased air and soil temperature in salvage-logged sites shortly after the disturbance induced heat stress and a large mortality of young trees (Mišíková & Škvarenina 2009). Extracted and mostly burnt sites (EXT, FIR) suffered from increased sensible heat flux in early stages after the disturbances (Matejka & Fleischer 2011). Soil moisture, a very variable indicator (Novák & Šurda 2010), was more sensitive to the type of land-use practice (bare soil in extracted sites) in the early stage of succession. A few years later, the new vegetation cover eliminated earlier extremes and differences (Fleischer & Koreň 2009). The difference in the air temperature during the growing season between REF and disturbed sites declined to 1.5 °C after 10 yr. Warmer soil stimulated more intensive soil respiration, which reduced RES. We found temporarily increased soil respiration in the salvage-logged sites, and a surprisingly large soil CO_2 efflux was measured close to the treeline. Canopy destruction caused by the bark beetle outbreak increased soil temperature mostly on steep, south-facing slopes. Such a change stimulated mineralisation of the thick soil organic layer and a large CO_2 efflux.

Excepting the effects of increased air temperature caused by the forest destruction, meteorological conditions also had a significant influence on ES. In 2015, the mean growing-season (May–August) air temperature above the treeline exceeded the long-term reference normal (1930–1960) by 2.0 °C. Such a temperature increase has been projected by regional climate change models for the time horizon 2050–2070 (Melo et al. 2013). Two, instead of the usual one, bark beetle *Ips typographus* generations developed in standing forest as a result of the high air temperature. Three generations developed in fallen trees with eliminated transpiration, which increased pressure on the remaining stands (Fleischer et al. 2016). Dead, frequently falling trees pose a high risk for tourists (Kajba et al. 2013) and preventive measures require large investments. The tree dieback has also increased the risk of falling rocks, mostly at the treeline damaged by bark beetle, which has strongly reduced RES. Warmer weather is correlated with fire risk in the Tatra region (Brunette et al. 2015) and thus reduced RES. Fire risk also significantly increased due to the large amounts of fuelwood that was mostly available in stands without salvage logging (NEX, IPS). Forestry evidence (www.lesytanap.sk) showed that more than 40 fires have occurred in the windthrow sites since 2005. Open, disturbed sites have boosted the spread of different plant and animal diseases (Turčeková et al. 2014), which was also interpreted as a decline of RES.

In contrast, the increased temperature positively stimulated the activity of soil microorganisms (Don et al. 2012), even though the differences in microorganism communities related to disturbance and land-use categories have remained rather limited even on ecologically contrasting sites (Gömöryová et al. 2008, 2011). The increased abundance of microorganisms, fungi (Hanajík & Šimonovičová 2015) and vegetation growth indicate the recovery of the ecosystem and related ES. High temperature and sufficient precipitation in recent years supported carbon sequestration. The disturbances immediately turned all the affected sites into large carbon sources. Ten years after the disturbance, the carbon balance was still slightly negative. The only case with a narrow positive balance was identified at the non-intervention site (NEX). This is a result of the highest LAI and a high proportion of naturally regenerated Norway spruce (Fleischer 2016).

Decline of the dangerous tropospheric ozone (O_3) concentration after the disturbances was observed. The previously high O_3 concentration above the treeline was understood to be one of the most important pressures on forest health (Bičárová & Fleischer 2006). The destruction of the forest reduced the natural precursors (volatile organic compounds) for ozone formation (Bičárová et al. 2015) produced by the conifers. Similarly, the reduced canopy led to a reduction of precipitation water dripping and stemflow, which would otherwise have channelled pollution deposition into the soil (Fleischer 2014).

3.3. CES

CES declined (Fig. 3) 5 yr after the disturbances to CI = 0.67 and this tendency continued (CI = 0.63 after 10 yr). In terms of CES, Božíková (2009) reported a decline in the clientele of the local climatic spas after the windstorm of 2004. Currently, allergens from grasses, birch and poplar are abundant in all sites. According to Vyskot et al. (2007), recreational and health-related forest CES have been reduced by 35–37 % due to forest destruction. Čekovská (2013) interviewed visitors (n = 138) and concluded that the images of vast wind forest destruction lowered the aesthetic appeal and thus also the recreational value of the area as perceived by the public. An increasing number of visitors passing the treeline induced a higher risk of injury through falling trees in the bark beetle-affected areas (Kajba et al. 2013). Most of the above indicators were below the pre-disturbance state and thus indicate the decline of CES provision. Education is one of the few indicators that has increased as the disturbance events have become an important source of information for research, education, practical forestry and nature conservation (Fleischer 2014, Pichler 2015).

PCA analysis revealed a temporal post-disturbance increase of disparity across the 5 study sites (Fig. 4). Before the disturbance in 2004, the environmental conditions at all sites were very similar, as indicated by small distances among the indicator values. The differences increased during the first 5 yr after the disturbances. In the following period (5–10 yr), the differences notably declined, especially among differently managed windthrow sites (EXT, NEX) and different disturbance factors (wind at EXT and fire at FIR). The development at the IPS site (damaged by bark beetle and unmanaged) was notably different; it documented a very sensitive response of the ecosystem at the tree line.

4. DISCUSSION

The DPSIR framework is usually presented as a linear chain or a circle with limited capability for in-depth analyses and is sometimes considered inappropriate as an analytical tool (Maxim et al. 2009). However, there are also positive examples that the DPSIR framework can reflect the complexity of many interacting factors, some of which may represent

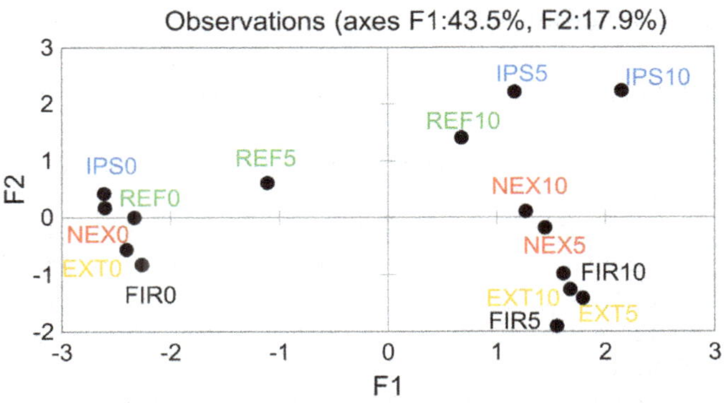

Fig. 4. Principal component analysis based on 44 ecosystem state indicators depicting temporal changes and differences among the study sites with different disturbance and land-use practices. See Fig. 3 for abbreviations and details

highly nonlinear dynamics (EEA 2005, Niemeijer & de Groot 2008, Wolfslehner & Vacik 2008).

The relevance of indicators used in the DPSIR frameworks for the whole system functioning is often expressed by its weight (Wolfslehner & Vacik 2008). We intentionally gave the same weight to each indicator to avoid a subjective judgement. The approach chosen has some limitations, e.g. occurrence of redundant information, which can bias the interpretation of ES calculated as an average from all indicators. Statistical selection of relevant key indicators often yielded sets with difficult or even unrealistic interpretations. When selecting the key ES indicators, we also considered local expert knowledge available at the Tatra National Park Research Station. Contrary to the key indicators selected by EFA, an expert estimation provided less correlated but more realistic relations. Such an approach based on the local expert knowledge was shown to reduce uncertainties in the estimation of future ES in the Alpine environment (Grêt-Regamey et al. 2015). Among the most informative indicators are those used for forest ES mapping in Europe e.g. risk protection, biomass production, carbon sequestration, habitat provision, recreation, etc. (Barredo et al. 2015), or for sustainable forest management e.g. growth and harvest, browsing, natural regeneration, tree species composition, woody debris, biodiversity loss, etc. (Wolfslehner & Vacik 2008). These indicators are more ES-specific and thus more suitable for further assessment, reporting and comparison of the ES status in the Tatra Mountains.

The concept offers an opportunity to affiliate desirable or consensus-based weight to any of the indicators used and to interpret the provisioning of ES not only from the viewpoint of economic gains or purely environmental protection, but from a much broader perspective. This includes sustained ability to yield a wide array of ES, which is important for the local population and Tatra National Park itself. We believe that this perspective is supported by the need to move away from area- and habitat-based assessment methods for both biodiversity and ecosystem services towards functional assessment at the landscape level, as highlighted by Tallis et al. (2015).

The set of indicators used in this study is based on published and available data dedicated to the ecosystem state before and after large-scale disturbances in the Tatra Mountains. As expected, not all the published data were temporally and spatially appropriate. A lot of data covered only a certain period (usually a short period after the disturbance), or focused only on limited disturbance type (e.g. fire), and therefore were excluded from the assessment.

The term 'impact' in the DPSIR framework usually implies negative environmental consequences of human activities. But there is substantial evidence that proactive management might successfully assist to sustain and maximise ES. Rounsevell et al. (2010) and Kelble et al. (2013) included ES directly into the DPSIR framework in lieu of impacts, thereby casting fresh light on how ecosystems function and produce ES, thus showing the capacity to provide more comprehensive information to decision-makers.

The response of the forest ecosystems in the Tatra Mountains to extensive disturbances exacerbated by climate and land-use changes has indicated a decline in ES provisioning. The biggest differences between the pre- and post-disturbance ecosystem capacity were recorded in the CES and RES in all disturbance and land-use categories.

It should be noted that ES also declined in the reference site. RES and CES declined in the first 5-yr period, whereas PES notably declined later. The decline of ES in the first 5 yr period resulted from individual tree dieback caused mostly by bark beetle. A steep decline in the second period was caused by the 2014 windthrow event. The ecosystem state and ES decline at the reference site are seen as consequences of synergies among the key driving forces (warmer climate, more frequent windstorms, impact of forest management and land-use changes) that affect the entire region of the Tatra Mountains.

The comparison of ES across the study sites showed that, as expected, the largest decline in the PES category was found in yield parameters. Ten years after disturbance, the CI values in all disturbed sites ranged from 0 to 0.11, indicating a pronounced decline in ES. The same negative effect is associated with canopy fragmentation. The differences in PES among the sites were observed mainly in the amount of deadwood, the intensity of natural regeneration and the diversity of tree species. Deadwood in non-intervention sites (NEX, IPS) exceeded that in reference and intervention sites (REF, EXT) 3-fold. Similarly, the CI for natural regeneration was highest at the NEX site. In contrast to the EXT and FIR sites, the NEX site showed the lowest tree species diversity. Surprisingly, little differences among the sites were found in photosynthesis rate (GPP) even in the first 5 yr after the windthrow disturbance (with the exception of the EXT site).

The decline in RES was the major concern when post-disturbance management strategies were discussed (Koreň 2015). Ten years after the windthrow disturbance, the wind-disturbed sites assimilated almost the same amount of carbon as the former mature forest. The NEX site assimilated even more carbon. The lowest assimilation rate was found at the IPS site. Fast recovery of the windthrow and fire sites was documented by a growing vegetation cover, expressed as LAI. The highest LAI was again found in the NEX site. IPS was the only site where LAI declined.

RES at the EXT and FIR sites 10 yr after the windthrow disturbance almost reached the pre-disturbance level. REF and IPS sites featured continuation of the decline caused by the windthrow and bark beetle mainly due to the subsequent windthrow event. The amount of respired CO_2 increased at all the study sites except at REF. It caused an obvious, large reduction in carbon sequestration. Except the NEX site, all the sites acted as carbon sources 10 yr after disturbance. The largest decline (CI = 0.04) was found at the IPS site. The CI at the salvage-logged sites (EXT, FIR) was influenced by change in natural composition and structure due to reforestation of the windthrow.

The energy balance (sensible heat) was found to be one of the most influential indicators for the overall CI of RES provision. The microclimate generated by fallen, unremoved wood at the NEX site reduced the heat impact 4 times more effectively than that at the reference site. Slight water erosion occurred during the first assessment period. The erosion was later eliminated by fast vegetation growth with the exception of the IPS site. The IPS site is located on steeper slopes and soil is more vulnerable to fine particle transport.

CES featured the lowest differences across the study sites. A relatively short distance between the study sites made it difficult to separate the number of visitors or their perception of attractiveness to distinct sites. The occurrence of allergens reduced CES most evidently (CI = 0.25) at the EXT and FIR sites.

With regard to biodiversity, available data on, for example, soil fauna or ungulates population, indicated only small differences among either disturbance or management types. Generally, biodiversity increased in disturbed ecosystems, which supports the idea of a so-called 'disturbance paradox' (Thom & Seidl 2016).

Significant and previously unknown changes occurred at the treeline. Forest ecosystems on the lee side of the Tatra Mountains are affected by significant disturbances caused by downslope winds in an approximately 60 yr cycle (Holeksa et al. 2016). Short periods between disturbances create conditions for the formation of fairly uniform forests, which are vulnerable to abiotic and biotic agents (Zielonka et al. 2010, Fleischer & Homolová 2011). Because the stands at the treeline are usually unaffected by this type of windstorm, late-stage succession forests were able to develop (Fleischer & Koreň 2009). Most of the treeline forests are more natural and show much higher structural complexity than other mountain forests in the Tatra Mountains. Despite a severe climate and long-lasting air pollution impact (Bowman et al. 2008, Fleischer 2014), the treeline had remained stable for a long time. The situation has changed in the last decade due to enormous bark beetle pressure generated by increasing air temperature (Fleischer et al. 2016, Økland et al. 2016) and dramatic changes in land-use practices (introduction of non-intervention management on one-third of the Tatra National Park forest area). A further massive impact on the remaining forest at the treeline is expected in the following years. These findings support the predictions of Hlásny & Turčáni (2013) and Hlásny et al. (2014) on the limited sustainability of Norway spruce forests in Central Europe associated with a climate change impact on spruce forest in the entire Carpathians.

Warm and wet summers have prevailed since 2004 and have probably facilitated the fast and successful recovery processes. Intense growth of young trees (positively interpreted from PES and RES due to carbon storage) was the reason for poor rooting and destruction of hundreds of hectares of 10 yr old stands by the windstorm in 2014 (Pajtík et al. 2015). This event suggested that positive experience with forest regeneration and certain adaptation to disturbance does not necessarily imply continuation of the same patterns and dynamics under changing conditions (Vido 2015).

Climate change is expected to cause additional changes in the Tatra Mountains. Melo et al. (2013) classified the Tatra Mountains as the most climate change-threatened region in the Carpathians. According to bioclimatological ranges in the 'envelope' models (Kölling 2007), the projected climate would displace the current key species (Norway spruce and European larch) outside their natural growing conditions (Fleischer 2014).

Increasing tree species diversity by planting mixed forests is one of the forest management responses to eliminate the predicted consequences of climate change. In contrast, a higher vertical and spatial heterogeneity of natural tree regeneration, which results from biological processes, might eventually lead to a higher stability of stands without forestry intervention (Bače et al. 2015, Michalová 2015).

5. CONCLUSIONS

The CES, PES and RES analysed in the study were below the pre-disturbance levels 10 yr after the windthrow disturbance from 2004. It should be noted that the ecosystem also changed at the reference site as indicated by the decreased CI.

The analysis of 44 indicators of ES in the study area influenced by the disturbances, climate extremes and land-use changes showed:

• fast recovery at mountain foothills and postponed recovery at high-elevation slopes

• natural regeneration provided a limited number of tree species, while reforestation in managed sites increased diversity

• that fast and intensive regeneration occurred, especially at the non-intervention sites, which served as carbon sinks 10 yr after the windthrow disturbance. However, the naturally dominant species *Picea abies* and *Larix decidua* are sensitive to climate change. In fact, this poses a risk for forest resilience under changing climate.

• no erosion or deviation from long-term hydrological conditions

• a high degree of variability in some indicators not only among the study sites, but also within them (soil moisture, soil respiration, carbon content, soil microorganisms, vegetation biomass, etc.) early after the disturbances. This spatial heterogeneity pattern supports forest resilience (Turner et al. 2013) and, together with fast and successful regeneration at disturbed sites, provides certain optimism with regard to the future uncertain conditions.

The analysis of indicators within the DPSIR framework proved useful for obtaining up-todate information on the ecosystem state needed by decision-makers. This is especially important in the TANAP area, because despite the enormous public response, no consensus on its future management strategy has been achieved to date (Fleischer et al. 2016).

Further research and monitoring should focus on the role of high slope, elevation, geology and soil conditions in maintaining ecosystem state and services. From the analysed indicators, we can conclude that sustainable provision of ES in the area affected by repeated natural disturbances and threatened by climate change might be constrained. The managed sites require more resource input but support more resilient ecosystems (lower biomass, higher tree diversity). In contrast, unmanaged forests require only very limited resources, but they prove to be more vulnerable to natural disturbances and to anticipated climate change.

Acknowledgements. This article is based on work conducted under COST Action ES1203 SENSFOR, supported by COST (European Cooperation in Science and Technology, www.cost.eu/COST_Actions/essem/ES1203) and prepared with the support from the Slovak Grant Agency, grant numbers APVV-0480-12, APVV-0744-12, APVV-15-0425, APVV-14-0086, APVV-15-0176, APVV-15-0270, VEGA 1/0589/15, VEGA 1/0783/15 and VEGA 2/0055/15, and grant NPU I LO1415 funded by the Ministry of Education Youth and Sports of Czech Republic. We appreciate the comments of Jozef Turok and 2 anonymous reviewers that helped to improve the manuscript.

LITERATURE CITED

Bače R, Svoboda M, Janda P, Morrissey RC and others (2015) Legacy of pre-disturbance spatial pattern determines early structural diversity following severe disturbance in montane spruce forests. PLOS ONE 10: e0139214

Barredo JI, Bastrup-Kirk A, Teller A, Onaindia M and others (2015) Mapping and assessment of forest ecosystem and their services. Joint Research Centre, Ispra

Bartík M, Sitko R, Ore ák M, Slovik J, Škvarenina J (2014) Snow accumulation and ablation in disturbed mountain spruce forest in West Tatra Mts. Biologia (Bratisl) 69: 1492–1501

Bičárová S, Fleischer P (2006) Windstorm effect on forest sources of biogenic volatile organic compound emissions in the High Tatras. Contrib Geophys Geod 36:269–282

Bičárová S, Holko L (2013) Changes of characteristics of daily precipitation and runoff in the High Tatra Mountains, Slovakia over the last fifty years. Contrib Geophys Geod 43:157–177

Bičárová S, Bilčík D, Pavlendová H, Janík R, Kellerová D (2015) Ground level ozone and windstorm. Stud Tan 11: 115–124 (in Slovak)

Bischoff WA, Mayer S, Schrumpf M, Freibauer A (2008) Nutrient leaching from soils affected by windfall in the High Tatras. In: Windstorm research 2008, State Forest of TANAP, Tatranská Lomnica, p 8–12

Bowman WD, Cleveland CC, Hakada L, Hreško J, Baron JS (2008) Negative impact of nitrogen deposition on soil bufferening capacity. Nat Geosci 1:767–770

Božíková J (2009) Climatic spas in Slovakia. In: Pribullová A, Bičárová S (eds) Sustainable development and bioclimate. Slovak Academy of Sciences, Stará Lesná, p 185–186

Brunette M, Holecy J, Sedliak M, Tucek J, Hanewinkel M (2015) An actuarial model of forest insurance against multiple natural hazards in fir (*Abies Alba* Mill.) stands in Slovakia. For Policy Econ 55:46–57

Bryce R, Irvine KN, Church A, Fish R, Ranger S, Kenter JO (2016) Subjective well-being indicators for large-scale assessment of cultural ecosystem services. Ecosyst Serv 21:258–269

Čekovská L (2013) Turism related problems in Tatranská Lomnica. PhD thesis, Comenius University, Bratislava

Dale VH (1997) The relationship between land-use change and climate change. Ecol Appl 7:753–769

Don A, Bärwolff M, Kalbitz K, Andruschkewitsch R, Jungkunst HF (2012) No rapid soil carbon loss after a windthrow event in the High Tatra. For Ecol Manage 276: 239–246

Don A, Walter K, Bauer A (2015) Impact of the Tatra windthrow on soil organic carbon stocks. Stud Tan 11:167–172

EEA (European Environment Agency) (2005) EEA core set of indicators: Guide. Tech rep No. 1/2005, EEA, Copenhagen

Fleischer P (2008) Windfall research and monitoring in the High Tatra Mts: objectives, principles, methods and current status. Contrib Geod Geophys 38:223–248

Fleischer P (2014) Impact of natural and antropogenic factors on state and changes in forest ecosystems in Tatra National Park. Hab thesis, Technical University in Zvolen

Fleischer P Jr (2016) Ecosystem exchange of CO_2 in Norway spruce ecosystems after natural disturbances. PhD thesis, Technical University in Zvolen (in Slovak)

Fleischer P, Homolová Z (2011) Long-term ecological research in larch-spruce forest community after natural disturbances in the Tatra Mts. For J 57:237–250

Fleischer P, Koreň M (2009) Selected forest soil properties after the 2004 windfall in the Tatra Mts. In: Pribullová A, Bičarová S (eds) Sustainable development and bioclimate. Slovak Academy of Sciences, Stará Lesná, p 77–78

Fleischer P, Matejka F, St elcová K, Jakubjak O (2012) Meteorological and microclimate conditions on windthrow in the High Tatra Mts in 2012. In: Fleischer P (ed) 5th Windstorm Proc. State Forest of TANAP, Tatranská Lomnica, p 22–23

Fleischer P, Fleischer P Jr, Homolová Z (2015) Carbon balance of early succession vegetation on managed sites ten years after large-scale windthrow. Stud Tan 11:113–122

Fleischer P Jr, Fleischer P, Ferenčík J, Hlaváč P, Kozánek M (2016) Elevated bark temperature in unremoved stumps after disturbances facilitates multivoltinism in *Ips typographus* population in mountainous forest. Lesn Cas For J 62:15–22

Fleischer P, Hreško J, Topercer J (2017) What is the progress with zoning of TANAP? Environment 50:251–253 (in Slovak)

Frič M, Škvarenina J (2008) Microclimate influence on hydrochemical conditions of lysimetric waters in TANAP calamity area. In: Rožnovský J, Litschmann T (eds) Bioclimatological assessment of landscape processes. Czech bioclimatological society, Mikulov, p 1–12

Gardiner B, Blennow K, Carnus J, Fleischer P and others (2010) Destructive storms in European forests: past and forthcoming impacts. Final report to European Commission - DG Environ, European Forest Institute, Bordeaux, p 138

Gömöryová E, Střelcová K, Škvarenina J, Bebej J (2008) The impact of windthrow and fire disturbances on selected soil properties in the Tatra National Park. Soil Water Res 3:74–80

Gömöryová E, Střelcová K, Fleischer P, Gömöry D (2011) Soil microbial characteristics at the monitoring plots on windthrow areas of the Tatra National Park (Slovakia): their assessment as environmental indicators. Environ Monit Assess 174:31–45

Gömöryová E, Fleischer P, Pichler V, Homolák M, Gere R, Gömöry D (2017) Soil microorganisms at the windthrow plots: the effect of postdisturbance management and the time since disturbance. iForest 10:515–521

Grêt-Regamey A, Bebi P, Brunner SH, Weibel B (2015) On the importance of integrating expert knowledge into mapping ecosystem services: Swiss Alps forests. In Barredo et al. (eds) Mapping and assessment of forest ecosystems and their services. JRS science for policy report, European Commission, Ispra, p 18–23

Gubka A, Pavlík Š, Vakula J, Galko J (2010) Natural enemies of bark beetle after windfall in the Tatra Mts. In: Konôpka B (ed) Research on spruce stands destabilized by disturbances. National Forest Center, Zvolen, p 128–140

Hanajík P, Šimonovičová A (2015) Monitoring of soil characteristics on the calamity sites in TANAP (2005–2009). Stud Tan 11:145–155 (in Slovak)

Havašová M, Ferenčík J, Jakuš R (2017) Interactions between windthrow, bark beetles and forest management in the Tatra national parks. For Ecol Manage 391: 349–361

Hlásny T, Turčáni M (2013) Persisting bark beetle outbreak indicates the unsustainability of secondary Norway spruce forest. Case study from Central Europe. Ann For Sci 70:481–491

Hlásny T, Nikolov C, Vakula J, Zúbrik M, Ferenčík J, Konôpka B (2010) Geostatistical analysis of *Ips typographus* population development after the 2004 windthrow in the High Tatra Mts. In: Konôpka B (ed) Research on spruce ecosystems destabilized by damaging factors. National Forest Center, Zvolen, p 86–95 (in Slovak)

Hlásny T, Mátyás C, Seidl R, Kulla L and others (2014) Climate change increases the drought risk in Central European forests: What are the options for adaptation? For J 60:5–18

Holeksa J, Zielonka T, Zywiec M, Fleischer P (2016) Identifying the disturbance history over a large area of larch-spruce mountain forest in Central Europe. For Ecol Manage 361:318–327

Holko L, Škoda P (2016) Assessment of runoff changes in selected catchments of the High Tatra Mountains ten years after the windthrow. Acta Hydrologica Slovaca 17: 43–50 (in Slovak)

Holko L, Hlavatá H, Kostka Z, Novák J (2009) Hydrological regimes of small catchments in the High Tatra Mountains before and after extraordinary wind-induced deforestation. Folia Geographica XI:33–44

Holko L, Fleischer P, Novák V, Kostka Z, Bičárová S, Novák J (2012) Hydrological effects of a large scale windfall degradation in the High Tatra Mts, Slovakia. In: Křeček J, Haigh M, Hofer T, Kubin E (eds) Management of mountain watersheds. Springer, Dordrecht, p 164–179

Homolová Z, Kyselová Z, Šoltés R (2015) Vegetation changes on windthrow sites in larch-spruce forest community. Stud Tan 11:191–200 (in Slovak)

Jonášová M, Vávrová E, Cudlín P (2010) Western Carpathian mountain spruce forest after a windthrow: natural regeneration in cleared and uncleared areas. For Ecol Manage 259:1127–1134

Kajba M, Nikolov C, Ferenčík J, Gubka A (2013) Bark beetle calamity and the risk for tourists in the Tatra Mts. In: Kunca A (ed) Problems in forest protection, National Forest Center, Zvolen, p 121–124

Kelble CR, Loomis DK, Lovelace S, Nuttle WK and others (2013) The EBM-DPSIR conceptual model: integrating ecosystem services into the DPSIR framework. PLOS ONE 8:e70766

Kňava K, Novák V, Orfánus T (2007) Canopy structure changes and potential evapotranspiration: possible influence of windthrow in High Tatra Mountains. In: Střelcová K, Matyas C, Kleidon A, Lapin M and others (eds) Bioclimatology and natural hazards. Slovak Bioclimatological Society, Zvolen, p 1–7

Kölling C (2007) Klimahüllen für 27 Waldbaumarten. AFZ-DerWald 23:1342–1345

Kopecká M, Nováček J (2009) Forest fragmentation in the Tatra Region in the period 2000–2006. Landf Anal 10: 58–63

Koreň M (2015) TANAP's forest development in historical context. Stud Tan 11:59–74 (in Slovak)

Kyselová Z, Homolová Z (2012) Vegetation succession and cellulose decomposition in disturbed sites. In: Fleischer P (ed) 5th Windstorm Proc, State Forest of TANAP, Tatranská Lomnica, p 41–42 (in Slovak)

Levin PS, Fogarty MJ, Matlock GC, Ernst M (2008) Integrated ecosystem assessment. NOAA Technical Memo, NMFS-MWFSC-92

Máliš F, Fábry R, Vodálová A (2015) Vegetation development in spruce forest with different management after a windthrow. Stud Tan 11:199–209 (in Slovak)

Matejka F, Fleischer P (2011) Energetic balance of the windthrow area in Tatra Mts. Stud Tan 10:77–86 (in Slovak)

Maxim L, Spangenberg JH, O'Connor M (2009) An analysis of risk for biodiversity assessment under the DPSIR framework. Ecol Econ 69:12–23

Melo M, Lapin M, Kapolková H, Pecho J (2013) Climate trends in the Slovak part of the Carpathians. In: Kozak J, Ostapowicz K, Bytnerowicz A, Wyżga B (eds) The Carpathians: integrating nature and society towards sustainability. Springer, Berlin, p 131–150

Michalová Z (2015) Post-disturbance study on vegetation composition and natural regeneration in Tatra National Park. Stud Tan 11:211–220 (in Slovak)

Mišíková N, Škvarenina J (2009) Microclimatological condition on the burnt windfall area in the Tatra National Park. Meteorol J 12:31–36

Niemeijer D, de Groot RS (2008) Framing environmental indicators: moving from causal chains to causal networks. Environ Dev Sustain 10:89–106

Nikolov C, Konôpka B, Kajba M, Galko J, Kunca A, Janský L (2014) Post-disaster forest management and bark beetle outbreak in Tatra National Park, Slovakia. Mt Res Dev 34:326–335

Novák V, Kňava K (2009) Infiltration of water into stony soil: To what extent is inlfiltration affected by stoniness? In: Pribullová A. Bičárová S (eds) Sustainable development and bioclimate. Slovak Academy of Sciences, Stará Lesná, p 111–112

Novák V, Šurda P (2010) The water retention of a granite rock fragments in High Tatras stony soils. J Hydrol Hydromech 58:181–187

Oesterwind D, Rau A, Zaiko A (2016) Drivers and pressures —untangling the terms commonly used in marine science and policy. J Environ Manage 181:8–15

Økland B, Nikolov C, Krokene P, Vakula J (2016) Transition from windfall-to patch-driven outbreak dynamics of the spruce bark beetle *Ips typographus*. For Ecol Manage 363:63–73

Pajtík J, Konôpka B, Šebeň V, Michelčík P, Fleischer P (2015) Biomass allocation of common larch in the first age class in the High Tatra Mts. Stud Tan 11:229–241 (in Slovak)

Pichler V (2015) Cooperation between Technical University in Zvolen and State Forest of TANAP in education for sustainable and adaptive forest management. Stud Tan 11:23–28 (in Slovak)

R Development Core Team (2008) R: a language and environment for statistical computing. R Foundation for Statistical Computing, Vienna

Rojan E (2015) Changes in unpaved forest roads of the windthrow area in the Slovak High Tatra Mts in years 2004–2014. Stud Tan 11:85–94

Rounsevell MDA, Dawson TD, Harrison PA (2010) A conceptual framework to assess the effects of environmental change on ecosystem services. Biodivers Conserv 19: 2823–2842

Schelhaas MJ, Nabuurs GJ, Schuck A (2003) Natural disturbances in the European forests in the 19th and 20th centuries. Glob Change Biol 9:1620–1633

Schroeder LM, Lindelöw Å (2002) Attacks on living spruce trees by the bark beetle *Ips typographus* (*Col. Scolytidae*) following a storm-felling: a comparison between stands with and without removal of wind-felled trees. Agric For Entomol 4:47–56

Šebeň V, Bošela M, Kula L (2011a) Monitoring network for studying revitalisation of the windthrow in the Tatra Mts. Stud Tan 10:13–24 (in Slovak)

Šebeň V, Homolová Z, Fleischer P (2011b) Forest regeneration on the windfall research sites. Stud Tan 10:187–199y (in Slovak)

Seidl R, Schelhaas MJ, Lindner M, Lexer MJ (2009) Modelling bark beetle disturbances in a large scale forest scenario model to assess climate change impacts and evaluate adaptive management strategies. Reg Environ Change 9:101–119

Seidl R, Schelhaas M, Rammer W, Verkerk P (2014) Increasing forest disturbances in Europe and their impact on carbon storage. Nat Clim Change 4:806–810

Šimkovič I, Dlapa P, Šimonovičová A, Ziegler W (2009) Water repellency of mountain forest soils in relation to impact of the katabatic windstorm and subsequent management practices. Pol J Environ Stud 18:443–454

Sing L, Ray D, Watts K (2015) Ecosystem services and forest management. Research Note, Forestry Commission, Edinburgh

Sitko R, Midriak R, Fleischer P, Goceliak T, Pavlarčík S (2011) Water erosion after disturbances in the High Tatra Mts. Stud Tan 10:115–130 (in Slovak)

Smeets E, Weterings R (1999) Environmental indicators: typology and overview. Technical report No. 25, European Environment Agency, Copenhagen

Tallis H, Kennedy CM, Ruckelshaus M, Goldstein J, Kiesecker JM (2015) Mitigation for one and and all. An integrated framework for mitigation of development impacts on biodiversity and ecosystem services. Environ Impact Assess Rev 55:21–34

TEEB (The economics of ecosystems and biodiversity) (2010) Mainstreaming the economics of nature: a synthesis report of the approach, conclusions and recommenadations of TEEB, UNEP, Malta

Thom D, Seidl R (2016) Natural disturbance impacts on ecosystem services and biodiversity in temperate and boreal forests. Biol Rev Camb Philos Soc 91:760–781

Turčeková Ľ, Hurniková Z, Spišák F, Miterpáková M, Chovancová B (2014) *Toxoplasmosa gondii* in protected wildlife in the Tatra National Park, Slovakia. Ann Agri and Envi Medic 21:235–238

Turner MG, Donato DC, Romme WH (2013) Consequences of spatial heterogeneity for ecosystem services in changing forest landscapes, priorities for future research. Landsc Ecol 28:1081–1097

Tutka J (2009) Enviromental and economic losses caussed by wind and bark beetle calamities. In: Kovalčík M (ed) Quantification of wind and bark beetle damage in TANAP. National Forest Center, Zvolen, p 145–148

Vido J (2015) Drought periods in TANAP analyzed by SPI in 1961–2010. Stud Tan 11:95–101 (in Slovak)
Vyskot I, Schneider J, Kupec P, Fialová J, Melicharová A, Smítka D (2007) Wind calamity damages to sanitary-hygienic and social-recreational functions of forest in Tatra National Park. In: Rožnovský J, Litschmann T, Vyskot I (eds) Forest climate. Czech Bioclimatological Society, Křtiny, p 5
Wang Z, Zhou J, Loaiciga H, Guo H, Hong S (2015) A DPSIR model for ecological security assessment through indicator screening: a case study at Dianchi Lake in China. PLOS ONE 10:e0131732
Wermelinger B (2004) Ecology and management of the spruce bark beetle *Ips typographus* — a review of recent research. For Ecol Manage 202:67–82
Wolfslehner B, Vacik H (2008) Evaluating sustainable forest management strategies with the analytic network process in a pressure-state-response framework. J Environ Manage 88:1–10
Ziegler W (2007) Windthrow 2004 effects on carbon and nitrogen cycling. In: Fleischer P (ed) Windstorm research 2007. Stae Forest TANAP, Tatranska Lomnica
Zielonka T, Holeksa J, Fleischer P, Kapusta P (2010) A tree-ring reconstruction of wind disturbances in a forest of the Slovakian Tatra Mountains, Western Carpathians. J Veg Sci 21:31–42

10

Soil properties as indicators of treeline dynamics in relation to anthropogenic pressure and climate change

M. Cristina Moscatelli[1], Eleonora Bonifacio[2], Tommaso Chiti[1], Pavel Cudlín[3], Lucian Dinca[4], Erika Gömöryova[5], Stefano Grego[6], Nicola La Porta[7], Leszek Karlinski[8], Guido Pellis[1], Maria Rudawska[8], Andrea Squartini[9], Miglena Zhiyanski[10], Gabriele Broll[11,*]

[1]Department of Innovation in Biological, Agrofood and Forest systems (DIBAF), University of Tuscia, Viterbo, Italy
[2]Department of Agricultural, Forest and Food Sciences (DISAFA), University of Torino, largo P. Braccini 2, 10095 Grugliasco, Italy
[3]Global Change Research Centre, Academy of Sciences of the Czech Republic, Lipová 1789/9, České Budějovice 370 05, Czech Republic
[4]National Forest Research-Development Institute I.N.C.D.S. Brasov, 13 Closca St., 500030 Brasov, Romania
[5]Technical University in Zvolen, Faculty of Forestry, T. G. Masaryka 24, 960 53 Zvolen, Slovakia
[6]Department of Science and Technology for Agriculture, Forestry, Nature and Energy (DAFNE), Tuscia University, Viterbo, Italy
[7]Research and Innovation Centre, Fondazione Edmund Mach (FEM) and MOUNTFOR Project Centre, European Forest Institute, Via E. Mach 1, 38010 San Michele all Adige (Trento), Italy
[8]Institute of Dendrology, ul. Parkowa 5, 62-035 Kórnik, Poland
[9]Department of Agricultural Biotechnology, University of Padua, Agripolis, Viale dell 'Università 16, 35020 Legnaro, Padua, Italy
[10]Forest Research Institute – Bulgarian Academy of Sciences, 132 'Kl. Ohridski' Blvd., 1756 Sofia, Bulgaria
[11]Institute of Geography, University of Osnabrueck, Seminarstr. 19, 49074 Osnabrueck, Germany

ABSTRACT: Mountain forests, treeline ecotones included, provide numerous ecosystem services. However, different drivers heavily impact the treeline areas, in particular anthropogenic pressure and climate change. Any change affecting the aboveground portion of terrestrial ecosystems automatically influences their belowground part, i.e. soil and soil organisms. Therefore, the focus of the present paper is on the soil resource that provides multiple ecosystem services, such as carbon storage, water filtration, food and biomass provisioning, biodiversity, maintenance, etc. Soil physical, chemical, and biological properties can be very helpful as indicators of ecosystem services in mountain regions. A selection and integration of appropriate indicators of soil quality is thus needed for soil monitoring and assessment in treeline areas. In this paper, results of case studies from mountain regions in Bulgaria, the Czech Republic, Italy, Romania, and Slovakia are presented. From these studies, it emerges that soil organic matter (content and quality), pH, and microbial parameters show significant changes in response to anthropogenic pressures and/or climate change. These indicators of soil quality, either in the short- or in the long-term, can thus be used as reliable and sensitive tools for monitoring actions. However, it is advisable to integrate this basic set with additional indicators that can be further selected in relation to specific conditions, such as geographical area, lithological substrate, land use, and management practices.

KEY WORDS: Ecosystem services · Forest resilience · Mountains

1. INTRODUCTION

Treeline ecotones are natural boundaries where plant species, plant productivity, and fauna change under the influence of global climate change along a small altitudinal gradient and at short distance (Broll & Keplin 2005, Holtmeier 2009, Körner 2012). Mountain forests, and thus treeline ecotones, provide sev-

*Corresponding author: gabriele.broll@uni-osnabrueck.de

eral ecosystem services, such as hydrological regulation and protection against avalanches and erosion. The upper treeline ecotone plays an important role as a potential indicator of environmental change. However, different drivers heavily impact treeline ecosystems. Among these are climate change and anthropogenic pressures such as overgrazing, abandonment, or reforestation of formerly grazed areas (e. g. Broll 2000, Holtmeier & Broll 2005, 2007, Palombo et al. 2014, Wieser et al. 2014). Climate change resulting in lasting drought periods, strong winds and snow storms, or modification of the snow regime, may potentially lead to physiological weakening, damage, or even destruction of forest stands, especially at the treeline (Holtmeier & Broll 2007, 2010). The soil in these areas may be affected in its structure, capacity to store carbon, cycling of mineral nutrients, and ultimately in the diversity of soil organisms. All these natural and anthropogenic disturbances (e.g. land use change, pollution, management practices), alone or combined in synergistic processes, induce changes that may impact the resilience and sustainability of these fragile ecosystems. The characteristics and properties of soils in treeline ecotones depend on specific climatic conditions and type of vegetation, as well as on the different land uses, silviculture, and management practices (Raev et al. 2011).

There is a strong interaction between plant cover and soil condition. Any change affecting the aboveground portion of terrestrial ecosystems automatically impacts the soil, and its complex multiphase system and biological network (Ponge 2013). Major changes affecting trees and vegetation, in general, may be expected to impact not only the soil organic matter content but also the entire soil profile (Bonifacio et al. 2008a). The opposite also occurs: e.g. root growth and seedling density are strongly influenced by the physical and chemical properties of the soil in terms of soil organic matter, nitrogen content, pH, temperature, porosity, structure, and moisture, which can change in response to natural and/or anthropogenic disturbances (land use changes, climate change) (e.g. Anschlag et al. 2008). Soil plays an important role in these ecosystems, and deserves special attention when studying the vulnerability of treeline ecotones. Therefore, a strong and increasing demand for effective monitoring of soils at local, regional and national scales is required. The soil security concept (McBratney et al. 2012) introduces a soil-centric framework that can be used as the basis of policies aimed to achieve ecological and human sustainable development.

2. SOIL INDICATORS

Soil quality indicators refer to measurable soil attributes that influence the capacity of soil to sustain crop production and perform several other environmental functions (Arshad & Martin 2002). The term soil 'indicators' was introduced together with the concepts of soil quality and soil health by Doran & Parkin (1994). However, there is still an ongoing scientific debate and lack of a shared consensus on the concepts of soil quality and soil health (Gil-Sotres et al. 2005) and on which properties of soils may be most appropriately monitored. As for the first issue, the term 'quality' seems to be more related to an anthropocentric view of soil, while the term 'health' encompasses soil functions and is focused on its ecological attributes (Gil-Sotres et al. 2005, Janvier et al. 2007). The difficulty in deciding which of the many properties of soil to choose is partly due to the wide range of goods and services that soils provide in diverse contexts, but also to their chemical, physical, and biological complexity linked to the strict interactions among the solid, liquid, and gaseous phases. There is currently a great variety of indicators that can potentially be used to monitor soil quality. It is, therefore, important and recommended to focus on those indicators that can act as early warning of any degradation process affecting the soil resource.

Among established soil chemical parameters like soil pH, soil organic matter (SOM) is the most important integrative indicator of soil quality. A strong interrelationship between SOM and other indicators such as soil aggregate stability, infiltration, bulk density, nutrient availability, microbial biomass, and activity has been widely demonstrated (e.g. Carletti et al. 2009); therefore several other soil characteristics can be inferred through SOM assessment. In particular, in relation to climate change, it is fundamental to evaluate soil carbon (C) sink potential. Therefore, the level of recalcitrance — the physical and chemical protection of SOM — is the focus of recent studies devoted to C cycling under global climate change (Schmidt et al. 2011, Cotrufo et al. 2015). In recent years, the attention of research into SOM has focused more on its quality than on its quantity, showing that this aspect is more relevant for ecological issues concerning C storage and C cycling in general (Rovira & Vallejo 2002, von Lutzow et al. 2007). Moreover, SOM content and quality are strictly related to soil stability and erosion processes, which are of major importance in mountainous areas (Egli & Poulenard 2017).

Soil biological properties can be successfully used as indicators of soil quality in addition to physical–chemical features (Ritz et al. 2009). Microorganisms are definitively the actors of almost all processes occurring in soils. Soil microbial communities enable the healthy functioning of terrestrial ecosystems through diversity in terms of community structure and/or activity (Nielsen & Winding 2002). Soil microbial processes include decomposition of organic residues, transformation of SOM, mineralization and immobilization of nutrients, and formation and stabilization of soil aggregates (Nannipieri et al. 2003). In forest ecosystems, many natural (e.g. climate) and anthropogenic disturbances (land use change, pollution, management practices) may modify soil microbial communities in terms of structure and functions (Moscatelli et al. 2007). Climate and land use changes may affect the microorganism–soil–plant root system, modifying many physical, chemical, and biological soil properties which can represent critical features of treeline ecotones. Changes in microbial populations, community structure, activity, and diversity have proved to be the most sensitive and reliable bioindicators in the monitoring of soil health and quality (Paz-Ferreiro & Fu 2016). Any change in plant cover modifies litter quality and quantity, alters root exudation, and changes the microclimate, and may lastingly affect soil biota and influence microbial community diversity. Microbial functional diversity was recently used to describe soil development under different moisture regimes and at different altitudes in mountain ecosystems. As soil evolution proceeded through an increasing niche separation, an interesting link between microbial functions and soil development was found (Marinari et al. 2013).

A healthy soil is defined as a stable system with high levels of biological diversity and activity, internal nutrient cycling, and resilience to disturbance. Soil biota (biomass, activity, and diversity) are primarily responsible for soil health and suppression of soil-borne pathogens (due to higher antagonism). Microbial fluctuations after a disturbance would dampen more quickly in a healthy than in a chronically damaged and biologically impoverished soil. The resistance and resilience of microbial communities in response to a disturbance can be considered quantitative indicators for soil health (van Bruggen & Semenov 2000). Due to this last, dynamic property of soil biota, using microbial properties as indicators of soil status in treeline ecotones can also take into account the resilience capacity of these ecosystems to face disturbances, as claimed in the SENSFOR concept note (Kyriazopoulos et al. 2014).

As far as is known, arctic and alpine vegetation of treeline regions are mycorrhizal (Haselwandter 2007) with several distinct mycorrhizal associations, such as arbuscular, ectomycorrhizal, ericoid, and dark septate fungi types. Mycorrhizal relationships are considered favorable, especially for plants in nutrient-stressed situations, and therefore may be particularly beneficial in treeline ecosystems characterized by harsh environmental condition like low temperatures, high rain or snowfall and low evaporation rates, nutrient-limited soils, and short-growing season (Haselwandter 2007, Schmidt et al. 2008). As a result, many alpine or tundra plants develop extensive root systems, often highly mycorrhizal, as the main route of nutrient supply to the host plants. Mycorrhizal fungi are also an important biotic factor in the survival of conifer seedlings under the stressful conditions of alpine or tundra habitats (Reithmeier & Kernaghan 2013). Alpine treeline ecosystems with widely distributed coniferous trees and shrubs are the main reservoir of ectomycorrhizal fungi. Their contribution as a symbiont decreases with increasing altitude (Gardes & Dahlberg 1996, Schmidt et al. 2008). In contrast to ectomycorrhizal fungi, ericoid symbionts dominate in heathlands and tend to increase root colonization at upper altitudes (Väre et al. 1997). Arbuscular mycorrhizas are ubiquitous in low alpine and arctic areas; however, their level of root colonization differs between species and habitats (Gardes & Dahlberg 1996). Host species appears to be a main factor influencing the community composition of arbuscular fungi (Becklin et al. 2012). An important group of fungi, common and abundant among plants of the treeline ecotone, is dark septate fungi, although their ecological role is still not sufficiently recognized. Since the report of Gardes & Dahlberg (1996), the number of studies exploring or characterizing fungal diversity in arctic and alpine ecosystems has gradually increased. It has been found that mycorrhizal diversity tends to be lower in arctic and alpine areas than in other regions, due to environmental constrains, dispersal barriers, and an increased number of facultative or non-mycorrhizal host species (e.g. Kernaghan & Harper 2001). The observed lower fungal diversity correlates well with observations of lower fungal biomass in soils of mountain forests stands, reflected by specific biomarkers (phospholipid fatty acid and ergosterol concentration) or content of extramatrical mycelium of ectomycorrhizal fungi in soil (Karliński et al. 2015). The biomass of extramatrical mycelium is largely influenced by the community structure of ectomycorrhizal fungi associated with the forest trees (Ekblad et al. 2013), as different fungal species produce variable amounts of extramatrical mycelium, which builds a mycelial network in the soil and may have variable tolerance to natural and anthropogenic stress factors (Karliński & Kieliszewska-Rokicka 2004). In contrast to the altitude gradient, seasonal fluctuations and changes in soil nutrient status play a significant role in microbial (including fungi) biomass and biodiversity of alpine and tundra forests (Buckeridge et al. 2013). Becklin et al. (2012) observed a stronger effect of host identity than habitat on the fungal community of alpine plant species. A high species diversity of fungi associated with alpine and arctic *Dryas octopetala* and a lack of influence of high altitude was demonstrated by Bjorbækmo et al. (2010). Schmidt et al. (2008) suggested that the altitudinal distributions of mycorrhizal fungi observed for European mountains do not necessarily apply to higher and drier mountains in other parts of the world. Thus, additional surveys spanning a greater number of sites, host species, and types of mycorrhiza are needed to fully understand the role of factors shaping the mycorrhizal community of the treeline ecotone.

The objective of this paper was therefore to identify reliable and strong indicators useful for monitoring soil properties in mountain forests and treeline ecotones, in particular. For this purpose, 7 case studies from mountains of different regions of Europe are presented to show the importance of soil characteristics as indicators of treeline dynamics in relation to anthropogenic pressures and/or climate change. When not specifically indicated, climatic and anthropogenic pressures were considered for their combined and/or synergic impact on soil properties (Oliver & Morecroft 2014). Within the EU project SENSFOR, the discussion of soil indicators was integrated into the discussion of indicators for changes in the treeline ecotone in general (Broll et al. 2016).

3. TREELINE ECOTONES AND SOIL INDICATORS

Within the numerous services that forested soil performs in treeline areas, key ecosystem services were selected for the aims of this paper: (1) protection of the area below treeline, i.e. prevent shallow landslide, erosion, and flooding, (2) sequestering of carbon, i.e. maintaining the SOM content and quality, and (3) acting as habitat for plants, animals, and microorganisms, and maintaining a high biodiversity (Adhikari & Hartemink 2016). Soil erosion, caused by landslide on steep slopes, represents a crucial problem affecting mountain soils, which are generally shallow, and their fertility is often concentrated in the uppermost layers (Stanchi et al. 2015, Egli & Poulenard 2017). Therefore, following soil loss, erosion provokes a loss of fertility, reduced carbon storage capacity, and ultimately a reduced level of biodiversity above- and belowground.

Once key ecosystem services performed by soil have been identified, the correct soil indicators to be used in a monitoring action can be selected. In this context, a promising way to analyze the impact of human–environment interactions on ecosystem services seems to be the application of the DPSIR (Drivers, Pressure, State, Impact, Response) framework (Kyriazopoulos et al. 2014). In order to assess the sustainability of treeline ecotones, it is important to use indicators that identify specific pressures (P) and impacts (I) within ecosystems that can also be monitored for their capacity to tolerate disturbances due to current and future changes. Many indicators can be used for monitoring changes in treeline ecotones. Ecological indicators related to plants, such as growth forms or regeneration processes (seed production, seedling mortality, or survival rate), are generally those aspects mostly studied. As for soils, physical, chemical, and biological properties can be selected as indicators of soil quality in treeline ecotones, in relation to their sensitivity to disturbance or stress. Most physical indicators are quite static, whereas chemical, and to a greater degree, biological characteristics may be very dynamic (Table 1). Integrating indicators (such as decomposition) can be very helpful for a complete soil-monitoring scheme.

An example of a useful ecological indicator for erosion resulting from grazing on sandy and dry soil would be the evaluation of areas of bare soil. Erosion removes soil cover, including most of the carbon contained in the upper layer. Loss of soil cover leads to fragmented treeline ecotones where the accumulation of organic matter is very low or restricted to small patches. In addition, aboveground and belowground biodiversity are reduced by erosion. Ecosystem services like decomposition and nutrient supply or carbon sequestration are also reduced. In the best case, recommendations based on the evidence from soil indicators, as summarized in this paper, will be accepted and implemented by policy makers in order to preserve carbon stocks and prevent soil erosion, and to protect biodiversity in the treeline ecotone, including belowground, as this is a precondition for the preservation of aboveground biodiversity.

Table 1. Soil properties as indicators in treeline ecotones. Highly dynamic, dynamic, static indicators: soil property change within days or weeks, years or decades, or centuries, respectively

	Highly dynamic	Dynamic	Static
Physical, chemical an biological indicators			
Physical	Porosity Temperature Water content	Area of bare soil Soil structure	Texture Soil depth
Chemical	Labile carbon Mineral nutrients (N,P,K) pH	Total carbon (TC) Total organic carbon (TOC) Total nitrogen (TN) C/N ratio	–
Biological	Soil organisms: • Microbial biomass, metabolism and diversity • Indicators of stress conditions (microbial quotients) • Enzyme activities • Root rot pathogens • Mycorrhiza	Rooting: • Depth • Biomass of fine roots • Architecture of roots	–
Integrating indicators			
Ecological	–	Humus forms and soil structure that result from the decomposition process	–

4. CASE STUDIES ON SOIL INDICATORS AT EUROPEAN TREELINES

In order to assess the importance of soil properties as indicators of natural and anthropogenic impacts, 7 case studies performed at the treeline in similar mountain regions were selected from a collaboration among the following European countries: Romania, Bulgaria, the Czech Republic, Slovakia, and Italy. Table 2 reports details for each case study, such as climate, soil type, vegetation, and the specific indicators used.

4.1. Romania (Case 1)

In Romania, Badea et al. (2012) studied air pollution, bulk precipitation, throughfall, soil condition, and foliar nutrients, as well as forest health and growth, from 2006 to 2009 at a study site belonging to the Long-Term Ecological Research (LTER) network in the Bucegi Mountains. They ascertained that bulk precipitation was generally acidic. Therefore rain could contribute to forest soil acidification with a possible negative effect on forest health. In fact, the pH of the soil solution was acidic (pH 4.4 to 4.8) in most of the studied sites. Concentrations of acidic ions such as sulfate were below critical limits in the bulk precipitation (CLRTAP 2004), but could have significant long-term cumulative effects on soils and forest vegetation, especially in the context of changing climate.

4.2. Romania (Case 2)

Similar research carried out in the Retezat Mountains showed that >90% of rain events were acidic with pH values <5.5, contributing to the high acidity of soils (Bytnerowicz et al. 2005). Drought that occurred in the southern Carpathians between fall 2000 and summer 2002, and frequent acidic rainfalls, could have caused the observed decline in forest condition, results confirmed at a national level for the 1992 to 2010 period (Badea et al. 2013). The very strongly acidic character of upper soil horizons (pH 3.4 to 4.2) can be partially attributed to the atmospheric deposition of sulfate, nitrate, and ammonium. Although soil acidification is common for this type of geological material (schist), it is possible that long-term deposition could further increase soil acidity in the Retezat Mountains.

4.3. Bulgaria

The Beklemeto region in the Central Balkan Mountains (Bulgaria) is representative of the relatively small mountains of the Balkan Peninsula, which share a similar land use history, where the upper

Table 2. Case studies: site characteristics of representative soils from different European countries. MAAT: mean annual air temperature, MAP: mean annual precipitation. Ranges are given for MAAT and MAP because of different altitudes. For definitions of soil type see IUSS Working Group WRB (2015)

Location of study sites	Elevation (m)	MAAT (°C)	MAP (mm)	Parent material chosen	Soil type	Tree cover	Key disturbance drivers	Indicators and rationale
Czech Republic								
Giant Mts	1185	2.6–3.5	1300–1414	Granite	Podzol, Histosol	Norway spruce	SO_2 pollution, acidic precipitation	Carbon content, proportion of fulvic and humic acids
Slovakia								
Tatra Mts	1150–1250	3.4–4.7	931–964	Moraine deposits	Dystric Cambisol	Norway spruce, larch	Storm, fire, land use change	Microbial biomass, activity and functional diversity
Romania								
Bucegi Mts	930–1720	2–6	800–1200	Conglomerate, sandstone	Dystric Cambisol, Eutric Cambisol, Entic Podzol	Norway spruce, silver fir, European beech	Acidic precipitation	Soil pH
Retezat Mts	800–1600	3–7	1300–1600	Schist	Dystric Cambisol, Entic Podzol	Norway spruce, European beech	Acidic precipitation	Soil pH
Bulgaria								
Central Balkan Mts	1050–1500	6.1	914	Schist	Dystric Cambisol	European beech, Norway spruce	Fires, forest conversion, land use change	Carbon content, soil pH
Italy								
Passo Brocon, Central-Eastern Alps	1700–2000	4.6	1286	Granite and limestone	Leptic Phaeozem	Norway spruce, larch	Pasture abandonment	Enzyme activity as indicators for microbial functional diversity, disturbances and stresses
Madonna di Campiglio	1730–1855	4.2	1148	Granite covered by moraine deposits	Umbrisol	Norway spruce, larch, Cembra pine	Pasture abandonment, ski resorts and tourism	Bacterial–fungal community molecular profiles shaped by aspect, parent material, and forest attributes

treeline is mainly formed by broadleaved forests. The treeline of the region is formed by 3 vegetation zones: broadleaved forests, areas of coniferous forests (plantations, natural), and a juniperus zone. In this case-study site, the combined effects of land use and climate change on soil properties was investigated. In the study region, altitudinal migration due to climate are expected for beech and occasionally for conifers. Changes in vegetation cover affect soil properties due to warming, changes in precipitation, increased transpiration, and poorer drainage. Among soil physical–chemical parameters, the focus was on SOM. Previous land use affects the potential for carbon sequestration in forest areas. Pastures are characterized by high carbon stocks and a greater density of roots in the surface layers of the mineral soil, thus conversions of grasslands to forests has clear consequences (Guo & Gifford 2002, Murty et al. 2002). Zhiyanski et al. (2008, 2016) and Zhiyanski & Sokolovska (2009) showed that land use change has affected the content of soil organic carbon in mountain ecosystems in the Central Balkan Mountains. The effects were associated with a reduction in soil organic carbon stocks when pastures were afforested to increase the treeline zone 40 yr ago. Nevertheless, a significant amount of carbon is stored in the forest floor, and thus the loss of soil is compensated. Zhiyanski et al. (2008) emphasized that the implementation of thinning activities in coniferous plantations is desirable to improve the microclimate conditions and the accumulation of carbon in the soil system. Results for other soil properties from the studied region showed variation in relation to forest tree vegetation (Doichinova & Zhiyanski 2009). Two groups of ecosystems could be distinguished according to soil properties in treeline zones with similar climatic characteristics: (1) beech and spruce forests, and (2) Scots pine forests. Clay and silt textural fractions, soil pH, and exchangeable

Ca were higher under Scots pine, while the soil under beech and spruce was characterized by higher total organic carbon (TOC) stocks, total N, and exchangeable Na, and lower pH (Zhiyanski et al. 2008). These differences could be related to differences in organic matter content caused by differences in forest vegetation and microclimate peculiarities of the sites. Soil physical–chemical properties and the SOM content could be used as composite indicators for assessing the effect of land use on soil systems and stability of ecosystems in treeline ecotones. The lower carbon content in the mineral soil under spruce plantations compared to that under natural beech forest is related to factors such as silvicultural practices and land use history. Overall, carbon stocks both on the forest floor and in soil were highest for the highland pasture, followed by the spruce plantation, while the beech forest and the pine plantation were characterized by comparatively lower carbon stocks. The conversion from natural pasture or beech forest to coniferous plantation in treeline ecotones of the Central Balkan Mountains has resulted in a decrease in TOC stock. Nevertheless, the large organic carbon storage in the forest floor in the spruce plantation compensates for the carbon leached out from the mineral soil after this land use change. In terms of stability, carbon sequestrated in the mineral soil is more desirable than carbon sequestrated on the forest floor, which is more vulnerable to decomposition following disturbances. Carbon stocks and SOM quality are very useful indicators for the detection of changes in mountain forests (Zhiyanski et al. 2008, Doichinova & Zhiyanski 2009) and their mapping could be used as an effective approach to identifying changes in supply of regulating ecosystem services (Zhiyanski et al. 2017).

4.4. Czech Republic

The research in the north-eastern part of the Czech Republic was carried out in the Mumlavska Hora area in the Giant Mountains National Park. The treeline in the Giant Mountains is at around 1200 to 1400 m above sea level (a.s.l.) (Treml & Chuman 2015), and the typical vegetation of the ecotone is represented by Norway spruce *Picea abies* intermixed with *Vaccinium* shrubs and graminoids communities. The forest stand of Norway spruce used for sampling was naturally established, and has been minimally influenced by forest management. The area was affected by a severe forest decline, which started >35 yr ago and was mostly caused by elevated levels of anthropogenic SO_2 deposition (Kopáček & Veselý 2005). Following the dieback of spruce trees, an increase in grass and shrub cover was observed. The aim of this study was to evaluate how this change in vegetation influences the amount and the quality of organic matter in the O horizon. The organic matter of Oa horizons was fractionated into humic and fulvic acids (HA and FA) to study differences related to the dominant ground vegetation, *Vaccinium myrtillus*, *Deschampsia flexuosa*, or *Molinia caerulea*. The results indicated that the dominant shrub or grass species could quantitatively influence the amount of carbon in organic horizons (organic C was higher under *M. caerulea* than under *V. myrtillus*). However, the humification rate was similar under all dominant plant species, probably because of a strong interaction with site condition, and differences appeared only in the quality of humic compounds. The proportion of fulvic and humic acids was in fact different according to vegetation type (a higher proportion of HA was found under *M. caerulea*). Regarding the importance of HA and FA in soil development, the findings suggest that, after a shift towards grass vegetation due to forest decline, major changes may be expected not only in the organic horizons, but also in the whole soil profile. Vegetation changes due to predicted climate change, probably leading to forest decline, could cause significant changes in quantitative and qualitative parameters of SOM. Thus, the quantity and quality of SOM in the organic layers are good indicators of changes in mountain forests (Bonifacio et al. 2008b).

4.5. Slovakia

Windthrow and fire belong to natural disturbances evoked by extreme climate events, such as strong wind, and extreme hot and dry periods, occurring as a consequence of ongoing climate changes. In 2004, forest stands near the treeline in the Tatra Mountains (Slovakia) were seriously damaged by strong wind and a part of this area was subsequently affected by fire. Four long-term research plots (intact reference forest, extracted windthrow, non-extracted windthrow, and extracted windthrow later affected by fire) were established in 2005 to monitor changes in the abiotic environment, including the soil microbial component. Based on the analysis of soil microbial biomass, activity, and structure of functional groups, Gömöryová et al. (2011) showed that several microbial community attributes assessed in windthrow areas appear suitable for the monitoring of long-term changes. In particular, catalase activity proved to be a useful indicator, as it exhibited clear interannual trends with low seasonal variability. N-mineralization also differed significantly among years, but varied considerably during the vegetation period. Richness and diversity of microbial functional groups, as determined by the Biolog assay (Garland & Mills 1991), did not differ between plots, but unidirectional temporal shifts qualify them as indicators of long-term change. In contrast, basal respiration and microbial biomass differed primarily among plots with different management practices. Basal respiration varied considerably among seasons as well as among years. Microbial biomass did not exhibit seasonal fluctuations, and inter-annual variability was much lower compared to other soil characteristics. It therefore proved to be a sensitive indicator of environmental change in disturbed plots.

4.6. Italy (Case 1)

This case study is part of a larger research program on the effects of woody plant encroachment on soil organic carbon dynamics, and in general on soil properties, along a latitudinal gradient in Italy. In this case, woody plant species recolonization after pasture abandonment was studied in the Central-East Italian Alps, in Castello Tesino (Trento province, Italy). Three stages of the recolonization process were selected: (1) a current cow pasture with annual grasses, (2) a heather *Calluna vulgaris* (L.) Hull and *Rhododendron hirsutum* L. shrubland, and (3) a Norway spruce *Picea abies* (L.) H. Karst. and larch *Larix decidua* Mill. forest. Three soil sample replicates were collected from each stage at both 0–5 and 5–15 cm depth intervals.

Two groups of soil indicators, chemical and biological, were used to describe the effect of the recolonization process: TOC and enzymatic activities of main soil nutrient cycles (C, N, P, S). TOC is considered the best indicator of soil quality, as soil carbon content provides information about soil degradation and influences all soil functions. On the other hand, enzyme activities are used to describe microbial functional diversity, and express the microbial community response to disturbances and stresses (Aon et al. 2001, Sowerby et al. 2005, Trasar-Cepeda et al. 2008). Fluorimetric analyses (Marx et al. 2001) were performed in order to evaluate the activity of cellulase, chitinase, α- and β-glucosidase, phosphatase, sulfatase, and xylosidase. The synthetic enzymatic index (SEI) (Dumontet et al. 2001) was then estimated in order to highlight variation in the sum of enzyme activities at both depth intervals.

When considering the first step of recolonization (from pasture to shrubland), we observed an increase in TOC content of +74 and +35% (calculated with respect to pasture for 0–5 and 5–15 cm, respectively). A similar trend was observed for the SEI (+38 and +156%). Then, from shrubland to forest, a general return to the original values of pasture was found both for TOC (+10 and –3%) and for the sum of the enzyme activities (–6 and +18%). When considering the enzyme specific activity (i.e. ratio of each enzyme activity to unit of total organic carbon), this increment was detected only in the upper intervals (+33 and +6%). A clear decrease in this ratio was observed for forest stand (–69, and –58% with respect to pasture). A change in land use, in this case pasture abandonment accompanied by a new vegetation cover, induced the microbial community to increase its decomposition activity during the first step of recolonization. An equilibrium level was then reached when the mature forest condition was established. In particular, the increase in enzymatic activities, which encompasses the increase in TOC, could reflect a modified functional diversity of microbial processes induced by the release of diverse carbon substrates by the new colonizing plant cover (shrubs etc.).

4.7. Italy (Case 2)

The aim of the last study was to determine whether a number of given state factors are important in shaping both chemical and microbial properties of the soil at the treeline in the Alps. Organic matter content and its characteristics, dissolved organic matter (DOM) and its phenolic content, bacterial–fungal community molecular profiles, and microbial biomass carbon and nitrogen content were evaluated (Carletti et al. 2009). Since microbiological processes regulate soil ecology and biogeochemistry, microbial community structure can be used as a dynamic indicator of forest ecosystem status after timber harvesting, revealing changes in nutrient and energy flow patterns measured under different aspects, parent material, and forest attributes. Aspect was considered as a proxy variable of microclimatic features. This topographic factor influences local site microclimate, mainly because it determines the amount of received solar radiation, snowmelt timing, soil freezing and thawing, and even water retention capacity and soil nutrient cycling (C and N), as well as other ecosystem processes.

In particular, parent material and forest coverage attributes were investigated in 6 Norway spruce *P. abies* forest sites located in the subalpine range of the eastern Italian Alps. Studying the relationships that stand age, site physical/topographic variables, and parent material have with soil carbon content and characteristics in mountain areas may be useful for predicting how environmental changes could affect soil carbon content and its accumulation. Stands were chosen so that they were consistent, as far as possible, with respect to elevation, macroclimatic, and dominant tree variables, while changing for parent material and aspect. The treeline was under pastoral use. In all sites, regeneration is natural and follows marginal cutting or canopy opening. Study areas were chosen according to local geological maps with different parent materials classified as acid, alkaline, and intermediate. For each parent material, A horizons of 2 sites were selected as north-facing and south-facing for a total of 6 sites. The geologic parent material had the strongest effect and was aspect-dependent. Microclimate features also played a distinct role in defining both soil chemistry and microbial community composition. In contrast, the composition of the deeper humus layers (OH, A) was stable and similar within a spruce tree canopy class. The most important variables in the construction of the discriminating models were soil pH, dissolved organic matter (DOM) content, and phenolic compounds. Bacterial communities appeared to be shaped first and foremost by the substratum, secondly by mountain slope orientation, and thirdly by forest stage, thus confirming the canonical discriminant analysis (CDA) model. Fungal communities indicated patterns guided strongly by pH, such as the highly diverse community of the acid/north, in which evenness was higher and differences related to 4 successional stages of the forest (in increasing order: gap [open patch, free of standing trees], juvenile, aggregate [slender tress reaching canopy height], and mature) were less evident, and did not lead to emerging dominance at mature stages, as occurred instead in each of the south-facing forests (La Porta et al. 2014). Rarefaction curves indicate a trend of higher diversity (less saturating) in acid/north conditions, while more saturating covered communities occur in southern-facing and mild bedrock pH conditions. With regard to the 4 forest stand age categories, the classes Archaeorhizomycetes and Sordariomycetes, and the genus *Archaeorhizomyces* presented with significantly higher values in gap samples, compared to the aggregate samples. Juvenile and mature forest samples presented intermediate quantities.

5. CONCLUSIONS

The case studies reported in this paper present an example of how the use of soil indicators at the treeline can help to describe the intrinsic properties of these fragile ecosystems. In particular, soil indicators can act as an early warning of degradation processes, leading to critical thresholds of sustainability in response to diverse types of pressure (Muscolo et al. 2015). Furthermore, certain soil properties can be chosen as indicators of the resilience of treeline ecosystems in response to stress, and be included in monitoring actions within a soil policy framework. The above case studies confirm SOM as one of the significant universal indicators (Lal 2004). In particular, SOM content fluctuations can account for modifications of soil quality due to land use change from pasture to forest (Bulgaria and Italy), while SOM quality assessment (humification rate, humic/fulvic acids ratio, etc.) accounts for modification of soil carbon stocks induced by plant cover changes due to climate and/or land use change (Bulgaria, Czech Republic, Italy).

Bioindicators focusing on soil biota, such as microbial biomass size, diversity, and activity, can inform on changes in soil biogeochemical processes (C or N cycling, mineralization, humification) and in food webs, or account for stress conditions (Slovakia, Italy). Moreover, due to their dynamic nature, bioindicators provide quick assessments of how the soil system reacts to the different pressures. This can help to prevent degradation processes by initiating correct policies and management strategies at the appropriate time.

In this study, several soil indicators were identified and used for the assessment of soil quality in relation to climatic and/or anthropogenic pressures in mountain forests. They were distinguished according to physical, chemical, and biological soil properties. Case studies in different mountain regions of Europe proved the importance of soil properties as indicators for the dynamics of mountain forest ecosystems and in the treeline ecotone in view of climate change and/or anthropogenic pressures. The case studies showed that SOM (in terms of quantity and quality), pH, and microbial biomass size, activity, and structural and functional diversity are reliable and sensitive indicators to be recommended for monitoring changes in the soil system. However, beyond this basic set of properties (and when possible), it is advisable to integrate the information obtained by using supplementary indicators to gain further insight and for a more holistic approach to the study of the soil resource.

Acknowledgements. This article is based upon work from COST ACTION ES 1203 SENSFOR, supported by COST (European Cooperation in Science and Technology, www.cost.eu).

LITERATURE CITED

Adhikari K, Hartemink AE (2016) Linking soils to ecosystem services — a global review. Geoderma 262:101–111

Anschlag K, Broll G, Holtmeier FK (2008) Mountain birch seedlings in the treeline ecotone, subarctic Finland: variation in above- and below-ground growth depending on microtopography. Arct Antarct Alp Res 40:609–616

Aon MA, Cabello MN, Sarea DE, Colaneri AC, Frenco MG, Burgos JL, Cortassa S (2001) I. Spatio-temporal patterns of soil microbial and enzymatic activities in an agricultural soil. Appl Soil Ecol 533:1–16

Arshad MA, Martin S (2002) Identifying critical limits for soil quality indicators in agro-ecosystems. Agric Ecosyst Environ 88:153–160

Badea O, Bytnerowicz A, Silaghi D, Neagu S and others (2012) Status of the Southern Carpathian forests in the long-term ecological research network. Environ Monit Assess 184:7491–7515

Badea O, Silaghi D, Taut I, Neagu S, Leca S (2013) Forest monitoring — assessment, analysis and warning system for forest ecosystem status. Not Bot Horti Agrobo 41: 613–625

Becklin KM, Hertweck KL, Jumpponen A (2012) Host identity impacts rhizosphere fungal communities associated with three alpine plant species. Microb Ecol 63:682–693

Bjorbækmo MFM, Carlsen T, Brystig A, Vrålstad T and others (2010) High diversity of root associated fungi in both alpine and arctic *Dryas octopetala*. BMC Plant Biol 10: 244

Bonifacio E, Caimi A, Falsone G, Trofimov S, Zanini E, Godbold DL (2008a) Soil properties under Norway spruce differ in spruce dominated and mixed broadleaf forests of the Southern Taiga. Plant Soil 308:149–159

Bonifacio E, Santoni S, Cudlín P, Zanini E (2008b) Effect of dominant ground vegetation on soil organic matter quality in a declining mountain forest of central Europe. Bor Environ Res 13:113–120

Broll G (2000) Influence of overgrazing by reindeer on soil organic matter and soil microclimate of well-drained soils in the Finnish Subarctic. In: Lal R, Kimble JM, Stewart BA (eds) Global climate change and cold region ecosystems. Advances in soil science. CRC Press, Bosa Roca, p 163–172

Broll G, Keplin B (2005) Mountain ecosystems. Studies in treeline ecology. Springer, Berlin

Broll G, Jokinen M, Aradottir AL, Cudlín P and others (2016) Indicators of changes in the treeline ecotone. Deliverable 5, ESSEM COST ACTION ES1203 — Enhancing the resilience capacity of sensitive mountain forest ecosystems under environmental change (SENSFOR). www.sensforcost.eu/index.php/deliverables

Buckeridge KM, Banerjee S, Siciliano SD, Grogan P (2013) The seasonal pattern of soil microbial community structure in mesic low arctic tundra. Soil Biol Biochem 65: 338–347

Bytnerowicz A, Badea O, Popescu F, Musselman R and others (2005) Air pollution, precipitation chemistry and forest health in the Retezat Mountains, Southern Carpathians, Romania. Environ Pollut 137:546–567

Carletti P, Vendramin E, Pizzeghello D, Concheri G, Zanella A, Nardi S, Squartini A (2009) Soil humic compounds and microbial communities in six spruce forests as function of parent material, slope aspect and stand age. Plant Soil 315:47–65

CLRTAP (Convention on Long-Range Transboundary Air Pollution) (2004) Manual on methodologies and criteria for modelling and mapping critical loads and levels and air pollution effects, risks and trends. Federal Environmental Agency, Berlin

Cotrufo MF, Soong JL, Horton AJ, Campbell EE, Haddix ML, Wall DH, Parton WJ (2015) Formation of soil organic matter via biochemical and physical pathways of litter mass loss. Nat Geosci 8:776–779

Doichinova V, Zhiyanski M (2009) Physical and chemical characteristics of forest soils from Beklemento region — Central Stara Planina Mountains. Nauka za Gorata – For Sci 2:45–52

Doran JW, Parkin TB (1994) Defining and assessing soil quality. In: Doran JW, Coleman DC, Bezdicek DF, Stewart BA (eds) Defining soil quality for a sustainable environment. SSSA, Madison, WI, p 3–21

Dumontet S, Mazzatura A, Casucci C, Perucci P (2001) Effectiveness of microbial indexes in discriminating interactive effects of tillage and crop rotations in a Vertic Ustorthens. Biol Fertil Soils 34:411–416

Egli M, Poulenard J (2017) Soils of mountainous landscapes. In: Richardson D, Castree N, Goodchild MF, Koboyashi A, Liu W, Marsten RA (eds) The international encyclopedia of geography. Wiley, New York, NY, p 1–10

Ekblad A, Wallander H, Godbold DL, Cruz C and others (2013) The production and turnover of extramatrical mycelium of ectomycorrhizal fungi in forest soils: role in carbon cycling. Plant Soil 366:1–27

Gardes M, Dahlberg A (1996) Mycorrhizal diversity in arctic and alpine tundra: an open question. New Phytol 133: 147–157

Garland JL, Mills AL (1991) Classification and characterization of heterotrophic microbial communities on the basis of patterns of community-level sole-carbon-source utilization. Appl Environ Microbiol 57:2351–2359

Gil-Sotres F, Trasar-Cepeda C, Leirós MC, Seoane S (2005) Different approaches to evaluating soil quality using biochemical properties. Soil Biol Biochem 37:877–887

Gömöryová E, Střelcová K, Fleischer P, Gömöry D (2011) Microbial characteristics at the monitoring plots on windthrow areas of the Tatra National Park (Slovakia): their assessment as environmental indicators. Environ Monit Assess 174:31–45

Guo LB, Gifford RM (2002) Soil carbon stocks and land use change: a meta analysis. Glob Change Biol 8:345–360

Haselwandter K (2007) Mycorrhiza in the alpine timberline ecotone: nutritional implications. In: Wieser G, Tausz M (eds) Trees at their upper limit. Plant ecophysiology, Vol 5. Springer, Dordrecht, p 57–66

Holtmeier FK (2009) Mountain timberlines. Ecology, patchiness, and dynamics, 2nd edn. Adv Global Change Res 36:1–437

Holtmeier FK, Broll G (2005) Sensitivity and response of northern hemisphere altitudinal and polar treelines to environmental change at landscape and local scales. Glob Ecol Biogeogr 14:395–410

Holtmeier FK, Broll G (2007) Treeline advance — driving processes and adverse factors. Landsc Online 1:1–33

Holtmeier FK, Broll G (2010) Wind as an ecological agent at treelines in North America, the Alps, and the European Subarctic. Phys Geogr 31:203–233

IUSS Working Group WRB (2015) World reference base for soil resources 2014, update 2015. International soil classification system for naming soils and creating legends for soil maps. World Soil Resources Reports No. 106. FAO, Rome

Janvier C, Villeneuve F, Alabouvette C, Edel-Hermann V, Mateille T, Steinberg C (2007) Soil health through soil disease suppression: which strategy from descriptors to indicators? Soil Biol Biochem 39:1–23

Karliński L, Kieliszewska-Rokicka B (2004) Diversity of spruce ectomycorrhizal morphotypes in four mature forest stands in Poland. Dendrobiology (Pozn) 51:25–35

Karliński L, Rudawska M, Leski T, Kieliszewska-Rokicka B (2015) Biomass of external mycelium of ectomycorrhizal fungi in Norway spruce stands in Poland. Acta Mycol 50: 1063

Kernaghan G, Harper KA (2001) Community structure of ectomycorrhizal fungi across an alpine/subalpine ecotone. Ecography 24:181–188

Kopáček J, Veselý J (2005) Sulfur and nitrogen emissions in the Czech Republic and Slovakia from 1850 till 2000. Atmos Environ 39:2179–2188

Körner C (2012) Alpine treelines. Functional ecology of the global high elevation tree limits. Springer, Basel

Kyriazopoulos A, Abraham E, Hofgaard A, Sarki S (2014) DPSIR for treeline ecosystems and their services. Deliverable 3, ESSEM COST ACTION ES1203—Enhancing the resilience capacity of sensitive mountain forest ecosystems under environmental change (SENSFOR). www.sensforcost.eu/index.php/deliverables

La Porta N, Sablok G, Pindo M, Squartini A (2014) 454 pyrosequencing analyses of mountain forest soils reveal a high fungal diversity and a rapid response to successional stages. In: Parrotta JA, Moser CF, Scherzer AJ, Koerth NE, Lederle DR (eds) Sustaining forest, sustaining people: the role of research. XXIV IUFRO World Congress, 5–11 October 2014, Salt Lake City. International Forest Review 16(Spec Issue):166

Lal R (2004) Soil carbon sequestration impacts on global climate change and food security. Science 304:1623–1627

Marinari S, Bonifacio E, Moscatelli MC, Falsone G, Antisari LV, Vianello G (2013) Soil development and microbial functional diversity: proposal for a methodological approach. Geoderma 192:437–445

Marx MC, Wood M, Jarvis SC (2001) A microplate fluorimetric assay for the study of enzyme diversity in soils. Soil Biol Biochem 33:1633–1640

McBratney AB, Minasny B, Wheeler I, Malone BP (2012) Frameworks for digital soil assessment. In: Minasny B, Malone BP, McBratney AB (eds) Digital soil assessment and beyond. Taylor and Francis, London, p 9–14

Moscatelli MC, Di Tizio A, Marinari S, Grego S (2007) Microbial indicators related to soil carbon in Mediterranean land use systems. Soil Tillage Res 97:51–59

Murty D, Kirschbaum MUF, McMurtrie RE, McGilvray H (2002) Does conversion of forest to agricultural land change soil carbon and nitrogen? A review of the literature. Glob Change Biol 8:105–123

Muscolo A, Settineri G, Attinà E (2015) Early warning indicators of changes in soil ecosystem functioning. Ecol Indic 48:542–549

Nannipieri P, Ascher J, Ceccherini MT, Landi L, Pietramellara G, Renella G (2003) Microbial diversity and soil functions. Eur J Soil Sci 54:655–670

Nielsen MN, Winding A (2002) Microorganisms as indicators of soil health. National Environmental Research Institute, Denmark. Technical Report No. 388

Oliver TH, Morecroft MD (2014) Interactions between climate change and land use change on biodiversity: attribution problems, risks, and opportunities. WIREs Clim Change 5:317–335

Palombo C, Marchetti M, Tognetti R (2014) Mountain vegetation at risk: current perspectives and research needs. Plant Biosyst 148:35–41

Paz-Ferreiro J, Fu S (2016) Biological indices for soil quality evaluation: perspectives and limitations. Land Degrad Dev 27:14–25

Ponge JF (2013) Plant-soil feedbacks mediated by humus forms: a review. Soil Biol Biochem 57:1048–1060

Raev I, Zhelev P, Grozeva M, Markov I and others (2011) Programme of measures for adaptation of forests in Bulgaria and mitigate the negative impact of climate change on them. Project FUTUREforest helping Europe tackle climate change. INTERREG IVC, ERDF, Sofia

Reithmeier L, Kernaghan G (2013) Availability of ectomycorrhizal fungi to black spruce above the present treeline in Eastern Labrador. PLOS ONE 8:e77527

Ritz K, Black HIJ, Campbell CD, Harris JA, Wood C (2009) Selecting biological indicators for monitoring soils: a framework for balancing scientific and technical opinion to assist policy development. Ecol Indic 9:1212–1221

Rovira P, Vallejo R (2002) Mineralization of carbon and nitrogen from plant debris, as affected by debris size and depth of burial. Soil Biol Biochem 34:327–339

Schmidt SK, Sobieniak-Wiseman LC, Kageyama SA, Halloy SRP, Schadt CW (2008) Mycorrhizal and dark-septate fungi in plant roots above 4270 meters elevation in the Andes and Rocky Mountains. Arct Antarct Alp Res 40: 576–583

Schmidt MWI, Torn MS, Abiven S, Dittmar T and others (2011) Persistence of soil organic matter as an ecosystem property. Nature 478:49–56

Sowerby A, Emmett B, Beier C, Tietema A and others (2005) Microbial community changes in heathland soil communities along a geographical gradient: interaction with climate changes manipulations. Soil Biol Biochem 37: 1805–1813

Stanchi S, Falsone G, Bonifacio E (2015) Soil aggregation, erodibility, and erosion rates in mountain soils (NW Alps, Italy). Solid Earth 6:403–414

Trasar-Cepeda C, Leirós MC, Gil-Sotres F (2008) Hydrolytic enzyme activities in agricultural and forest soils. Some implications for their use as indicators of soil quality. Soil Biol Biochem 40:2146–2155

Treml V, Chuman T (2015) Ecotonal dynamics of the altitudinal forest limit are affected by terrain and vegetation structure variables: an example from the Sudetes Mountains in Central Europe. Arct Antarct Alp Res 47:133–146

van Bruggen AHC, Semenov AM (2000) In search of biological indicators for soil health and disease suppression. Appl Soil Ecol 15:13–24

Väre H, Vestberg M, Ohtonen R (1997) Shifts in mycorrhiza and microbial activity along an oroarctic altitudinal gradient in northern Fennoscandia. Arct Alp Res 29: 93–104

von Lutzow M, Koegel-Knabner I, Ekschmitt K, Flessa H, Guggenberger G, Matzner E, Marschner B (2007) SOM fractionation methods: relevance to functional pools and to stabilization mechanisms. Soil Biol Biochem 39: 2183–2207

Wieser G, Holtmeier FK, Smith WK (2014) Treelines in a changing global environment. In: Tausz M, Grulke N (eds) Trees in a changing environment. Springer, Dordrecht, p 221–263

Zhiyanski M, Sokolovska M (2009) Macro- and microelements in soils under spruce and beech forests in Central Stara Planina Mountain. Proc Seminar of Ecology, 23–24 April 2009, Sofia. Central Laboratory of General Ecology BAS, p 132–143

Zhiyanski M, Kolev K, Sokolovska M, Hursthouse A (2008) Tree species effect on soils in Central Stara Planina Mountains. Nauka za Gorata – For Sci 4:65–82

Zhiyanski M, Gikov A, Nedkov S, Dimitrov P, Naydenova L (2016) Mapping carbon storage using land cover/land use data in the area of Beklemeto, Central Balkan. In: Koulov B, Zhelezov G (eds) Sustainable mountain regions: challenges and perspectives in Southwestern Europe. Springer, Basel, p 53–67

Zhiyanski M, Glushkova M, Kirova L, Filcheva E (2017) Quantitative and qualitative features of soil humus in mountain treeline ecosystems. Silva Balcanica 18:5–23

11

Feedback effects of clonal groups and tree clusters on site conditions at the treeline: implications for treeline dynamics

Friedrich-Karl Holtmeier[1,*], Gabriele Broll[2]

[1]Institute of Landscape Ecology, Heisenbergstraße 2, Westfälische Wilhelms-Universität, 48149 Münster, Germany
[2]Institute of Geography, University of Osnabrück, Seminarstr. 19 a/b, 49074 Osnabrück, Germany

ABSTRACT: The article examines interactions between closely spaced clonal tree groups (spruce, fir, mountain birch) and dense tree clusters and their environment with special regard to Northern Europe, the Alps, and the Rocky Mountains. Modification of wind velocity and direction, trapping blowing snow and eolian dust, ground-shading and the release of sensible heat to the immediate vicinity ('black-body effect') are the main direct feedback effects of grouped trees on site conditions. Under long-lived clonal groups, soil physical and chemical properties gradually change. Clonal groups also leave their marks in geomorphic micro-structures. Compact groups may facilitate establishment of tree seedlings by providing shelter. Under global warming, the existing clonal groups above the forest limit may act as 'new' sources of viable seed that facilitate infilling of gaps in the treeline ecotone and the establishment of new tree populations above the present tree limit. On the other hand, increased deposition of snow as a result of the presence of tree groups may affect seedling establishment, and thereby overrule the positive effects of climate warming. If climate warming continues, current spatial structures, surface roughness, albedo, and snow distribution pattern, along with the resulting effects of these factors, will change across the entire ecotone. The feedback effects of closely spaced clonal groups will spatially overlap. Thus, it will be hard to distinguish between the effects of the single groups when new trees become established within the existing gaps. The northward shift of the polar treeline ecotone might contribute to heating of the lower atmosphere. On steep high-mountain slopes, however, no similar effects can be expected because of the limited size of the lower alpine area that has the potential to become covered with new forest and tree stands.

KEY WORDS: Clonal groups · Spruce · Fir · Mountain birch · Black-body effect · Wind · Snowpack · Treeline pattern

1. INTRODUCTION

Feedback effects of trees, and tree groups in particular, on site conditions in the treeline ecotone have been little studied, even though these effects may exert a strong influence on soil temperatures, soil moisture, nutrients, soil forming processes and, not least, microtopography. Since the end of the last century, however, this aspect has increasingly attracted attention (e.g. Marr 1967, Holtmeier 1980, Daly 1984, Callaway et al. 2002, Alftine & Malanson 2004, Bekker 2005, Holtmeier 2005, 2009, Resler 2006, Zeng & Malanson 2006, Humphries et al. 2008, Butler et al. 2009, Malanson et al. 2009, Holt-

*Corresponding author: fkholtmeier@arcor.de

§Advance View was available online January 17, 2017

meier & Broll 2010, Weiss et al. 2015). The objective of the present study is to examine interactions between closely spaced tree groups and the treeline environment, with special regard to the role of tree groups' architecture and distribution patterns. Our own previous studies were carried out in treeline ecotones in northern Fennoscandia, the Alps (Europe), and the Rocky Mountains (North America). This study summarizes current knowledge. It is a timely contribution to the debate on treeline dynamics and their driving factors (e.g. Harsch et al. 2009).

Fig. 1. Tree-line ecotone with clonal groups influencing snow pattern on a wind-exposed slope. Niwot Ridge, Colorado Front Range, USA, at about 3450 m. Photo by F.-K. Holtmeier, 2 March 1990

Treeline ecotones are often characterized by tree groups of different ages and sizes, surrounded by alpine grass and/or dwarf shrub vegetation. Clustering of trees may be a result of windborne seeds that, for example, germinate from seedbeds close to each other or from the scattered distribution of suitable seedbeds and/or safe sites (Holtmeier 2009, Holtmeier & Broll 2011, Holtmeier et al. 2003). Clustered trees may also emerge from seed caches of animals, such as nutcrackers (*Nucifraga caryocatactes, N. columbiana*), for example. Compact tree groups with a closed canopy are usually formed by species that are able to reproduce and propagate by layering (i.e. formation of adventitious roots on lateral branches embedded in the litter layer). Vegetative reproduction and propagation still continues at low temperatures that would otherwise preclude any seed-based regeneration (Larcher 1980, Holtmeier 1985a). Hereafter we call tree groups originating from layering 'clonal groups' and other closely spaced groups 'tree clusters'.

Clonal groups can be found at the altitudinal tree line of many high mountains and in the polar forest-tundra ecotone of Eurasia and North America. Clonal groups are particularly common in genera of spruce (e.g. *Picea abies*, Europe; *P. engelmannii, P. mariana*, North America) and fir (e.g. *Abies lasiocarpa*, North America; *A. sibirica*, northern Eurasia). In several Rocky Mountain areas, clonal groups prevail and give the treeline ecotone a distinctive physiognomy (Fig. 1). In the Colorado Front Range, >70 % of all tree groups in the treeline ecotone originated from layering spruce *P. engelmannii* and subalpine fir *A. lasiocarpa* (Holtmeier 1999). On level terrain and in the absence of a one-sided disturbance by wind, almost circular clonal groups (plan view) with the mother trees in the center may occur. In extremely windy places, mats, as well as wedge-like and low hedge-like growth forms, prevail. The orientation of the hedges is parallel to the prevailing wind direction and to each other (cf. Fig. 1). Tree clusters typically occur for example in zoochoric pines (*Pinus cembra*, European Alps; *P. flexilis, P. albicaulis*, Rocky Mountains).

In the European Alps and in many other Eurasian mountain areas already settled in prehistoric times, the natural treeline zone where layering in spruce probably prevailed has now largely disappeared mainly due to pastoral use and fire (Holtmeier 1974). Wildfires also occur under present conditions. At the treeline, where trees grow very slowly, wildfires may cause serious damage, in particular to dense tree clusters and clonal groups (e.g. Holtmeier 2009, Holtmeier & Broll 2010). Wildfires will probably be more frequent under warmer and drier conditions (e.g. Schumacher & Bugmann 2006).

At the mountain treeline in northern Finland, south of the polar limit of spruce, and in the Swedish Scandes, clonal groups of Norway spruce are well represented (Fig. 2; e.g. Kullman 2007, 2012, Öberg & Kullman 2011). In many areas of subarctic Northern Europe, compact groups of multistemmed mountain birch *Betula pubescens* ssp. *tortuosa* (Fig. 3) prevail, that sprouted from the rootstock after birch decline due to episodic defoliation by leaf-eating insects (*Epirrita autumnata, Operophtera brumata*; Bylund 1999, Neuvonen & Wielgolaski 2005).

Fig. 2. Clonal Norway spruce *Picea abies* in the forest-alpine tundra ecotone on Yllästunturi, a fell in Finnish Lapland, at about 480 m. Photo by F.-K. Holtmeier, 1 September 2000

Fig. 3. Clonal groups formed by mountain birch *Betula tortuosa* and willow *Salix caprea* (with darker stems, at the right hand side) on Korkea Jeähkkraš, a mountain in the Kilpisjärvi area, NW Finland. Photo by F.-K. Holtmeier, 10 September 2000

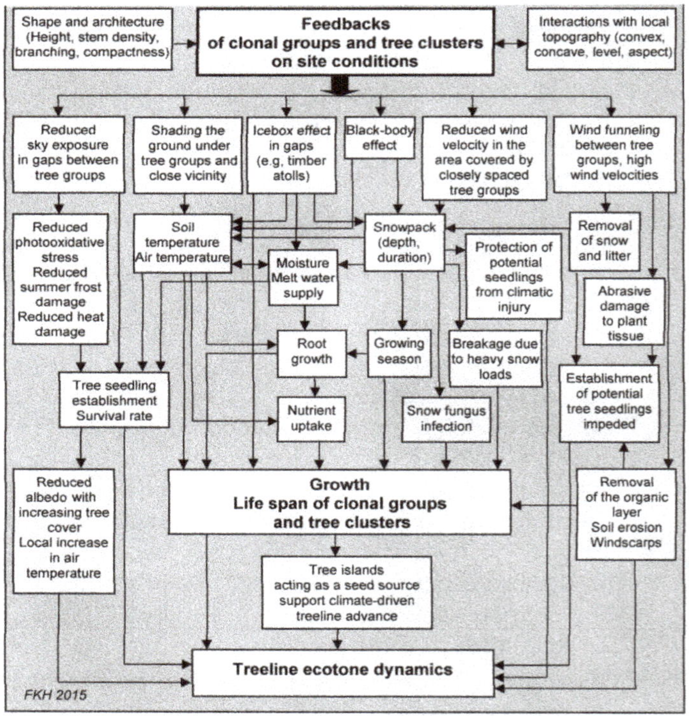

Fig. 4. Possible feedbacks of clonal tree islands and tree clusters on site conditions. Considerably modified from Holtmeier (2009)

2. FEEDBACKS EFFECTS OF TREE GROUPS ON THEIR ENVIRONMENT

Trees start to influence their close environment when projecting above the ground vegetation (e.g. dwarf shrubs) and the winter snowpack. The feedbacks from trees and tree groups depend largely on tree stature, stem density, architecture and species as well as on the interactions with local topography (Fig. 4; see also Holtmeier 2009, Holtmeier & Broll 2010).

2.1. Influence on direct solar radiation, and resultant effects

Transformation of direct solar radiation into sensible heat and subsequent release to the environment by long-wave radiation and turbulent flux — all subject to modification by wind velocity and direction — are important feedback effects of grouped trees (e.g. Holtmeier 1987, Oke 1987) (cf. Fig. 4). Due to the low albedo of conifers (0.05 to 0.15) their sun-exposed sides rapidly warm up to >10°C (maximum ~20°C) above the temperature of ambient air (e.g. Tranquillini 1957, 1979, Havranek & Tanquillini 1995, Hadley & Smith 1986). Thus, they act as a source of longwave radiation which is immediately absorbed by the surrounding snow and may cause early snowmelt at the directly south-facing sun-exposed side of the trees. In addition, the ground (with lower albedo) may become exposed to solar radiation and warm up rapidly (Oke 1987). This effect of trees on snowmelt has been called the 'black-body effect' in treeline literature (e.g. Brink 1959, Habeck 1969, Franklin & Dyrness 1973). The temperature of the exposed ground may rise and accelerate snowmelt. Nevertheless, wind and evaporation of melt water may delay the increase in soil temperature (e.g. Holtmeier 1987). The black-body effect can facilitate the establishment of tree seedlings (Fig. 5). Blowing snow accumulating at the sun-facing leeward side of tree groups may override the black body effect completely (Holtmeier 2009, Holtmeier & Broll 2010).

Shading of the ground by compact tree groups plays an even bigger role in the treeline ecotone. Soil temperature under the canopy of tree groups is relatively low during the growing season and relatively high in winter, and moisture is high compared to open terrain (Holtmeier & Broll 1992). In the open, temperatures rapidly rise in spring, with a positive influence on fine root development, nutrient uptake and seedling growth, provided that water supply is sufficient (Anschlag et al. 2008). In ecotones sparsely covered with low tree clusters and clonal groups, the total shaded area is relatively small. Thus, the uppermost soil layers and the air next to the ground surface may warm up during the day far above ambient temperature, unless prevented by wind and evaporation. In narrow gaps between high tree groups, however, most of the ground is usually in shade, and temperatures during the growing season are relatively low, compared to gaps of the same size surrounded, for example, by low krummholz (i.e. trees of which height growth is suppressed due to the adverse climatic conditions) or willow scrub. However, high trees surrounding narrow gaps minimize sky exposure and thus may mitigate photooxidative stress due to high daytime radiation loads combined with low night temperatures (e.g. Germino et al. 2002, Brodersen et al. 2006, Maher & Germino 2006). In addition, outgoing long-wave radiation and nocturnal summer frosts are reduced (Germino & Smith 1999, Smith et al. 2003, Johnson et al. 2004). Winter snowpack on the ground, however, often lingers into summer, even on southern aspects (see also Elliott & Kipfmueller 2010, Elliott 2012, Elliott & Cowell 2015). This 'icebox effect' also occurs in gaps resulting from death of the initial trees or older clone members in

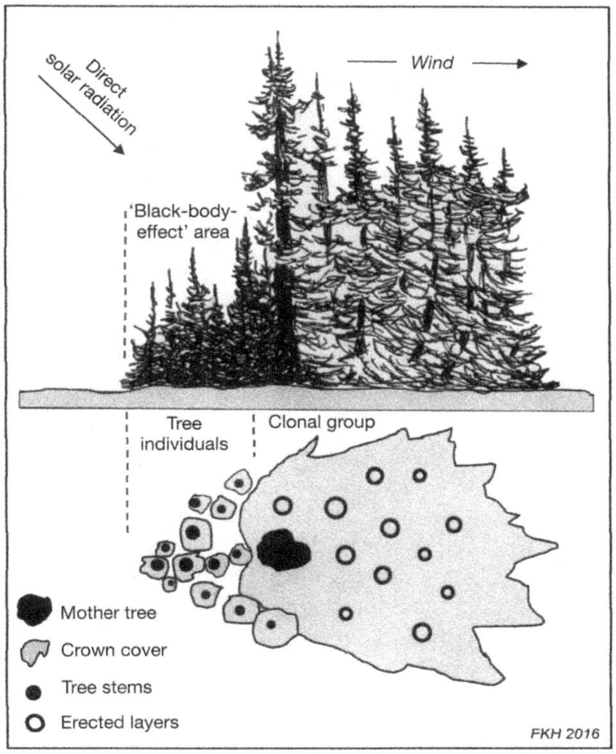

Fig. 5. Clonal group with younger trees that established themselves facilitated by the 'black-body effect'. Drawing by F.-K. Holtmeier after a field sketch. Erected layers are lateral branches that form adventitious roots, after which their distal end turns to a vertical position

more circular clonal groups ('timber atolls'; cf. Holtmeier 2005). Late-lying snow curtails the growing season. Tree seedlings are often destroyed by mechanical damage (settling snow) and/or infection by snow fungi (*Herpotrichia juniperi*, *H. coulteri*, *Phacidium infestans*) (Roll-Hansen 1989, Nierhaus-Wunderwald 1996, Senn 1999, Holtmeier 2005, 2009, Cunningham et al. 2006, Streule & Häsler 2006, Butin 2011, Holtmeier & Broll 2011, Barbeito et al. 2013).

2.2. Influence of tree groups on wind and resultant effects

Most treeline ecotones are in a windy environment (e.g. Holtmeier 1978, 2005, Hiemstra et al. 2002, 2006, Alftine & Malanson 2004, Bekker 2005, Kullman 2007, 2012). Trees, tree clusters and compact clonal groups in particular, reduce wind velocity and modify wind flow pattern near the ground surface. Compact tree groups and open tree stands also trap more blowing snow compared to treeless mountain and northern tundra (cf. Fig. 1; Holtmeier 1996, Liston et al. 2002, Tape et al. 2006). Relatively open tree clusters usually collect snow inside, whereas a deep snowdrift piles up behind the compact clonal groups' leeward edge. Its length is often several times the height of the windward tree group. At the windward front of clonal groups and tree clusters, and also along their lateral edges, blowouts occur, where intense scouring effects (eddies) remove the winter snowpack. Closely spaced clonal groups may increase deposition of blowing snow to such an extent that even on wind-exposed slopes, where the winter snow cover would normally be very patchy or even absent, relatively large areas get buried under a deep and late-lying snowpack (cf. Fig. 1).

The effects of clonal groups and terrain on relocation of snow and its consequences often overlap (Holtmeier 2005). Snowdrifts at the downwind edge of clonal groups are usually shorter on slopes with prevailing upslope winds than those with downslope winds, as supply of blowing snow from the closed forest on windward slopes is limited. In contrast, on lee slopes, the alpine zone is usually a large source of blowing snow that is partly deposited as snowdrifts behind the downwind edge of clonal groups, in tree clusters and also within more open tree stands. The amount of accumulated drifting snow may be up to 20 times (e.g. Dyunin 1967, Kobayashi 1972), in extreme cases even 100 times (Blanken 2009) greater than that deposited by direct snowfall (see also Holtmeier 1987). Clonal groups and tree clusters on top of convex topography increase deposition of snow on the lee slope, which in any case is normally rich in snow (Holtmeier 1996, 2005).

Soil temperatures in leeward snowdrift areas and also inside clonal groups do not usually drop much below the freezing point due to the deep insulating snowpack (Aulitzky 1961, Brooks et al. 1995). However, deep and late-lying snowdrifts may keep soil temperatures low far into the growing season (Holtmeier & Broll 1992, Hinkel & Hurd 2006). Consequently, the onset of root growth is delayed (e.g. Turner & Streule 1983). Root growth, nutrient uptake, photosynthesis, and carbon acquisition are reduced (e.g. Kramer 1982, Karlsson & Nordell 1996). On the other hand, nonstructural carbon reserves (e.g. Sveinbjörnsson et al. 1992, Skre 1993, Grace et al. 2002, Hoch & Körner 2003, Shi et al. 2006) may compensate for the effects of low soil temperature during the early growing season. In dry climates and during dry summers, moisture supply from melting snow drifts may mitigate moisture stress in late spring and early summer. As a result, comparatively luxuriant ground vegetation can be found in such places. Tree seedlings may also profit from extended moisture supply. On the other hand, at the treeline in Northern Europe and North America we found multi-stemmed clonal groups of spruce that locally increase deposition of blowing snow even on wind-swept convexities to an extent that needles and shoots of the lower branches encased in late-laying snow are destroyed by snow fungi. In the European Alps, Swiss stone pine *Pinus cembra* clusters that originated from seed caches of the European nutcracker *Nucifraga caryocatactes* on convex topography in the treeline ecotone may also act as a snow trap and increase the risk of snow fungus damage (Holtmeier 1974, 1985b, 2005). While the older trees lose all their lower needles, the younger pines and seedlings buried under the snowpack are completely destroyed (e.g. Holtmeier 1974, 2005, Stöcklin & Körner 1999).

Clonal groups have a lasting influence not only on soil physical and chemical properties (Holtmeier & Broll 1992) but also on geomorphic micro-structures, such as wind-scarps and small 10 to 20 cm high 'platforms' (Broll & Holtmeier 1994). Dieback at their windward front and elongation by layering on their leeward side (ca. 2 cm yr^{-1}; Benedict 1984) can bring about a slow downwind migration of clonal groups (Fig. 6). Inside the clonal groups a deep litter layer

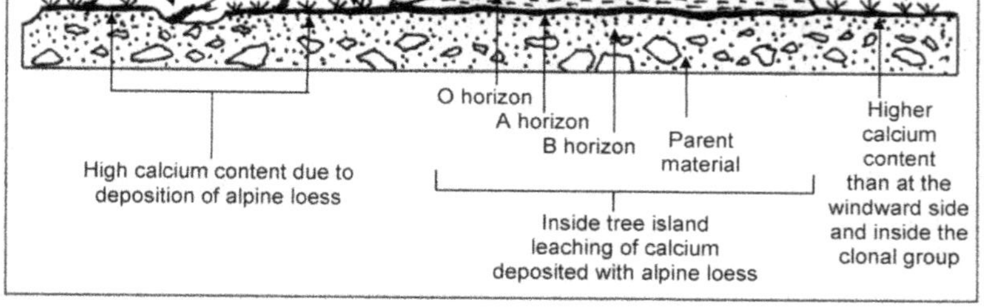

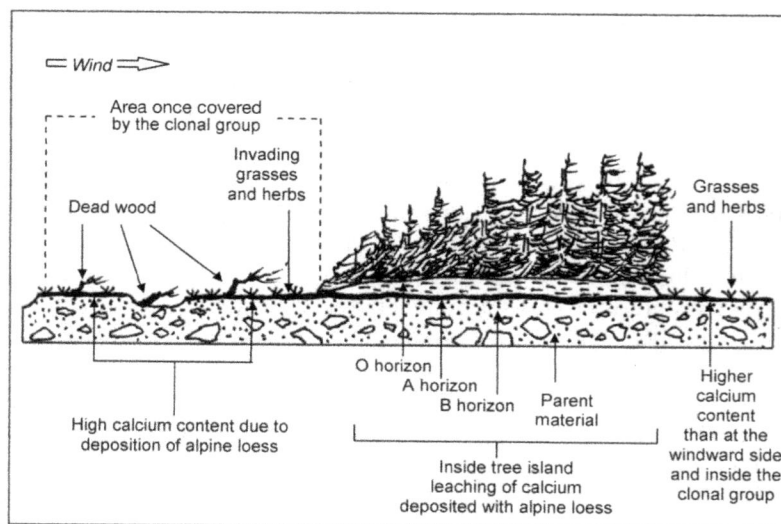

Fig. 6. Effects of a downwind 'migrating' clonal group. Modified from Broll & Holtmeier (1994)

accumulates. When the windward portions of the clonal groups are destroyed and the ground surface becomes exposed, the litter layer and the slightly decomposed organic matter (O_f) horizon are the first to erode. At least a part of the highly decomposed organic matter (O_h) horizon may be left intact, however. Almost simultaneously, alpine vegetation invades. This not only prevents further deflation and erosion for a while, but also increases deposition of calcium-rich eolian dust ('alpine loess'; Thorn & Darmody 1980, Litaor 1987, Holtmeier & Broll 1992, Broll & Holtmeier 1994). Eolian dust influences soil nutrient content, soil acidity, water-storing capacity, heat conductivity and decomposition. Nitrogen-rich litter produced by invading herbs and grasses, together with increased cation exchange capacity and relatively favorable soil moisture conditions, enhance humification in the topsoil. However, the wind-exposed edge of the tree island is partly eroded by wind. Needle ice formation often reinforces erosion. The resultant wind scarp gradually recedes in the downwind direction (Holtmeier & Broll 1992, Broll & Holtmeier 1994). Lateral erosion separates it into isolated small 10 to 20 cm high 'plateaus', with different physical and chemical soil conditions compared to the scarps still connected with the downwind migrating clonal groups (cf. Broll & Holtmeier 1994). Additional evidence of tree-island migration is provided by dead wood remains in the soil and on the surface. The remaining humus profile slowly decomposes in the long term due to higher temperatures at the formerly shaded tree island site. However, as suggested by Benedict (1984) the E-horizon and well developed B-horizons can record tree island passage long after wood remains have rotted away. Comparable lasting effects of mountain birch groups on geomorphic micro-structures were also documented at treeline in northernmost Finland (Holtmeier et al. 2004).

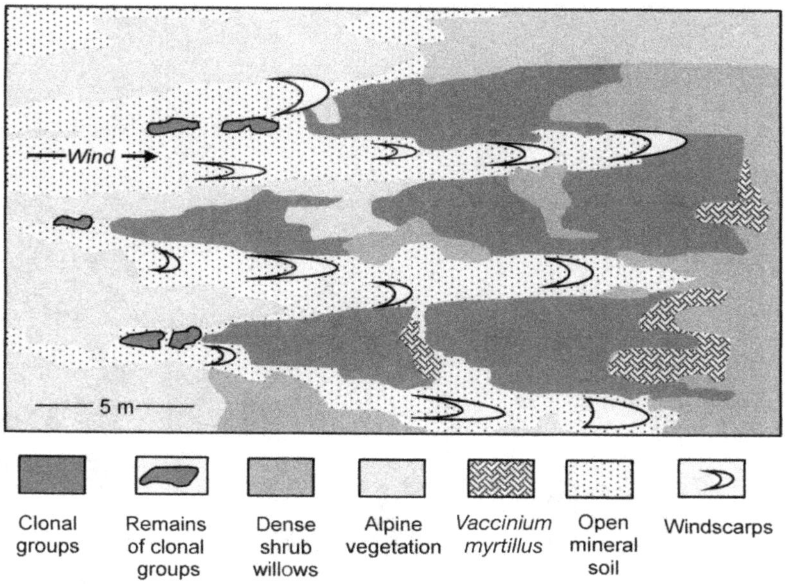

Fig. 7. Effects of wind funneling between compact clonal groups

Narrow 'corridors' between closely spaced 'hedges' or wedge-like clonal groups act as wind channels. In such places, strong winds often expose the mineral soil and create wind scarps (Fig. 7; Holtmeier 1978, 2005). Needle ice reinforces deflation and may impede or even prevent colonization by vegetation. At the leeward end of these corridors, however, where the wind funneling effect dwindles away, herbs, grasses, willow shrubs *Salix* spp., and blueberry *Vaccinium myrtillus* become established. Low willow shrubs often also line the lateral sides of hedges if snowpack is deep enough in such places to provide shelter from climatic injuries. Such interactions are characteristic traits of the wind-swept tree island areas in the entire treeline ecotone on the Colorado Front Range and neighboring mountain ranges (e.g. Hiemstra et al. 2002). They are particularly conspicuous on gently sloping or level uplifted old land surfaces, whereas on steep and rugged terrain, effects of geomorphic structures and mass-wasting are often of major importance (Holtmeier 2009, Holtmeier & Broll 2012). The same patterns have been documented by other treeline studies (e.g. Bekker 2005, Resler et al. 2005, Elliott 2012, Kharuk et al. 2010).

The influence of clonal groups and clusters of deciduous trees such as mountain birch or aspen on wind and snow relocation during winter is usually less conspicuous than that caused by evergreen conifer groups. However, in the treeline ecotone in Northern Europe, compact mountain-birch krummholz thickets, although without leaves in winter, collect drifting snow, even on wind-swept topography. Wind-trimmed wedge-like birches growing on top of convex topography or near the crest of the upper lee slope of convexities often produce a snowdrift at their downwind edge that enhances snow deposition on the lee slope rich in snow anyway. This is reflected in the occurrence and relatively luxuriant growth of

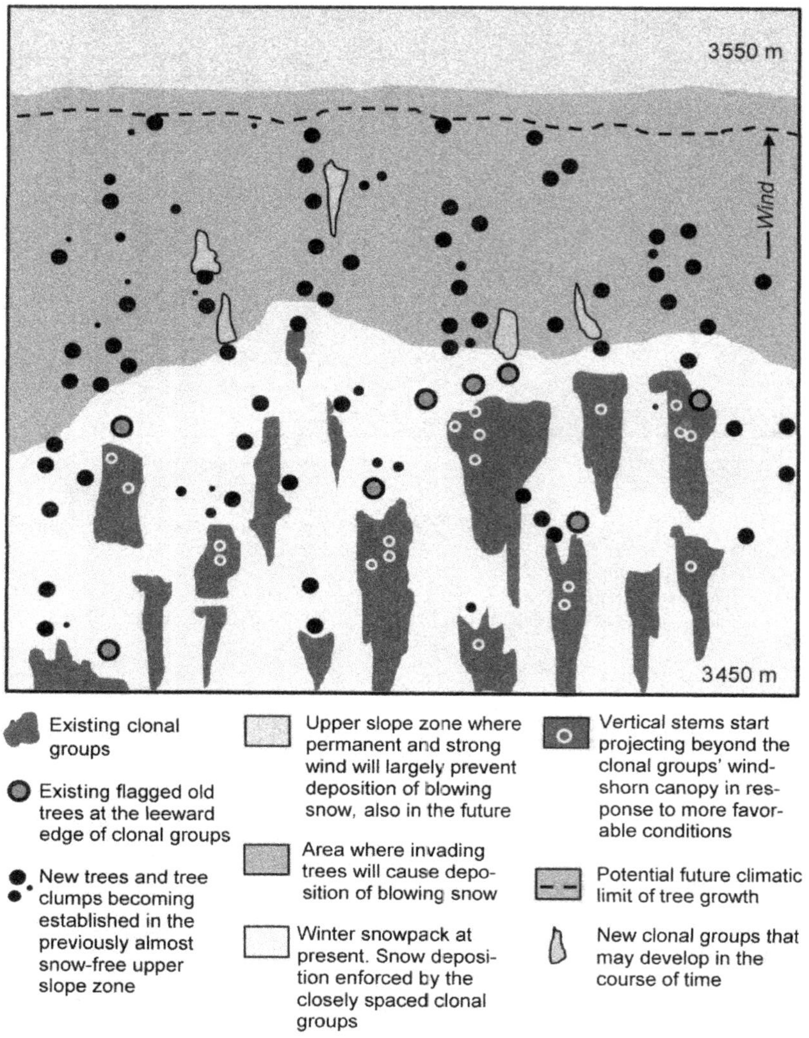

Fig. 8. Schematic representation of possible spatial dynamics in the tree-line ecotone with special regard to snow pattern. On wind-exposed slopes, where the winter snow cover would normally be very patchy or even absent, relatively large areas are buried under a deep and late-lying snowpack as existing tree groups and new trees reduce wind velocity and increase surface roughness

species that are intolerant of severe wind impact, deep frost and desiccation, such as *Vaccinium myrtillus*. On the other hand, dieback of snow-trapping multi-stemmed birch clusters exposes plants or parts of plants in the understory to severe winter injuries, in particular if terrain is wind-exposed. Thus, after mountain birch had declined due to mass-outbreak of the autumnal moth *Epirrita autumnata* in the mid-1960s (Kallio & Lehtonen 1973, 1975, Holtmeier 1974) common juniper *Juniperus communis*, previously well protected by the winter snowpack in the birch stands, was exposed to severe damage by frost, desiccation and ice particle abrasion (Holtmeier 2005). In general, plants buried under snow are less tolerant of winter-desiccation than plants projecting above the winter snowpack (e.g. Havranek 1993, Neuner 2007).

3. CLIMATE CHANGE AND FEEDBACK EFFECTS OF TREE GROUPS

Climate warming may promote establishment of trees in the treeline ecotone possibly up to a new upper physiological limit of tree growth (Fig. 8) (see also Mathisen 2013, Hofgaard et al. 2013). However setbacks due to extreme events may occur. For example, strong frosts early in the growing season and in summer (e.g. Wardle 1968, Tranquillini 1979, Kauhanen 1987), lasting summer drought periods, extremely snow-rich winters or lack of snow could occur that can adversely affect seedling establishment and performance. Seedling recruitment may also fail, for example due to scarce availability of safe microsites (e.g. Resler et al. 2005, Ninot et al. 2008, Batllori et al. 2009).

When considering feedback effects of tree groups, changes in winter climatic conditions are of particular interest. Observations and model-based predictions, however, show great regional variation due to climate regime and elevation (Brown & Mote 2009, Räisänen & Eklund 2012, Mountain Research Initiative EDW Working Group 2015). Thus, in Northern Europe, the number of days with fresh snow is projected to decrease most in the coastal regions and least in the mountain areas. Winter precipitation in northern Europe is projected to decrease by 20 to 30% (Lehtonen et al. 2013). The snowpack season decreased by ~10 d between 2002 and 2011. In the central Alps, precipitation is not expected to change a great deal at high elevations during the coming decades. The number of days with fresh snow, however, may be reduced by about 3 wk (MeteoSchweiz 2013). An increase in snowfall is possible in the North American cordillera, while there is little evidence of increasing snow in northern

Canada (Brown & Mote 2009). On the other hand, in the high Colorado Rocky Mountains, the timing of snowmelt at high elevations has shifted toward earlier in the year (in March or April) (Clow 2010). Under these changing climatic conditions, the feedback mechanisms themselves will not change fundamentally; however, the effects on the treeline environment may be modified.

On wind-swept topography, infilling of gaps between existing clonal groups and increasing release of vertical stems from their wind-shorn compact surface (e.g. Caccianiga & Payette 2006, Devi et al. 2008, Kullman 2015) will enhance deposition of blowing snow (cf. Fig. 4). This might be the most important effect of climate change in the ecotone. More wet snow, however, would reduce wind-mediated snow relocation.

More snow is expected to protect potential young growth from adverse climatic factors. It has been argued that at the alpine treeline of white spruce *Picea glauca*, winter conditions rather than summer temperatures are critical for seedling establishment (Renard et al. 2016). Moisture supply from melting snow might facilitate germination of seeds and support tree seedlings as well as ground vegetation. This may be of particular importance on dry southern exposures (see also Elliott & Kipfmueller 2010, Elliott & Cowell 2015). Clonal groups and tree clusters as well may also facilitate seed-based establishment of other species such as pine, spruce, fir and mountain birch (in northern Europe), at their lee side where these are protected from prevailing winds. On the other hand, late-lying snowpack will shorten the growing season, delay rise of soil temperature and increase the risk of potential seedlings being destroyed by settling of wet snow and/or by parasitic snow fungi (cf. Fig. 4). Moreover, increase of stem breakage and tension cracks due to heavy snow loads, snow creep and wind pressure is likely, while layering will not be affected. Some clonal groups, however, may die back, mainly due to recurrent damage by heavy snow loads. Seedling establishment, however, is not always linked to the presence of clonal and other compact tree groups. Microgeomorphic structures such as solifluction-terrace risers, large stones and boulders, shallow depressions, grass tussocks, cushion plants and dwarf shrubs may also facilitate seedlings (Holtmeier & Broll 1992, 2010, 2012, Holtmeier 1996, Butler et al. 2009, Malanson et al. 2009, Elliott & Kipfmueller 2010).

Geomorphic micro-structures, in particular wind scarps that have developed at the wind-exposed edge and lateral sides of migrating clonal groups may gradually become disguised by expanding grasses, shrubs and herbaceous vegetation (cf. Fig. 8). Moreover, the feedbacks from a single clonal group will become less obvious when new trees become established within the existing gaps between clonal groups. Clonal groups may also expand and increase in height, mainly due to emergence of vertical stems (e.g. Kullman 1979, 1986, 2007, Holtmeier 1985b, Lescop-Sinclair & Payette 1995).

Due to the changing spatial structures and surface properties (roughness, albedo) the entire ecotone will change. However, heating of the lower atmosphere, suggested as a consequence of the northward migration of the polar treeline ecotone (e.g. Sturm et al. 2005), has been less than expected (Chapin et al. 2005). In rugged high-mountains, the lower alpine area with the potential to become covered with forest and tree stands is usually too small for such an effect to occur (Holtmeier 2009, Holtmeier & Broll 2009, 2012, Mathisen 2013, Mathisen et al. 2014), compared to rolling subarctic/arctic landscapes where the polar treeline ecotone may shift northward hundreds of kilometers (Callaghan et al. 2002a, Callaghan et al. 2002b, Harding et al. 2002).

However, positive feedback effects of expanding mountain birch forest on warming as a result of reduced snow cover and lower albedo were found in south-central Norway (de Wit et al. 2014). In the Swiss Alps, the upward shift of the climatic snowline is expected to cause a significant heating as the exposed ground will absorb incoming solar radiation and warm up during a longer snow-free season (Pepin et al. 2015).

Infilling of the gaps between clonal groups and tree clusters will also bring about changes in biodiversity (see also Grace et al. 2002, Hofgaard et al. 2012). While subalpine species may profit from deeper snowpack, snow-intolerant species will only survive on wind-swept topography where snowpack is absent or episodic in winter. In better wind-protected sites, diversity generally decreases in response to increasing tree cover (e.g. Malanson et al. 2007). In the long term, the new trees that can establish themselves beyond the present upper limit of clonal groups (cf. Fig. 8) might reproduce by layering and give rise to clonal groups that may persist for centuries or longer, even if climate deteriorates again. Thus, these groups would contribute to stabilization of the future treeline ecotone. New clonal groups of mountain birch and multi-stemmed mountain birch sprouting from the rootstock (basal shoots) could have a similar effect.

4. CONCLUSIONS

Compact clonal groups and tree clusters influence their sites and those in close vicinity mainly by their effects on direct solar radiation and on ground-level wind-flow. The black-body effect occasionally facilitates establishment of seedlings at the sun-exposed side of trees and tree groups.

Small-scale wind-mediated relocation of snow is a key factor in the treeline ecotones considered. Tree clusters can accumulate large quantities of drifting snow. Compact clonal groups produce a long and deep snowdrift behind their leeward edge. Increased snow accumulation may have both positive and negative effects on site conditions, seedling establishment and survival of young trees. Deep snow prevents soils from freezing, provided that the soil was not frozen already before the snow accumulated. Increased soil moisture mitigates drought stress at the beginning of the growing season. On the other hand, infiltrating meltwater and evaporation delay warming of the soil. Late-laying snowdrifts shorten the growing season and may impair tree seedlings that establish themselves in the snowdrift site. In addition they may be affected by snow fungi and mechanical damage caused by settling snow.

The relative effects of clonal groups and compact tree clusters vary depending on local conditions. Thus, in dry climates and on southern aspects, increased moisture supply as a result of snow deposition caused by the tree groups appears to be more important than on northern exposures, which are usually rich in snow anyway (see also Elliott & Kipfmueller 2010).

Locally, trapped eolian dust (alpine loess) and wind-blown organic particles both play an important role. Eolian dust influences soil nutrient content, soil acidity, water-storing capacity, heat conductivity and decomposition. Downwind 'migrating' clonal groups may lastingly influence soils and leave micro-geomorphic structures in the ecotone. No comparable effects of tree clusters can be found.

Climate change will bring about far-reaching changes in treeline spatial structures, dynamics, and in biodiversity. On terrain currently covered with clonal groups, the infilling of existing gaps with trees and tree clusters might be the most important effect (e.g. Harding et al. 2002). Development of new clonal groups at higher elevation and beyond the northern treeline will increase the persistence of the future treeline as they reduce wind velocity and increase deposition of drifting snow and may also act as a new source of viable seeds. On the other hand, wildfires, which are likely to increase under warmer and drier conditions, may adversely affect clonal groups, tree clusters and young trees occupying gaps in the present treeline ecotone, even at sites where wildfires are absent under current conditions.

Warming of the lower atmosphere might result from a poleward shift of the northern coniferous treeline. However such an effect cannot be expected in rugged high-mountains, because on steep slopes the lower alpine area with potential to become covered with forest and tree stands is usually of too small a scale to have this effect. The retreating climatic snowline is more likely to contribute to a warming of the lower atmosphere. Conditions may be different on undulating smooth topography as, for example, in northern Finland and in many areas of the Rocky Mountains, where the treeline ecotone often extends over old uplifted land surfaces. However, since a multitude of regionally and locally varying factors are involved, prognoses of the possible influence of increasing tree populations on warming of the lower atmosphere are uncertain.

Acknowledgements. We are grateful to the COST action SENSFOR (Enhancing the resilience capacity of SENSitive mountain FORest ecosystems under environmental change) for support. Our thanks go to Professor R. M. M. Crawford (St. Andrews University) for revising the English text.

LITERATURE CITED

Alftine KJ, Malanson GP (2004) Directional positive feedback and pattern at an alpine tree line. J Veg Sci 15:3–12
Anschlag K, Broll G, Holtmeier FK (2008) Mountain birch seedlings in the treeline ecotone, Subarctic Finland. Variation in above- and belowground growth in relation to microtopography. Arct Antarct Alp Res 40:609–616
Aulitzky H (1961) Die Bodentemperaturen in der Kampfzone oberhalb der Waldgrenze und im subalpinen Zirben-Lärchenwald. Mitteilungen der Forstlichen Bundes-Versuchsanstalt Mariabrunn 59:155–208
Barbeito I, Brücker RL, Rixen C, Bebi P (2013) Snow fungi-induced mortality of *Pinus cembra* at the alpine treeline: evidence from plantations. Arct Antarct Alp Res 45: 455–470
Batllori E, Camarero JJ, Ninot JM, Gutiérrez E (2009) Seedling recruitment, survival and facilitation in alpine *Pinus uncinata* tree line ecotones. Implications and potential responses to climate warming. Glob Ecol Biogeogr 18:460–472
Bekker MF (2005) Positive feedbacks between tree establishment and patterns of subalpine forest advancement, Glacier National Park, Montana, USA. Arct Antarct Alp Res 37:97–107
Benedict JB (1984) Rates of tree-island migration, Colorado Rocky Mountains. Ecology 65:820–823
Blanken PD (2009) Designing a living snow fence for snow drift control. Arct Antarct Alp Res 41:418–425

Brink VC (1959) A directional change in the subalpine forest-heath ecotone in Garibaldi-Park, British Columbia. Ecology 40:10–16

Brodersen CR, Germino MJ, Smith WK (2006) Photosynthesis during an episodic drought in *Abies lasiocarpa* and *Picea engelmannii* across an alpine treeline. Arct Antarct Alp Res 38:34–41

Broll G, Holtmeier FK (1994) Die Entwicklung von Kleinreliefstrukturen im Waldgrenzökoton der Front Range (Colorado, USA) unter dem Einfluß leewärts wandernder Ablegergruppen (*Picea engelmannii* und *Abies lasiocarpa*). Erdkunde 48:48–59

Brooks PD, Williams MW, Walker DA, Schmidt SK (1995) The Niwot Ridge snow fence experiment: biogeochemical response to changes in seasonal snowpack. Biogeochemistry of seasonally snow-covered catchments. In: Tonessen KA, Williams MW, Tranter M (eds) Biogeochemistry of seasonally snow-covered catchments. Proc Symp, Boulder 1995. Publication 228, International Association of Hydrological Sciences (IAHS), Wallingford, p 293–302

Brown RD, Mote PW (2009) The response of northern hemisphere snow cover to a changing climate. J Clim 22: 2124–2145

Butin H (2011) Krankheiten der Wald- und Parkbäume. Diagnose – Biologie – Bekämpfung, 4th edn. Verlag Eugen Ulmer, Stuttgart

Butler DR, Malanson GP, Resler LM, Walsh SJ, Wilkerson FD, Schmid GL, Sayer CF (2009) Geomorphic patterns and processes at alpine treeline. In: Butler DR, Malanson GP, Walsh SF, Fagre DB (eds) The changing alpine timberline: the example of Glacier National Park, MT, USA. Elsevier, Amsterdam, p 62–84

Bylund H (1999) Climate and the population dynamics of two insect outbreak species in the north. In: Hofgaard A, Ball JP, Danell K, Callaghan T (eds) Animal responses to global change in the north. Ecol Bull 47: 54-62

Caccianiga M, Payette S (2006) Recent advance of white spruce (*Picea glauca*) in the coastal tundra of the eastern shore of Hudson Bay (Québec, Canada). J Biogeogr 33: 2120–2135

Callaghan TV, Crawford RMM, Eronen M, Hofgaard A and others (2002a) The dynamics of the tundra-taiga boundary: an overview and suggested coordination and integrated approach to research. Ambio 12:3–5

Callaghan TV, Werkman BR, Crawford RMM (2002b) The tundra–taiga interface and its dynamics: concepts and applications. Ambio 12:6–14

Callaway RM, Brooker RW, Choler P, Kikvidzer Z and others (2002) Positive interactions among alpine plants increase with stress. Nature 417:844–848

Chapin FS, Sturm M, Serreze MC, McFadden JP and others (2005) Role of land-surface changes in arctic summer warming. Science 310:657–660

Clow DW (2010) Changes in timing of snowmelt and streamflow in Colorado: a response to recent warming. J Clim 23:2293–2306

Cunningham C, Zimmermann NE, Stoeckli V, Bugmann H (2006) Growth response of Norway spruce saplings in two forest gaps in the Swiss Alps to artificial browsing, infection with black snow mold and competition by ground vegetation. Can J Forest Res 36:2782–2793

Daly C (1984) Snow distribution pattern in the alpine krummholz zone. Prog Phys Geogr 8:157–175

de Wit HA, Bryn A, Hofgaard A, Karstensen J, Kvalevåg MM, Peters GP (2014) Climate warming feedback from mountain birch forests expansion: reduced albedo dominates carbon uptake. Glob Change Biol 20:2344–2355

Devi N, Hagedorn F, Moiseev P, Bugmann H, Shiyatov S, Mazepa V, Rigling A (2008) Expanding forest and changing growth forms of Siberian larch at the polar Urals treeline during the 20th century. Glob Change Biol 14: 1581–1591

Dyunin AK (1967) Fundamentals of the mechanics of snow storms. In: Ohmura (ed) Physics of snow and ice. Proc Conf Physics of Snow and Ice. II. Cryobiology, Sapporo, 14–19 Aug 1966. Institute of Low Temperature Science, Hokkaido University, p 1965–1173

Elliott GP (2012) The role of thresholds and fine-scale processes in driving upper treeline dynamics in the Bighorn Mountains, Wyoming. Phys Geogr 33:129–145

Elliott GP & Cowell CM (2015) Slope aspect mediates fine-scale tree establishment patterns at upper treeline during wet and dry periods of the 20th Century. Arct Antarct Alp Res 478:679–690

Elliott GP, Kipfmueller KF (2010) Multi-scale influences of slope aspect and spatial pattern on ecotonal dynamics at upper treeline in the Southern Rocky Mountains, USA. Arct Antarct Alp Res 42:45–56

Franklin JF, Dyrness CT (1973) Natural vegetation of Oregon and Washington. General Technical Report PNW-8, USDA Forest Service, Portland, OR

Germino MJ, Smith WK (1999) Sky exposure, crown architecture, and low-temperature photoinhibition in conifer seedlings at alpine treeline. Plant Cell Environ 22: 407–415

Germino MJ, Smith WK, Resor AC (2002) Conifer seedling distribution and survival in an alpine treeline ectone. Plant Ecol 162:157–168

Grace J, Berninger F, Nagy L (2002) Impacts of climate change on tree line. Ann Bot 90:537–544

Habeck JR (1969) A gradient analysis of a timberline zone at Logan Pass, Glacier Park, Montana. Northwest Sci 43: 65–73

Hadley JL, Smith WK (1986) Wind effects on needles of timberline conifers: seasonal influences on mortality. Ecology 67:12–19

Harding R, Khury P, Christensen TR, Sykes MT, Dankers R, Van der Linden S (2002) Climate feedbacks at the tundra–taiga interface. Ambio 12:47–55

Harsch MA, Hulme PE, McGlone MS, Duncan RP (2009) Are treelines advancing? A global meta-analysis of treeline response to climate warming. Ecol Lett 12:1040–1049

Havranek WM (1993) The significance of frost and frost-drought for the alpine timberline. In: Anfodillo T, Urbinati C (eds) Ecologia delle Foreste di Alta Quota. Proc XXX Corso di Cultura in Ecologia. Università di Padova, p 115–117

Havranek W, Tranquillini W (1995) Physiological processes during winter dormancy and their ecological significance. In Smith WK, Hinckley TM (eds) Ecophysiology of forests. Academic Press, New York, NY, p 95–124

Hiemstra CA, Liston GE, Reiners WA (2002) Snow redistribution by wind and interactions with vegetation at upper treeline in the Medicine Bow Mountains, Wyoming, USA. Arct Antarct Alp Res 34:262–273

Hiemstra CA, Liston GE, Reiners WA (2006) Observing, modelling and validating snow redistribution in a Wyoming upper treeline landscape. Ecol Modell 197:35–51

Hinkel KM, Hurd JK (2006) Permafrost destabilization and thermokarst following snow fence installation, Barrow, Alaska, USA. Arctic Antarct Alp Res 38:530–539

Hoch G, Körner C (2003) The carbon charging of pines at the climatic treeline: a global comparison. Oecologia 135:

10–21
Hofgaard A, Harper A, Golubeva E (2012) The role of circum-arctic forest–tundra ecotone for Arctic biodiversity. Biodiversity 13:174–181
Hofgaard A, Tömmervik H, Rees G, Hanssen F (2013) Latitudinal forests advance in northernmost Norway since the early 20th century. J Biogeogr 40:938–949
Holtmeier FK (1978) Die bodennnahen Winde in den Hochlagen der Indian Peaks Section, Colorado Front Range. Münstersche Geographische Arbeiten 3:3–47
Holtmeier FK (1974) Geoökologische Beobachtungen und Studien an der subarktischen und alpinen Waldgrenze in vergleichender Sicht (nördliches Fennoskandien/Zentralalpen). Erdwissenschaftliche Forschung 8, Steiner Verlag, Wiesbaden
Holtmeier FK (1980) Influence of wind on tree-physiognomy at the upper timberline in the Colorado Front Range. In Benecke U, Davis MR (eds) Mountain environments and subalpine tree growth. Forest Research Insitute Technical Paper 70, New Zealand Forest Service, Wellington, p 247–251
Holtmeier FK (1985a) Die klimatische Waldgrenze — Linie oder Übergangssaum (Ökoton)? Ein Diskussionsbeitrag unter besonderer Berücksichtigung der Waldgrenze in den mittleren und hohen Breiten der Nordhalbkugel. Erdkunde 39:271–285
Holtmeier FK (1985b) Climatic stress influencing the physiognomy of trees at the polar and mountain timberline. Eidgenössische Anstalt für das forstliche Versuchswesen, Berichte 270:31–40
Holtmeier FK (1987) Beobachtungen und Untersuchungen über den Ausaperungsverlauf und einige Folgewirkungen in 'ribbon-forests' und an der oberen Waldgrenze in der Front Range, Colorado. Phytocoenologia 15:373–396
Holtmeier F-K (1996) Die Wirkungen des Windes in der subalpinen und alpinen Stufe der Front Range, Colorado, USA. Arbeiten aus dem Institut für Landschaftsökologie 1:15–45
Holtmeier FK (1999) Ablegerbildung im Hochlagenwald und an der oberen Waldgrenze in der Front Range, Colorado. Mitt Deutsch Dendrol Ges 84:39–61
Holtmeier F-K, Broll G, Müterthies A, Anschlag K (2003) Regeneration of trees in the treeline ecotone, northern Finnish Lapland. Fennia 181:103–128
Holtmeier FK (2005) Relocation of snow and its effects in the treeline ecotone — with special regard to the Rocky Mountains, the Alps and Northern Europe. Erde 136:343–373
Holtmeier FK (2009) Mountain timberlines: ecology, patchiness, and dynamics, 2nd edn. Advances in Global Change Research 36, Springer, Dordrecht
Holtmeier FK, Broll G (1992) The influence of tree islands on microtopography and pedoecological conditions in the forest-alpine tundra on Niwot Ridge, Colorado Front Range, USA. Arct Alp Res 24:216–228
Holtmeier FK, Broll G (2009) Altitudinal and polar treelines in the northern hemisphere — causes and response to climate change. Polarforschung 79:139–153
Holtmeier FK, Broll G (2010) Wind as an ecological agent at treelines in North America, the Alps, and the European Subarctic. Phys Geogr 31:203–233
Holtmeier F-K, Broll G (2011) Response of Scots Pine (*Pinus sylvestris*) to warming climate at its altitudinal limit in northernmost subarctic Finland. Arctic 64: 269–280
Holtmeier FK, Broll G (2012) Landform influences on treeline patchiness and dynamics in a changing climate. Phys Geogr 33:403–437
Holtmeier FK, Broll G, Anschlag K (2004) Winderosion und ihre Folgen im Waldgrenzbereich und in der alpinen Stufe einiger nordfinnischer Fjelle. Geo-öko 25:203–224
Humphries HC, Bourgeon PS, Mujica-Crapanzano LR (2008) The spatial patterns and environmental relationships in the forest-alpine tundra ecotone at Niwot Ridge, Colorado, USA. Ecol Res 23:589–605
Johnson DM, Germino MJ, Smith WK (2004) Abiotic factors limiting photosynthesis in *Abies lasiocarpa* and *Picea engelmannii* seedlings below and above the alpine timberline. Tree Physiol 24:377–386
Kallio P, Lehtonen J (1973) Birch forest damage by *Oporinia autumnata* (Bkh.) in 1965–1966 in Utsjoki, N Finland. Rep Kevo Subarct Res Stn 10:55–69
Kallio P, Lehtonen J (1975) On the ecocatastrophe of birch forest caused by *Oporinia autumnata* (Bkh.) and the problem of reforestation. In: Wielgolaski FE (ed) Fennoscandian tundra ecosystems. 2. Animals and system analysis. Ecological Studies 17, Springer, Berlin, p 174–180
Karlsson PS, Nordell O (1996) Effects of soil temperature on the nitrogen economy and growth of mountain birch seedlings near its presumed low temperature distribution limit. Ecoscience 3:183–189
Kauhanen H (1987) On growth problems of mountain birch near its distributional limits. In: Alexanderson H, Holmgren B (eds) Climatological extremes in mountains. UNGI (Uppsala universitet, Naturgeogrfiska institutionen) 65:183–190
Kharuk VI, Ranson KJ, Im ST, Vdovin AS (2010) Spatial distribution and temporal dynamics of high elevation forest stands in southern Siberia. Glob Ecol Biogeogr 19:822–830
Kobayashi D (1972) Studies of transport in low level drifting snow. Contributions from the Institute of Low Temperature Science, Serie A 24:1–58
Kramer PJ (1982) Water relations of plants. Academic press, New York, NY
Kullman L (1979) Change and stability in the altitude of the birch tree-limit in the southern Swedish Scandes 1915–1975. Acta Phytogeographica Suecica 65, Uppsala
Kullman L (1986) Recent tree-limit history of *Picea abies* in the southern Swedish Scandes. Can J For Res 16: 761–771
Kullman L (2007) Modern climate change and shifting ecological states of the subalpine/alpine landscape in the Swedish Scandes. Geo-öko 28:187–221
Kullman L (2012) The alpine treeline ecotone in the southwestern Swedish Scandes: dynamics on different scales. In Myster RW (ed) Ecotones between forest and grassland. Springer, New York, NY, p 271–298
Kullman L (2015) Norway spruce (*Picea abies* (L.) Karst.). Treeline performance since the mid-1970s in the Swedish Scandes — evidence of stability and minor change from repat surveys and photography. Geoöko 36: 25-53
Larcher W (1980) Klimastress im Gebirge — Adaptionstraining und Selektionsfilter für Pflanzen. Rheinisch-Westfälische Akademie der Wissenschaften. Rheinisch-Westfälische Akad Wiss Vorträge 291:49–80
Lehtonen I, Venäläinen A, Ikonen J, Puttonen N, Gregory H (2013) Some features of winter climate in northern Fennoscandia. Finnish Meteorological Institute Reports 2013 (3), Unigrafia, Helsinki
Lescop-Sinclair PS, Payette S (1995) Recent advance of the arctic treeline along the eastern coast of Hudson Bay. J Ecol 83:929–936
Liston GE, McFadden JP, Sturm N, Pielke RA Sr (2002) Modelled changes in arctic tundra snow, energy and moisture fluxes due to increased shrubs. Glob Change Biol 8: 17–32

Litaor MI (1987) The influence of eolian dust on the genesis of alpine soils in the Front Range, Colorado. Soil Sci Soc Am J 51:142–147

Maher EL, Germino MJ (2006) Microsite differentiation among conifer species during seedling establishment at alpine treeline. Bioscience 13:334–341

Malanson GP, Butler DR, Fagre DB, Walsh SJ and others (2007) Alpine treeline of western North America: linking organism-to-landscape dynamics. Phys Geogr 28: 378–396

Malanson GP, Brown DG, Butler DR, Cairns DM, Fagre DB, Walsh SJ (2009) Ecotone dynamics: invisibility of alpine tundra. In Butler DR, Malanson GP, Walsh SF, Fagre DB (eds) The changing alpine timberline: the example of Glacier National Park, MT, USA. Elsevier, Amsterdam, p 35–61

Marr JW (1966) The development and movement of tree islands near the upper limit of tree growth in the southern Rocky Mountains. Ecology 58:1159–1164

Mathisen EI (2013) Structure, dynamic sand regeneration capacity at the subarctic forest-tundra ecotone of northern Norway and Kola Peninsula, NW Russia. Doctoral thesis, Norwegian University of Science and Technology, Trondheim

MeteoSchweiz (2013) Klimaszenarien Schweiz: eine regionale Übersicht. Fachbericht 243, MeteoSchweiz, Zürich

Mountain Research Initiative EDW Working Group (2015) Elevation-dependent warming in mountain regions of the world. Nat Clim Change 5:424–430

Neuner G (2007) Frost resistance at the upper timberline. In: Wieser G, Tausz M (eds) Trees at their upper limit: treelife limitation at the upper timberline. Plant Ecophysiology 5, Springer, Dordrecht, p 171–180

Neuvonen S, Wielgolaski FE (2005) Herbivory in northern birch forests. In: Wielgolaski FE (ed) Plant ecology, herbivory, and human impact in Nordic mountain birch forests. Springer, Berlin p 183–189

Nierhaus-Wunderwald D (1996) Pilzkrankheiten in Hochlagen, Biologie und Befallsmerkmale. Wald und Holz 10: 18–24

Ninot JM, Batllori E, Carillo E, Carreras J, Ferré A, Gutiérrez E (2008) Timberline structure and limited tree recruitment in the Catalan Pyrenees. Plant Ecol Divers 1:47–51

Öberg L, Kullman L (2011) Ancient subalpine clonal spruces (*Picea abies*): sources of postglacial vegetation history in the Swedish Scandes. Arctic 64:183–196

Oke TR (1987) Boundary layer climates, 2nd edn. Routledge, London

Pepin N, Bradley RS, Diaz HF, Baraer M and others (2015) Elevation-dependent warming in mountain regions of the world. Nat Clim Change 5:424–430

Räisänen J, Eklund J (2012) 21st century changes in snow climate in Northern Europe: a high resolution view from ENSEMBLES regional climate models. Clim Dyn 38:2575

Renard SM, McIntire EJB, Fajrado A (2016) Winter conditions — not summer temperature — influence establishment of seedlings at white spruce alpine treeline in Eastern Quebec. J Veg Sci 27:29–39

Resler LM (2006) Geomorphic controls of spatial pattern and process at alpine treeline. Prof Geogr 58:124–138

Resler LM, Butler DR, Malanson GP (2005) Topographic shelter and conifer establishment and mortality in an alpine environment, Glacier National Park, Montana. Phys Geogr 26:112–125

Roll-Hansen F (1989) *Phacidium infestans*: a literature review. Eur J Forest Pathol 19:237–250

Schumacher S, Bugmann H (2006) The relative importance of climatic effects, wildfires and the management for future forest landscape dynamics in the Swiss Alps. Glob Change Biol 12:1435–1450

Senn J (1999) Tree mortality caused by *Gremeniella abietina* in subalpine afforestations in the central Alps and its relationship with duration of snow cover. Eur J Forest Pathol 29:65–75

Shi P, Körner C, Hoch G (2006) End of season carbon supply status of woody species near the treeline in western China. Basic Appl Ecol 7:370–377

Skre O (1993) Growth of mountain birch (*Betula pubescens* Ehrh.) in response to changing temperature. In Alden J, Mastrantonio JL, Ødum S (eds) Forest development in cold climates. Plenum Press, New York, NY, p 65–78

Smith WK, Germino MJ, Hancock TE, Johnson DM (2003) Another perspective on altitudinal limits of alpine timberlines. Tree Physiol 23:1101–1112

Stöcklin J, Körner C (1999) Recruitment and mortality of *Pinus sylvestris* near the Nordic treeline: the role of climatic change and herbivory. Ecol Bull 47:168–177

Streule A, Häsler R (2006) Windschutz für junge Bäume in subalpinen Aufforstungen an stark windexponierten Wuchsorten 'Pru dal Vent' (Alp Grüm, GR). Eidgenössische Anstalt für Wald, Schnee und Landschaft WSL, Birmensdorf

Sturm K, Douglas T, Racine R, Liston G (2005) Changing snow and shrub conditions affect albedo with global implications. J Geophys Res 110:1–3

Sveinbjörnsson B, Nordell O, Hauhanen H (1992) Nutrient relations of mountain birch growth at and below the elevational tree line in Swedish Lapland. Funct Ecol 6: 213–220

Tape K, Sturm M, Racine C (2006) The evidence for shrub expansion in Northern Alaska and the Pan-Arctic. Glob Change Biol 12:686–702

Thorn CE, Darmody RG (1985) Grain-size distribution of the insoluble compontents of contemporary eolian deposits in the alpine zone, Front Range, Colorado. Arct Alp Res 17:433–442

Tranquillini W (1957) Standortsklima, Wasserbilanz und CO_2-Gaswechsel junger Zirben (*Pinus cembra* L.) an der alpinen Waldgrenze. Planta 49:612–661

Tranquillini W (1979) Physiological ecology of the alpine timberline: tree existence at high altitudes with special reference to the European Alps. Ecological Studies 31, Springer, Berlin

Turner H, Streule A (1983) Wurzelwachstum und Sprossentwicklung junger Koniferen im Klimastress der alpinen Waldgrenze, mit Berücksichtigung von Mikroklima, Photosynthese und Stoffproduktion. In: Böhm W et al. (eds) Wurzelökologie und ihre Anwendung. Internationales Symposium Gumpenstein 1982. Bundesanstalt Gumpenstein, Irdning, p 617–635

Wardle P (1968) Engelmann spruce (*Picea Engelmannii* ENGEL) at its upper limit in the Front Range, Colorado. Ecology 49:483–495

Weiss DJ, Malanson GP, Walsh SJ (2015) Multiscale relationships between alpine treeline elevation and hypothesized environmental controls in the western United States. Ann Assoc Am Geogr 105:437–453

Zeng Y, Malanson GP (2006) Endogenous fractal dynamics at alpine treeline ecotones. Geogr Anal 38:271–287

12

Forests dynamics in the montane–alpine boundary: a comparative study using satellite imagery and climate data

Lucian Dinca[1], Mihai Daniel Nita[2,*], Annika Hofgaard[3], Concepcion L. Alados[4], Gabriele Broll[5], Stelian Alexandru Borz[2], Bogdan Wertz[6], Antonio T. Monteiro[7]

[1]National Forest Research-Development Institute, Cloşca 13, Braşov, 500040, Romania
[2]Faculty of Silviculture and Forest Engineering, Transilvania University of Brasov, Ludwig van Beethoven 1, Braşov, 500123, Romania
[3]Norwegian Institute for Nature Research, Høgskoleringen 9, 7034 Trondheim, Norway
[4]Pyrenean Institute of Ecology (CSIC), Av. Ntra. Sra. de la Victoria, Huesca, 22700, Spain
[5]University of Osnabrück, Neuer Graben, 49074 Osnabrück, Germany
[6]University of Agriculture, Faculty of Forestry, Adama Mickiewicza 21, Kraków, 30-001, Poland
[7]Predictive Ecology Group, Research Center on Biodiversity and Genetic Resources, CIBIO-InBIO, Rua Padre Armando Quintas 7, Vairão, 4485-661, Portugal

ABSTRACT: Over the past decades, the altitudinal and latitudinal advance of forest lines has increased due to global warming and the abandonment of less productive areas previously subject to agricultural activities. The intensity and speed of the forest line advance also depend on numerous physical, biological and human factors that are region-specific. It is important to fully understand the mechanisms behind forest line behaviour, as existing studies do not report global figures. We selected 4 study areas in which to analyse the temporal and spatial behaviour of the forest line and of forest cover based on selection criteria such as minimal human interference and maximal representativeness at the European level. The sites were located in national parks that were evenly spread across some of the dominant European mountain ranges such as the Pyrenees, Alps and Carpathians, at comparable altitudes and latitudes, and with similar land cover proportions in the year 1970. Methodologically, we used cloud-free Landsat satellite images that were acquired in the same month during the growing season. A post-classification comparison technique, using all bands but the thermal one, was implemented to evaluate forest line behaviour, while the accuracy of image classification was evaluated by random sampling. Four time frames were used to evaluate forest cover behaviour in relation to the non-forested areas: 1971–1980, 1981–1990, 1991–2000 and 2001–2014. Also, climate and topography data were included in this study, which enabled comparison and computation of dependence relations. Our results indicate significant differences between the analysed areas. For instance, for the same reference period (1981–1990), the greatest differences in terms of forest cover change were specific to the Austrian Alps (28%), whereas the lowest differences were those from the Spanish Pyrenees (1%). Similar forest line shifts were found in the Austrian Alps and in the Romanian Carpathians, whereas the lowest altitudinal advancement was specific to the Spanish Pyrenees. According to this study, the temperature trend could have significantly influenced tree line behaviour.

KEY WORDS: Remote sensing · Tree lines · Mountains · National park

1. INTRODUCTION

Forested areas are increasing in many developed countries due to the combined effects of more intensive agriculture, abandonment of less productive areas and increasing awareness regarding the environmental importance of forests (Müller et al. 2009, Kozak 2010, Lambin & Meyfroidt 2010, Baumann et

*Corresponding author: mihai.nita@unitbv.ro

§Advance View was available online April 13, 2017

al. 2011). In mountainous afforested areas, the dynamics in the montane–alpine belts, namely in the tree and forest lines (limit of uppermost >2 m tall trees, and limit of continuous forest [canopy cover of >10%], respectively), are particularly relevant. Here, climate influences ecological processes governing dynamics in forest stands, but human land use has also been a control factor for centuries (Gehrig-Fasel et al. 2007). Owing to the climatic influence on the tree and forest lines, they are regarded as environmental change descriptors. In Europe, changes in these boundaries have been observed (Gehrig-Fasel et al. 2007) and the discussion on whether these changes are related to human land use or climate is ongoing. Therefore, it is in the public interest to monitor forest changes within the montane–alpine belts, and evaluate their relationship with altered climatic features, particularly if multiple sites are considered, allowing comparisons on a continental scale.

In general, information on changes in mountain forest cover is not always publicly accessible. There is still a lack of comprehensive knowledge of spatially and temporally explicit forest cover dynamics, especially across large areas and at sufficient spatial detail to resolve the full range of forest change processes (Griffiths et al. 2014). However, the technology that enables full global monitoring and better comprehension of the various properties of forest resources is evolving, for instance the increasing availability of remote sensing data. Satellite remote sensing is frequently used to assess land ecosystem dynamics, as it provides consistent measurements of bio-geophysical processes, including natural and anthropogenic disturbances (Jin & Sader 2005, Linderman et al. 2005). The synoptic and regular coverage at short intervals of consistent remote sensing data provides a valuable source for updated land cover information necessary to monitor the type and extent of environmental changes (Mas 1999). The Landsat archive is among the most used data sources for studying land cover changes, including deforestation, agricultural expansion and intensification, urban growth, and wetland loss (Coppin & Bauer 1996, Woodcock et al. 2001, Seto et al. 2002, Galford et al. 2008, Gartzia et al. 2014), due to its long record of continuous measurement, spatial resolution and near-nadir observations (Pflugmacher et al. 2012, Wulder et al. 2012). For forests, these data have been providing spatial and spectral details, allowing the capture of forest attributes at adequate scales for analysis since the 1970s (Griffiths et al. 2014). More recently, continental or even global data analyses with Landsat are feasible, due to advances in automated imagery pre-processing and processing methods, and data storage capabilities (Townshend et al. 2012). For instance, Hansen et al. (2008) produced 2 regional Landsat composites that aimed to detect forest changes in the Congo Basin, also integrating the MODIS Vegetation Continuous Field product (Hansen et al. 2003) for classified training purposes. Potapov et al. (2011) focused on boreal forest changes between 2000 and 2005 in European Russia, using composited Landsat data, and independently obtained indicators of forest change. For mountain tree line and forest line monitoring, Wang et al. (2006) proposed a Landsat satellite-based monitoring that considers the temporal response of continuous vegetation indices (e.g. normalised difference vegetation index), while Klinge et al. (2015) combined Landsat image classification and spatial modelling to understand the distribution patterns. Both studies used multi-temporal image analysis. Dynamics of mountain tree lines and forest lines vary across the globe. Previous studies reveal an average northward advance of 156 and 71 m for birch and pine forest lines in Norway during the 20th century, respectively (Hofgaard et al. 2013). Shifts in tree lines were more pronounced in birch (340 m yr^{-1}), whereas pine advance was very limited (10 m yr^{-1}). In Russia's Khibiny Mountains, the mean tree line advance of birch and pine was 29 and 27 m in altitude (0.6 and 0.5 m yr^{-1}), respectively, between 1958 and 2008 (Mathisen et al. 2014). In the Italian Alps, tree lines shifted upwards by 115 m between 1901 and 2000 (Leonelli et al. 2011). Overall, analysis of historical and recent forest delineation data shows a very restricted advance rate compared to the predictions of dynamic global vegetation models (ACIA 2005, Kaplan & New 2006).

Complex interactions between factors affects the patterns of mountain tree lines (e.g. exposure to wind, snow depth, animals, mass outbreak of insects) and forest lines. Tree line responses to climate change are dependent on a multitude of interacting abiotic and biotic drivers in a site-specific manner (Holtmeier & Broll 2005, Holtmeier 2009, Hofgaard et al. 2012), with characteristics determined by multiple variables (Holtmeier & Broll 2007, Hofgaard et al. 2012), dominated by temperature, precipitation, wind and herbivory (Sveinbjörnsson et al. 2002, Cairns & Moen 2004, Holtmeier & Broll 2005). A change in the regeneration capacity in the tree line ecotone has caused its main features, forest line, tree line and species line (limit of tree saplings and seedlings) to move back and forth over time in accordance with long- and short-term climate changes

(Dalen & Hofgaard 2005, Payette 2007, Shiyatov et al. 2007, Holtmeier & Broll 2010, Chhetri & Cairns 2015).

Ameztegui et al. (2015) parameterised and used an individual based, spatially explicit model of forest dynamics (SORTIE-ND) to investigate the role of species-specific differences in juvenile performance induced by climate change (juvenile growth and recruitment ability) in the dynamics of mixed forests located in the montane–subalpine ecotone of the Pyrenees. Batllori & Gutiérrez (2008) found that past and recent synchronous recruitment trends (mid-19th century, second half of the 20th century) were apparent at the tree line over the studied area of the Pyrenean range. Altitudinal treeline ecotone, growth and establishment has been occurring since the 1950s in the context of climatic warming and substantial land-use abandonment; both gradual and step-like transition patterns in tree age and size along the ecotone have been observed. Gehring-Fasel et al. (2007) found a significant increase in forest cover between altitudes of 1650 and 2450 m in the Swiss Alps. Above 1650 m, 10% of the new forest areas were identified as true upward shifts, whereas 90% represented in-growth, and both land use and climate change were identified as likely drivers. Most upward shift activities occurred within a band of 300 m below the potential regional tree line, indicating land use as the most likely driver. Only 4% of the upward shifts were identified to rise above the potential regional tree line, thus indicating climate change. In this context, the present study aimed to (1) quantify forest dynamics in the montane–alpine belts over the past 40 yr, using 4 protected European representatives of mountain forest ecosystems; (2) investigate whether there were significant changes in forest line and forest cover; and (3) understand the relationship between these forest changes and climate modification over the past 40 to 50 yr. For this purpose, multi-temporal analysis with Landsat satellite data was performed. Regression models between forest line and cover, using baseline climate data from 1901 to 1970, were established.

2. MATERIALS AND METHODS

2.1. Study areas

The boundaries between the montane and alpine belts in 4 European mountain areas (Fig. 1) were considered for analysis. The areas shared homogeneous minimum human influence and protected status (in the core areas of national parks), and were located in the Austrian Alps, Slovak and Romanian Carpathians, and Spanish Pyrenees at similar altitudes and comparable un-forested surfaces in 1970 (pastures, alpine hollows, etc.): 19.945 ha (Ordesa, Spain), 18.945 ha (Nockberge, Austria), 10.436 ha (Tatra, Slovakia) and 15.175 ha (Retezat, Romania). The latitudinal range of the selected areas was 4° (between 43 and 47° N), while the general aspect (orientation) was eastern–western.

The Ordesa and Monte Perdido National Park, located in the central Pyrenees (Spain), covers 15 608 ha and was established in 1918. Altitude ranges between 700 and 3355 m above sea level, and Monte Perdido is Europe's highest calcareous peak. Average annual rainfall is 1688 mm, and average maximum and minimum temperatures are 8.7 and 1.5°C, respectively (Góriz Refuge Meteorological Station at 2200 m between 1992 and 2012, both included). Natural vegetation is dominated by coniferous forest *Pinus sylvestris*, beech *Fagus sylvatica* and several oak species at lower altitudes. At higher altitudes, *P. uncinata* forests have been replaced by *Buxus sempervirens*, *Echinospartum horridum* and *Juniperus communis* shrublands, followed by alpine grasslands. At altitudes below 2100 m, grasslands were created by humans to obtain pastures for livestock during summer. Since the Middle Ages, the Pyrenees forest line has been artificially lowered to increase grassland availability for livestock summer pastures (Monserrat-Martí 1992). But since 1930, a marked reduction in agro-pastoral activities in mountain areas has taken place (Alados et al. 2014, Gartzia et al. 2016b), which has also affected the Ordesa and Monte Perdido National Park, where the grazing activity was maintained as part of the conservation programme (Gartzia et al. 2016a). Consequently, we are facing important changes in the land cover in the national park, mainly below the 2100 m, where approximately 30% of shrubland has turned to forest and 12% of grasslands have become shrubland. Above 2100 m, the woody encroachment of grasslands is low (Gartzia et al. 2014).

The Nockberge Biosphere Park (Austria) was established initially as national park in 1987. It covers 18 300 ha and is among the oldest and most interesting upland formations in Europe. Altitude ranges between 600 and 2441 m, with the highest peak at Mt. Eisenhut in Styria (2441 m). The climate is continental dry, due to its central location in the European land mass. Average annual rainfall and temperature are 1100 mm and 7°C, respectively. The region boasts the Eastern Alps' largest pine *P. sylvestris*, larch *Larix decidua* and spruce *Picea abies* forests. Wind throw is

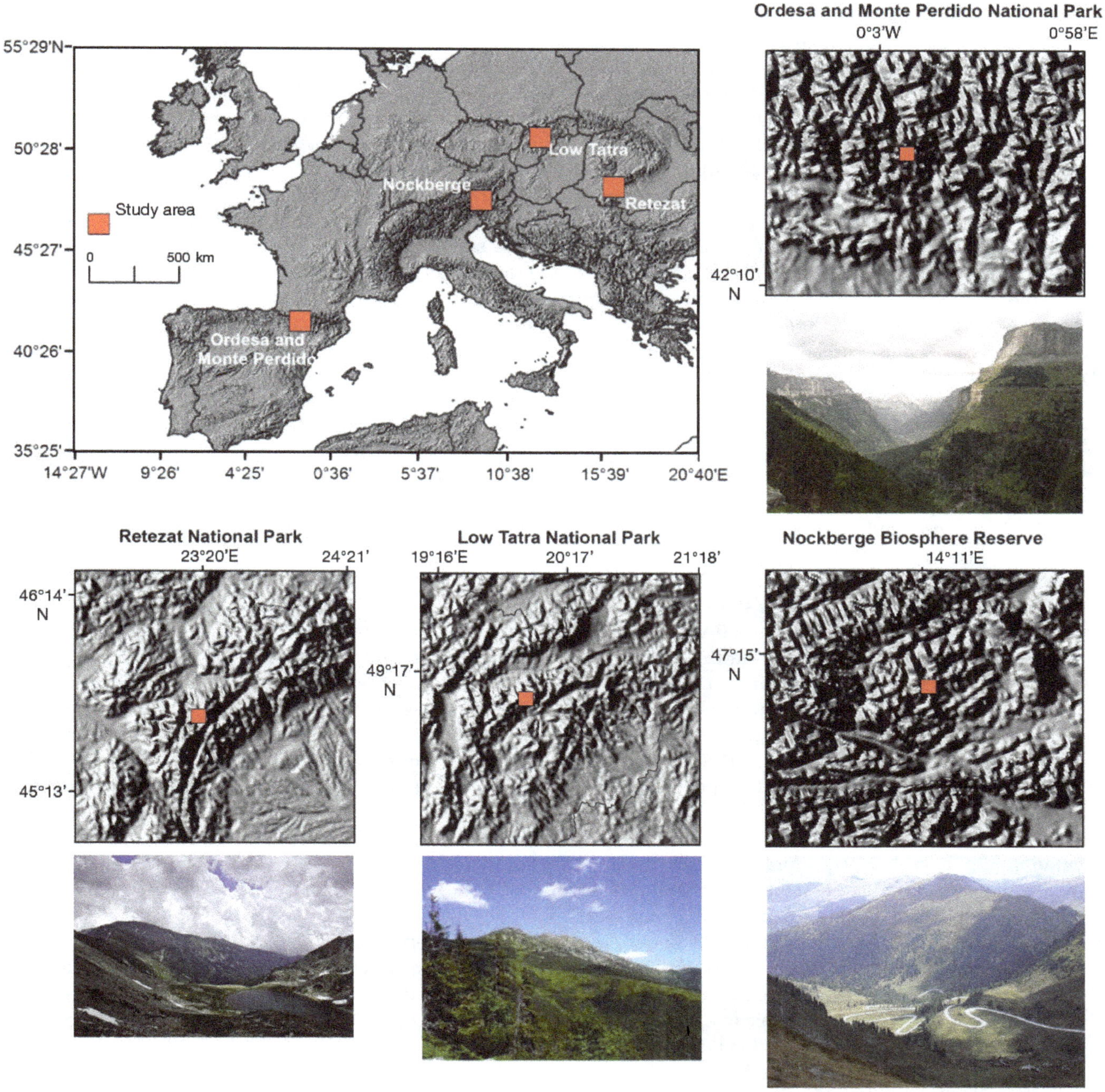

Fig. 1. Location of the mountain study sites considered in this study: Ordesa and Monte Perdido, Spanish Pyrenees; Retezat and Low Tatra in the Slovak and Romanian Carpathians; and Nockberge, Austrian Alps

the main factor that influences tree lines and their advance.

The Low Tatra National Park (Slovakia) was created in 1978 and covers 72 842 ha. The first attempt to create this national park was between 1918 and 1921. Altitude ranges between 700 and 2043 m, with the highest peak at Dumbier (2043 m). The mountains are characterised by continental mountain climate on the slopes and a slightly cold climate at the foothills. Average annual temperature is 4°C and average annual precipitation 1200 mm. Dwarfed pines *P. mugo*, mountain ashes *Sorbus aucuparia* and small pines *P. sylvestris* occupy the park's highest areas. The constant interaction between natural forces and human effects (mainly destructive activities) has significantly influenced the tree line of the Tatra Mountains, (as well as of other important mountains regions of Slovakia). As a general rule, tree lines in Slovakian mountains are not situated at their natural elevation. For example, nowadays, the tree line is situated on average at altitudes of 1185 to 1430 m. Furthermore, strong winds affected the forest stands near the tree line from the Tatra Mountains in 2004, which was followed by a damaging fire.

The Retezat National Park (Romania) was created in 1935. With an area of 38 138 ha, it shelters one of Europe's last remaining intact old-growth forests and the continent's largest single area of pristine mixed forest. Altitude ranges between 600 and 2509 m, with the highest peak at Peleaga (2509 m). Average annual temperature is 6°C and average annual precipitation is 1400 mm. The main tree line species are Norway spruce *P. abies*, mountain pine *P. mugo* and alpine stone pine *Pinus cembra*. Wind throw is the main factor that influences tree lines and their advance.

2.2. Forest line and forest cover quantification using satellite imagery data

The tree line was defined in this study as the pixels classified as forest (according to the FAO definition) near the non-forest pixels located in the upper area of the massif. Assessing the expansion of forest line over time is a typical problem of change detection, and many approaches can be found in the literature (Singh 1989, Almutairi & Warner 2010, Canty 2010, Hecheltjen et al. 2014), each with specific advantages and disadvantages.

Change detection methods can be grouped into 6 categories (Lu et al. 2004): algebra (which includes image differencing, vegetation indexes, change vector analysis); transformation (e.g. principal component analysis); classification; advanced models (where image reflectance values are often converted to physically based parameters); GIS approaches; and visual analysis. Generally, it is not possible to establish *a priori* which method of change detection is the most convenient, so the choice is often made on a pragmatic and application-driven basis (Coppin et al. 2004).

2.2.1. Imagery pre-processing

The Landsat Data Continuity Mission (LDCM) and the launch of the Landsat 8 platform in 2013 offer a unique opportunity to perform monitoring by remote sensing on a regional scale (Mandanici & Bitelli 2015). Owing to the large temporal extent (1972–2014), remote sensing data were obtained from different Landsat sensors: Landsat Multispectral Scanner (MSS) sensor for 1972 and 1980; Landsat Thematic Mapper (TM) sensor for 1990; Landsat Enhanced Thematic Mapper Plus (ETM+) for 2000; and Operational Land Imager (OLI) for 2014.

Sixteen Landsat scenes acquired during the vegetation season (May to September between 1973 and 2014) were used to evaluate temporal dynamics in forest lines and forest covers across the study sites. The scenes were obtained from the United States Geological Survey (USGS) repository (http://earthexplorer.usgs.gov/). In order to reduce noise caused by different acquisition angles, seasons and reflectance features, scenes captured in different years, but within the same vegetation season, were chosen for each study area. In this study, we used L1T products which, according to producer, provide a high radiometric and geodetic accuracy by incorporating ground control points while employing a digital elevation model for topographic displacement (Table 1). The spectral coverage of band images is 0.42–2.29 μm for OLI, 0.45–2.35 μm for ETM, 0.45–2.35 μm for TM and 0.5–1.1 μm for MSS.

As reference cartographic material we used the Soviet Topographic Map, a declassified map that was created based on intelligence information gathered in the Cold War period (Oberman & Mazhitova 2003). The map was georeferenced based on the original Gauss-Kruger grid reprojected in Universal Transverse Mercator (UTM), World Geodetic System 1984 (WGS84) projection, with an average root mean square error of 5 m.

2.2.2. Image processing

Atmospheric correction. Images acquired on different dates are affected to some extent by the presence of haze and dust in the atmosphere, a fact that could mask real changes in a given territory. In order to reduce this effect, images were corrected atmospherically using the European Cooperation in Science and Technology (COST) model developed by Chavez (1996), which employs the cosine of the solar zenith angle, representing a good approximation of atmospheric transmittance.

Topographic correction. To remove the relief effects, images were normalised using the Shuttle Radar Topographic Mission (SRTM) digital terrain model downloaded from http://earthexplorer.usgs.gov/. SRTM is a key breakthrough in digital mapping and provides a major advance in the accessibility of high-quality elevation data worldwide.

Table 1. Summary of satellite data used for multi-temporal mapping of forest dynamics in the mountain study sites. NIR: near-infrared band; SWIR: shortwave infrared band

Year	Ordesa	Nockberge	Low Tatra	Retezat	Metadata: Band-to-band registration accuracy	Metadata: Geometric accuracy and registration success[a]
1970	Soviet Topographic Map	Soviet Topographic Map	Soviet Topographic Map	Soviet Topographic Map	–	5 m
1980	1975-07-26 GLS 1975	1979-05-22 Landsat 2	1979-09-03 Landsat 2	1980-09-21 Landsat 3	0.2 pixel (90 %)	<40 m (average 30.6 m)
1990	1989-07-17 GLS 1990	1988-08-07 Landsat 5	1990-07-16 Landsat 5	1988-08-29 Landsat 5	0.2 pixel (90 %)	<30 m (average 22 m)
2000	2001-08-1 Landsat 7	2003-06-30 Landsat 7	2001-05-26 Landsat 7	2000-08-22 Landsat 7	0.2 pixel (90 %)	<30 m (average 4.6 m)
2014	2014-07-22 Landsat 8	2014-09-18 Landsat 8	2014-08-03 Landsat 8	2014-07-04 Landsat 8	0.2 pixel (90 %)	<30 m (average 8.3 m)
Spatial resolution (m)	30	30	30	30		
Spectral resolution (μm)	Visible and NIR	Visible, NIR, SWIR	Visible, NIR, SWIR	Visible, NIR, SWIR		

[a]According to Landsat metadata files

The sun-canopy-sensor (SCS) topographic correction method was used to remove topographic effects. SCS correction (Gu & Gillespie 1998) is based on SCS geometry and it can be expressed as:

$$L_m = L \cdot \left(\frac{\cos\theta \cdot \cos\alpha}{\cos i} \right) \quad (1)$$

where L_m is the normalised radiance, L is the uncorrected radiance, θ is the solar zenith angle, i is the incident angle and α is the slope of the surface.

All images have been thus co-registered to UTM, WGS84, and the accuracy of image registration was assessed using topographic plans and local cartographic products based on terrestrial measurements and aerial surveys, according to the acquisition period (Fig. 2).

Owing to the different characteristics of spectral sensors (i.e. TM and ETM+) in the Landsat image series, we also corrected the spectral reflectance between images acquired by different sensors (MSS, TM, ETM+ and OLI8). The empirical line approach for reflectance factor retrieval from Landsat-5 TM and Landsat-7 ETM+ was used for this purpose (Moran et al. 2001). All operations were performed in ENVI 5.0.

2.2.3. Mapping and change detection in forest line and cover

Temporal and spatial dynamics in forest covers and lines within the montane–alpine elevation belts (>900 m) were measured using post-classification comparison (PCC) change detection with independently classified images. This method compares, pixel by pixel, 2 independent classified images acquired on different dates, using a change detection matrix (Jensen 2004). PCC minimises the influence of sensor variation in the detection of change. Results depend on the accuracy of initial classifications (Coppin & Bauer 1996). The method locates changes and provides 'from–to' change information. Here, the source for PCC change detection was land cover data created for each Landsat scene. Land cover classification was performed using a supervised classification with maximum likelihood algorithm supported by forest management data that provided information on the spatial distribution of forest and other additional variables (e.g. stand age, height or diameter). Two land cover classes were considered (forest and pasture), and all bands were used for classification, except the thermal band. The approach included the 3 steps: training site selection, classification and assessment of results (Lillesand et al. 2008).

Reference data for training and validation were collected based on high-resolution satellite images or air photos available in Google Earth that cover the complete study area (Knorn et al. 2009, Baudron et al. 2011). We sampled 200 random training areas and classified those as either forest or non-forest, based on visual interpretation. Areas were considered forested if tree cover exceeded 60 % and forest

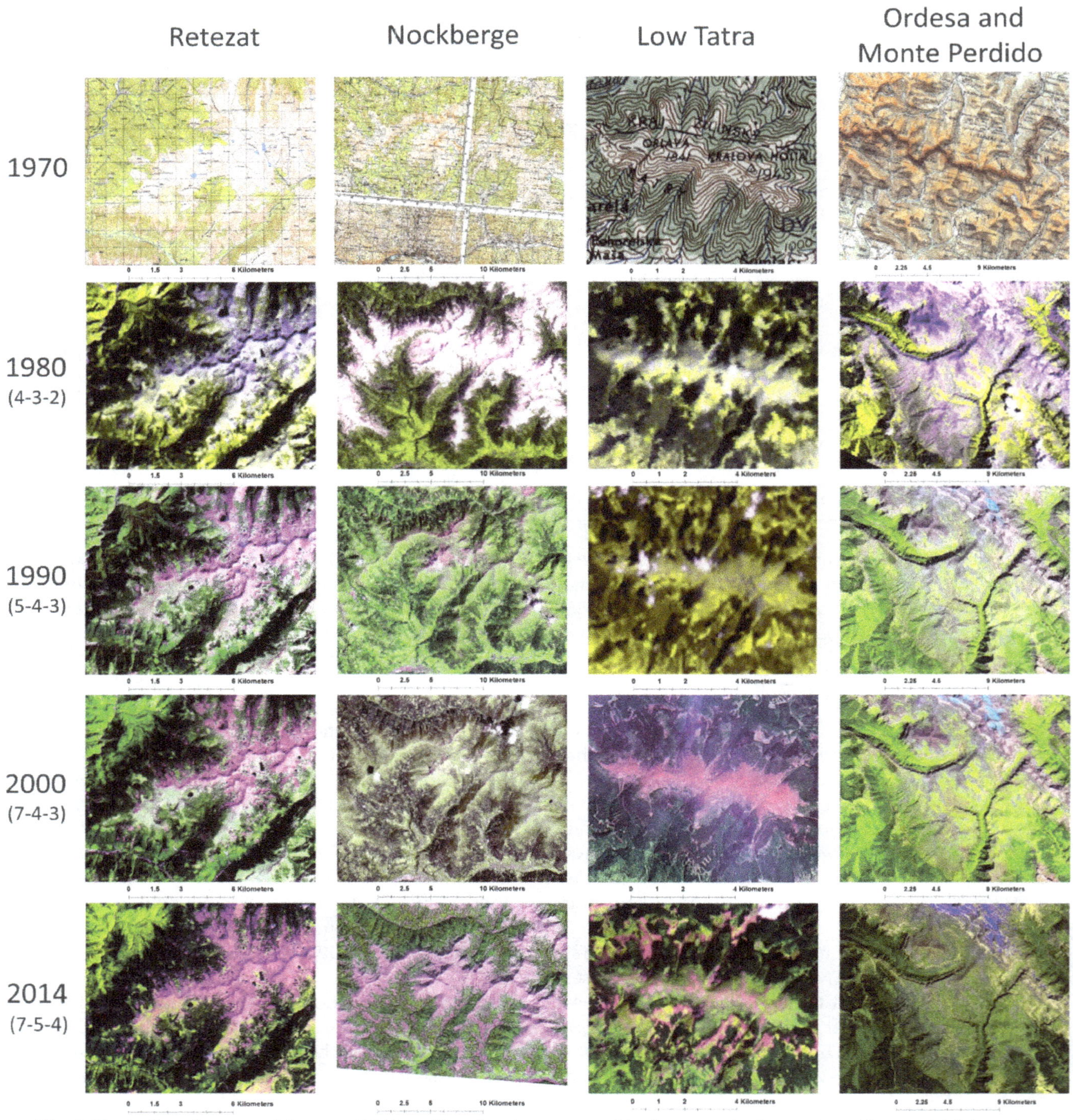

Fig. 2. Illustration of topographic maps (1970) and Landsat imagery used in the supervised mapping of forest with false colour composites: 1980 (Landsat MSS 4-3-2), 1990 (Landsat TM 5-4-3), 2000 (Landsat ETM+ 7-4-3) and 2014 (Landsat OLI 7-5-4)

patches were larger than 1 Landsat pixel (900 m^2) (Kuemmerle et al. 2009).

Classification accuracy was evaluated through a confusion matrix based on a minimum of 100 ground truth sites for each image, other than the training sites, established through a random sampling strategy based on field recommendations (Congalton & Green 2009, Vorovencii 2014). In order to emphasise the changes in land cover classes over the 1970–2014 period, the classified images were compared by cross-tabulation, which resulted in the change matrix that estimates quantitative change (Fig. 3). Using

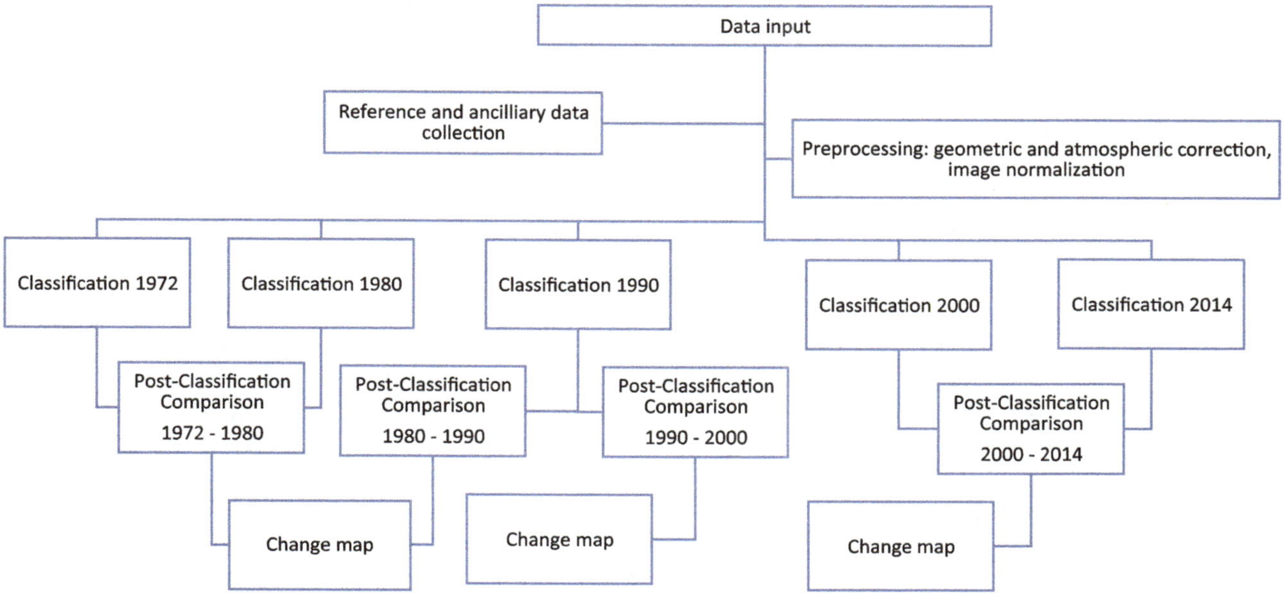

Fig. 3. Analytical framework used for multi-temporal land cover analysis across the study sites with Landsat satellite scenes

these data, forest line and forest cover dynamics were quantified (Fig. 4).

2.3. Historical climate data

Climate data retrieved from Climate Explorer Utility (http://climexp.knmi.nl) were used to pattern climate conditions across each study area from 1900 to 2014 (we took monthly station data, introducing the coordinates of the national parks, from the 10 nearest meteorological stations to that point). To reconstruct climate condition trends during the analysed tree line period (1970–2014), climate data between 1901 and 1970 were taken as a baseline. Climate conditions were represented by average temperature, precipitation and de Martonne aridity index (de Martonne 1926). However, preliminary correlation analysis showed an insignificant statistical relationship between growth in forest area and the last 2 variables, which led to their exclusion from the analysis. This was expected, as precipitation in forest line ecosystems is not a limiting factor (Leal et al. 2007, Grytnes et al. 2014)

2.4. Statistical analysis

Temperature evolution was analysed using simple graphical trends for the studied regions. The reference mean temperature values calculated as simple arithmetic means of periods covering 10 to 15 yr each were used to plot the trends. A clear growth trend was observed, so further analyses were carried out to estimate the relationship between temperature increase and both forest cover growth and altitudinal forest line migration. The percentage of forested cover growth and the forest line altitudinal migration were plotted versus the relative temperature increment during the studied period. Furthermore, the dependence between the percentage of forest cover growth and forest line altitude migration, with respect to temperature variation in the studied period, were modelled using least-squares simple regression.

3. RESULTS

3.1. Forest line and forest cover dynamics across the montane–alpine boundary

The error matrix obtained for each land cover classification is presented in Table 2. Overall accuracy values reached the minimum average standard of 85% recommended by the USGS classification scheme (Anderson et al. 1976), which indicates a strong link between map classification and ground reference data.

The accuracy values can be grouped (Congalton & Green 1999) into (1) strong link (>0.80; >80%); (2)

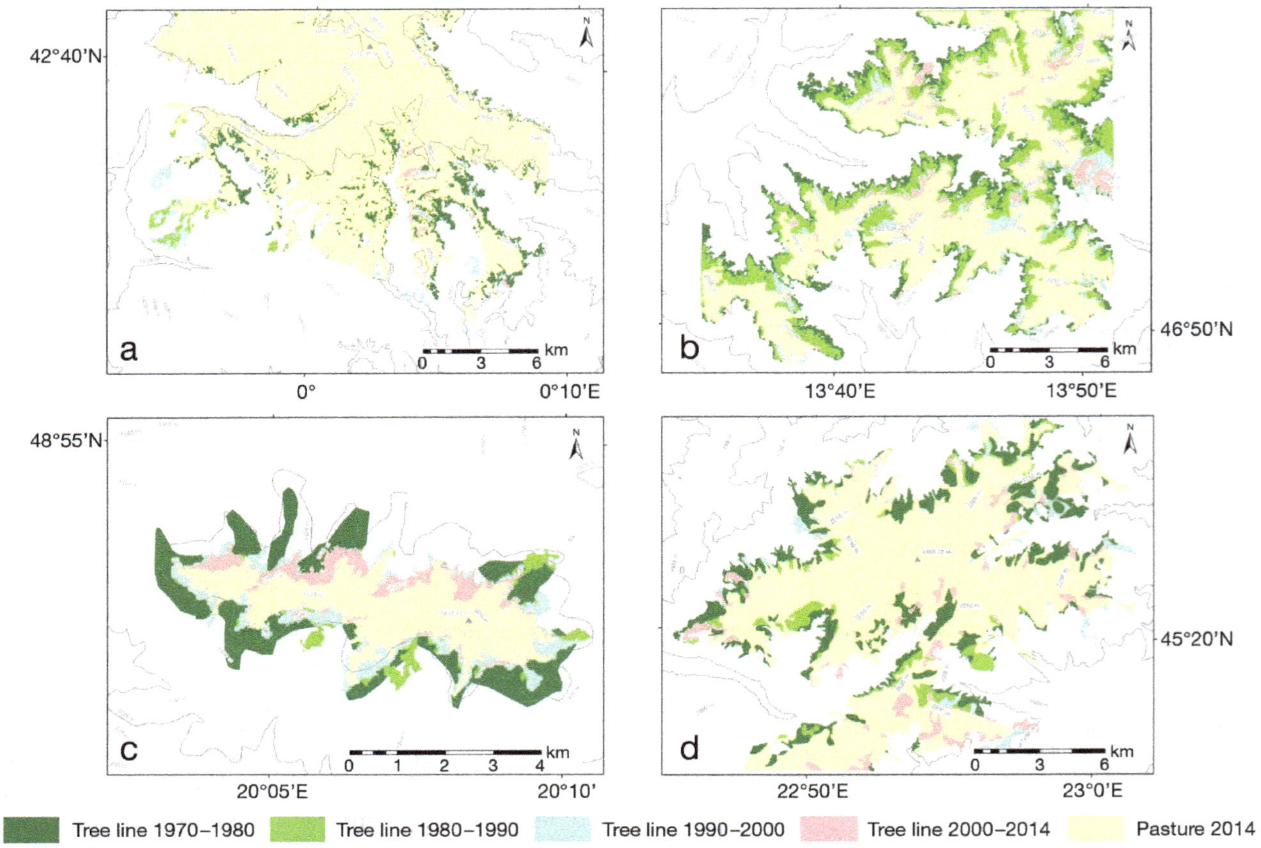

Fig. 4. Spatial and temporal dynamics of forest line and cover in 4 European mountain sites obtained from supervised land cover classification of Landsat satellite data between 1970 and 2015. Dynamics are represented using a 10 yr time-step interval for (a) Ordesa and Monte Perdido National Park, Spain; (b) Nockberge Biopshere Park, Austria; (c) Low Tatra Park, Slovakia; and (d) Retezat National Park, Romania

moderate link (0.40–0.80; 40–80%); and (3) weak link (<0.40; <40%).

All study sites showed changes in forest line and forest cover. Highest and lowest change in elevation of the upper forest limit (altitude growing) occurred in the Alps (Nockberge) and Pyrenees (Ordesa), respectively. Surface forest growth indicated differences between sites. The largest forest expansion was registered in the Alps (28%; Nockberge) and the lowest in the Pyrenees (1%; Ordesa) for the period 1981–1990. The largest forest line advance was recorded in the Alps (Nockberge) and the Carpathians (Retezat), while the lowest was found in the Pyrenees (Ordesa) (Table 3).

Table 2. Summary of forest class accuracies (%) of land cover maps (see Fig. 4) obtained with maximum likelihood supervised classification of Landsat satellite data (years 1980, 1990, 2000 and 2014), along with Kappa statistics. PA: producers' accuracy; UA: users' accuracy

	Ordesa (ES)		Nockberge (AU)		Low Tatra (SK)		Retezat (RO)	
	PA	UA	PA	UA	PA	UA	PA	UA
1980	90	85.60	86.64	82.10	86.00	81.30	92.40	88.30
1990	85	78.40	84.70	79.40	89.50	83.40	90.30	84.20
2000	94	89.40	97.50	94.30	98.20	97.10	96.30	95.30
2014	95	91.50	88.12	98.70	96.30	95.20	96.40	96.20
Mean	86.23		88.63		89.25		91.00	
Kappa statistics	0.84		0.87		0.88		0.90	

Table 3. Summary of changes in forest line and forest cover across the study areas between 1970 and 2014, and average annual temperatures registered during this period

	Forest expansion (ha)	Forest expansion (%)	Altitude gain (m)	Temp-erature (°C)
Ordesa				
1901–1970	0	0	0	4.45
1971–1980	243	1.22	5	4.63
1981–1990	195	0.99	4	5.25
1991–2000	874	0.48	18	5.79
2001–2014	3034	1.28	5	6
Nockberge				
1901–1970	0	0	0	4.1
1971–1980	3204	16.91	90	4.29
1981–1990	4430	28.14	44	4.73
1991–2000	1601	14.15	34	5.29
2001–2014	1539	10.03	2	5.51
Low Tatra				
1901–1970	0	0	0	2.22
1971–1980	181	12.6	43	2.41
1981–1990	5	0.4	10	2.81
1991–2000	271	21.68	40	2.91
2001–2014	183	18.69	1	3.62
Retezat				
1901–1970	0	0	0	4.36
1971–1980	1997	13.16	61	4.08
1981–1990	737	5.67	13	4.24
1991–2000	414	3.38	5	4.54
2001–2014	962	8.12	17	5.25

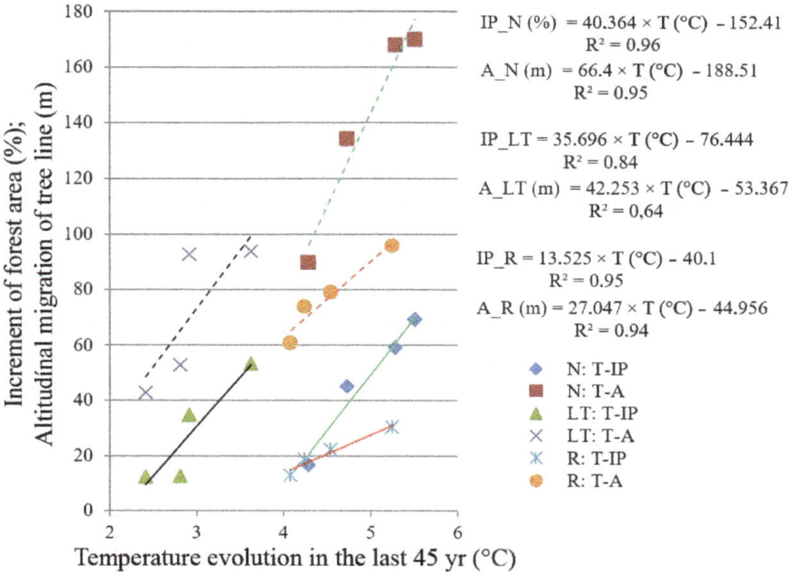

Fig. 5. Estimation equations indicating trends, relationships and intensity of dependence between the studied parameters for the 3 areas (N: Nockberge; LT: Low Tatra; R: Retezat). IP: increment percentage of forest; A: altitude; T: temperature

Table 4. Spatial distribution of the expansion in forest cover (%) across topographic exposure classes in the study areas

Exposition	Forest area growth (%) Nockberge	Retezat	Low Tatra
N	18	14	15
NE	18	12	12
E	11	10	7
SE	6	11	20
S	5	9	22
SW	8	11	10
W	16	14	3
NW	19	20	9

Forest expansion also varied across exposure classes and study sites. In the Austrian Alps (Nockberge) and the Romanian Carpathians (Retezat), expansion was strongly skewed to northerly slopes, while in the Slovakian Carpathians (Low Tatra), it occurred mostly on southerly slopes (Table 4).

3.2. Relationship between average annual temperature and forest line and cover dynamics

Historical climate data evidence an increase in average annual temperature between 1950 and 2014 across all sites. Overall, the average increase was 1.1°C, with the Carpathian sites showing both the highest (Low Tatra; +1.4°C) and the lowest (Retezat; +0.89°C) increases (Fig. 5). In 3 of the studied areas, mean temperatures of the past 50 yr showed a continuously growing trend (Table 3, Fig. 5). Regression models showed strong linear relationships between forest cover increment and altitudinal migration of forest line across all sites with respect to mean annual temperature rise (Fig. 5). The increment of dependent variables had a lower magnitude in the Romanian Carpathians (Retezat), significantly contrasting with the magnitude in the Alps (Nockberge), which was the highest. Fig. 5 illustrates the magnitude of the dependencies between dependent and independent variables.

4. DISCUSSION

In this study, we combined multi-temporal satellite image analysis and observed the impacts of climate on forest cover and forest line dynamics in the montane–alpine boundary across European mountains between 1970 and 2014. Four protected areas, distributed along a west–east longitudinal gradient and enclosed in 3 emblematic European mountain ranges (Pyrenees, Alps and Carpathians), were considered. These areas were located in national parks that were created before the oldest satellite images considered for analysis. Therefore, human influence (cattle grazing, tree cutting, etc.) in the period of analysis was minimised, and changes in forest cover and line were mostly driven by natural processes. The influence of natural disruptive factors (windfalls, fires, insect attacks, etc.) was not accounted for in this study. The accuracy values of imagery classification were predominantly above the average standard (85%) suggested by the USGS classification scheme (Anderson et al. 1976) and in line with previous studies using Landsat data in mountain regions (Kharuk et al. 2010, Hagedorn et. al 2014). Lower accuracy values in some scenes, which can slightly influence the estimations, are related to common factors influencing multi-temporal satellite studies in mountain areas (e.g. spatial resolution, atmospheric anomalies and cloud cover), including in tree line studies (Wang et al. 2006).

Forest change analysis revealed a considerable expansion in forest cover and altitudinal migration of forest lines in the montane–alpine boundary across all sites, particularly in the central eastern mountains. The expansion of forest cover and increase in altitude of the upper forest line are 2 common patterns in Northern Hemisphere mountains in the past decades (Bolli et al. 2006, Gehring-Fasel et al. 2007, Harsch et al. 2009, Schickhoff et al. 2015). Both forest cover and forest line were below their potential ranges (Gehring-Fasel et al. 2007), especially in mountains where humans have social and economic interests (Motta et al. 2006), as in the case of our study sites. The combination of atmospheric warming and decreased human activities at high altitudes promoted forest cover expansion and forest line shifts (Motta et al. 2006, Gehring-Fasel et al. 2007, Leonelli et al. 2011). Our patterns of upward shifts of the forest line were not uniform and varied across sites (Harsch et al. 2009). Registered upward shifts of forest lines were below (Low Tatra), in line with (Retezat) or above (Nockberge) the 70 to 100 m belt previously proposed (Moiseev et al. 2010). Forest cover expansion also differed among sites and topographic exposure. While the western forest (Pyrenees) was slightly altered, central and eastern sites showed a considerable forest cover expansion of up to 28%. Dynamics in the forests of central and eastern Europe have been widely discussed in the past few years (e.g. Gherig-Fasel et al. 2007, Hartl-Meier et al. 2014, Pretzsch et al. 2014, Munteanu et al. 2016).

The anthropogenic effect on forest expansion is reinforced by asymmetric historical distributions of the largest expansion periods across the study areas. While in the Romanian Carpathians (Retezat), large expansion occurred until the 1980s, in the Austrian Alps (Nockberg) and the Slovak Carpathians (Low Tatra), the high expansion periods were registered in and after the 1990s. Moreover, large tree line upward shifts occurred in the early 1980s. This asymmetry within study sites under the same trend (increasing average temperatures during the last century) suggests that average temperature is not the only factor impacting forest and forest line expansion. Indeed, anthropogenic activity determines European mountain tree lines (Dirnböck et al. 2003, Kuemmerle et al. 2008).

Several alpine tree line studies document altitudinal shifts and tree density increase during the 20th century (Kullman 1979, MacDonald et al. 1998), although only 51% out of 166 sites reviewed by Harsch et al. (2009) showed tree line advance. The main factors responsible for those changes were climate warming and land-use modification. In the Pyrenees, increases in temperatures between 1882 and 1970 were observed at the Pic du Midi meteorological station (Bücher & Dessens 1991). In parallel, grazing pressure has been declining drastically since the 1950s (Alados et al. 2014). In spite of these changes, we did not observe the expected tree line upward shift. Previous studies (Camarero & Gutiérrez 2004, Camarero et al. 2015) also show that the tree line remained static in the central Pyrenees, while tree density increased within the ecotone.

A careful reading of the overall effects of mean temperature increase on forest expansion (1.31°C; 1970–2014) is therefore required (Körner 1998), as alpine vegetation can tolerate temperature increases only between 1 and 2°C without major changes (Theurillat & Guisan 2001), and the response of the tree line to observed climate warming is still globally inconsistent (Harsch et al. 2009, Schickhoff et al. 2015). Nonetheless, a strong statistical relationship between tree line expansion and annual increased average temperatures was found, with sites experiencing increased temperatures showing a large advance in tree line. The temperature trend is consistent with the general constant increasing trend in average temperatures across European mountain ranges (Kullman 2007, Leonelli et al. 2011, González de Andrés et al. 2015). The impacts of temperature on tree lines are well known (Körner 1998), but the significant responses are mostly related to early or late growing seasons (Rammig et al. 2010, González de Andrés et al. 2015) or even winter temperatures (Kullman 2007). Our findings suggest that increasing annual average temperatures favour the expansion of tree cover and line in the Austrian Alps, Low Tatra and the Romanian Carpathians. Annual average temperatures may influence tree cover and line due (to a lesser effect) to factors affecting plant growth (e.g. wind, snow cover), improvement of microsite conditions (Leonelli et al. 2011, Pardo et al. 2013) and the favouring of seed viability and seedling emergence at higher temperatures (Kullman 2007).

In conclusion, rising mean annual temperatures in mountainous European areas have influenced tree line advance, especially in areas with reduced human intervention (national parks). This mechanism was stronger in the mountainous areas located in central Europe (Alps, Carpathians) and weaker in the warmer European areas (Pyrenees). However, uncertainty in the definition of line position suggests that interpretations should be carried out carefully. The accuracy with which the location of forest lines can be measured in historical data is influenced by the georeferencing accuracy of the data sources and the accuracy of the mapping technique in the definition of line position (Hofgaard et al. 2013). Our data include estimated uncertainties of 0.2 pixels band-to-band, which indicate a relative misregistration of less than 1/3 pixel in the position of forest lines due to misresgistration of Landsat imagery. In addition, uncertainty due to misclassification of forest class has also occurred. Mean uncertainty due to misclassification amounted to 11.2%, with the Pyrenees (Ordesa) presenting the higher values. Interpretation in sites with limited change in forest line (e.g. Ordesa) and high uncertainty therefore requires caution.

Overall, open access to historical forest maps and multispectral satellite imagery archives combined with accurate pre-processing and classification is well suited to multi-site comparative analysis on forest line and forest cover dynamics. Comparative analysis of forest lines in protected areas can be a good strategy to better understand the response of natural systems to changes in climate conditions. The consideration of novel variables can further enhance the benefits of understanding mountain forest lines.

Acknowledgements. This article is based on work from COST Action ES1203 SensFor, supported by COST (European Cooperation in Science and Technology, www.cost.eu).

LITERATURE CITED

ACIA (2005) Arctic climate impact assessment. Cambridge University Press, Cambridge

Alados CL, Errea P, Gartzia M, Saiz H, Escós J (2014) Positive and negative feedbacks and free-scale pattern distribution in rural-population dynamics. PLOS ONE 9:e114561

Almutairi A, Warner TA (2010) Change detection accuracy and image properties: a study using simulated data. Remote Sens 2:1508–1529

Ameztegui A, Coll L, Messier C (2015) Modelling the effect of climate-induced changes in recruitment and juvenile growth on mixed-forest dynamics: the case of montane–subalpine Pyrenean ecotones. Ecol Modell 313:84–93

Anderson JR, Hardy EE, Roach JT, Witmer RE (1976) A land use and land cover classification system for use with remote sensor data. US Geological Survey Professional Paper 964. United States Government Printing Office, Washington, DC

Batllori E, Gutiérrez E (2008) Regional tree line dynamics in response to global change in the Pyrenees. J Ecol 96: 1275–1288

Baudron F, Corbeels M, Andersson JA, Sibanda M, Giller KE (2011) Delineating the drivers of waning wildlife habitat: the predominance of cotton farming on the fringe of protected areas in the mid-Zambezi valley, Zimbabwe. Biol Conserv 144:1481–1493

Baumann M, Kuemmerle T, Elbakidze M, Ozdogan M and others (2011) Patterns and drivers of post-socialist farmland abandonment in western Ukraine. Land Use Policy 28:552–562

Bolli JC, Rigling A, Bugmann H (2006) The influence of changes in climate and land-use on regeneration dynamics of Norway spruce at the tree line in the Swiss Alps. Silva Fenn 41:55–70

Bücher A, Dessens J (1991) Secular trend of surface temperature at an elevated observatory in the Pyrenees. J Climatol 4:859–868

Cairns DM, Moen J (2004) Herbivory influences tree lines. J Ecol 92:1019–1024

Camarero JJ, Gutiérrez E (2004) Pace and pattern of recent treeline dynamics: response of ecotones to climatic variability in the Spanish Pyrenees. Clim Change 63: 181–200

Camarero JJ, García-Ruiz JM, Sangüesa-Barreda G, Galván JD and others (2015) Recent and intense dynamics in a formerly static Pyrenean tree line. Arct Antarct Alp Res 47:773–783

Canty MJ (2010) Image analysis, classification, and change detection in remote sensing: with algorithms for ENVI/IDL, 2nd edn. CRC Press, Boca Raton, FL

Chavez PS Jr. (1996) Image-based atmospheric corrections —revisited and improved. Photogramm Eng Remote Sensing 62:1025–1036

Chhetri PK, Cairns DM (2015) Contemporary and historic population structure of *Abies spectabilis* at treeline in Barun valley, eastern Nepal Himalaya. J Mt Sci Engl 12: 558–570

Congalton RG, Green K (1999) Assessing the accuracy of remotely sensed data: principles and practices. Lewis Publishers, Boca Raton, FL

Congalton RG, Green K (2009) Assessing the accuracy of remotely sensed data: principles and practices, 2nd edn. CRC Press Taylor & Francis Group, London

Coppin PR, Bauer ME (1996) Digital change detection in forest ecosystems with remote sensing imagery. Remote Sens Rev 13:207–234

Coppin P, Jonckheere I, Nackaerts K, Muys B, Lambin E (2004) Digital change detection methods in ecosystem monitoring: a review. Int J Remote Sens 25:1565–1596

Dalen L, Hofgaard A (2005) Differential regional treeline dynamics in the Scandes Mountains. Arct Antarct Alp Res 37:284–296

de Martonne E (1926) Areisme et indice aridite. C R Acad Sci 182:1395–1398

Dirnböck T, Dullinger S, Grabherr G (2003) A regional impact assessment of climate and land use change on alpine vegetation. J Biogeogr 30:401–417

Galford GL, Mustard JF, Melillo J, Gendrin A, Cerri CC, Cerri CEP (2008) Wavelet analysis of MODIS time series to detect expansion and intensification of row-crop agriculture in Brazil. Remote Sens Environ 112:576–587

Gartzia M, Alados CL, Pérez-Cabello F (2014) Assessment of the effects of biophysical and anthropogenic factors on woody plant encroachment in dense and sparse mountain grasslands based on remote sensing data. Prog Phys Geogr 38:201–217

Gartzia M, Fillat F, Pérez-Cabello F, Alados CL (2016a) Influence of agropastoral system components on mountain grassland vulnerability estimated by connectivity loss. PLOS ONE 11:e0155193

Gartzia M, Pérez-Cabello F, Bueno CG, Alados CL (2016b) Physiognomic and physiologic changes in mountain grasslands in response to environmental and anthropogenic factors. Appl Geogr 66:1–11

Gehrig-Fasel J, Guisan A, Zimmermann NE (2007) Tree line shifts in the Swiss Alps: climate change or land abandonment? J Veg Sci 18:571–582

González de Andrés E, Camarero J, Buntgen U (2015) Complex climate constraints of upper treeline formation in the Pyrenees. Trees (Berl) 29:941–952

Griffiths P, Kuemmerle T, Baumann M, Radeloff V and others (2014) Forest disturbances, forest recovery and changes in forest types across the Carpathian ecoregion from 1985 to 2010 based on Landsat image composites. Remote Sens Environ 151:72–88

Grytnes JA, Kapfer J, Jurasinski G, Birks HH and others (2014) Identifying the driving factors behind observed elevation range shifts on European mountains. Glob Ecol Biogeogr 23:876–884

Gu D, Gillespie A (1998) Topographic normalization of Landsat TM images of forest based on subpixel sun-canopy-sensor geometry. Remote Sens Environ 64:166–175

Hagedorn F, Shiyatov SG, Mazepa VS, Devi NM and others (2014) Treeline advances along the Urals mountain range — driven by improved winter conditions? Global Change Biol 20:3530–3543

Hansen MC, DeFries RS, Townshend JRG, Carroll M, Dimiceli C, Sohlberg RA (2003) Global percent tree cover at a spatial resolution of 500 meters: first results of the MODIS Vegetation Continuous Fields algorithm. Earth Interact 7:1–15

Hansen MC, Roy DP, Lindquist E, Adusei B, Justice CO, Altstatt A (2008) A method for integrating MODIS and Landsat data for systematic monitoring of forest cover and change in the Congo Basin. Remote Sens Environ 112:2495–2513

Harsch MA, Hulme PE, McGlone MS, Duncan RP (2009) Are tree lines advancing? A global meta-analysis of tree line response to climate warming. Ecol Lett 12:1040–1049

Hartl-Meier C, Dittmar C, Zang C, Rothe A (2014) Mountain forest growth response to climate change in the Northern Limestone Alps. Trees 28:819–829

Hecheltjen A, Thonfeld F, Menz G (2014) Recent advances in remote sensing change detection—a review. In: Manakos I, Braun M (eds) Land use and land cover mapping in Europe. Springer, Dordrecht, p 145–178

Hofgaard A, Harper KA, Golubeva E (2012) The role of the circumarctic forest–tundra ecotone for arctic biodiversity. Biodiversity 13:174–181

Hofgaard A, Tømmervik H, Rees G, Hanssen F (2013) Latitudinal forest advance in northernmost Norway since the early 20th century. J Biogeogr 40:938–949

Holtmeier FK (2009) Mountain timberlines: ecology, patchiness, and dynamics. Springer, Dordrecht

Holtmeier FK, Broll G (2005) Sensitivity and response of Northern Hemisphere altitudinal and polar treelines to environmental change at landscape and local scales. Glob Ecol Biogeogr 14:395–410

Holtmeier FK, Broll G (2007) Treeline advance — driving processes and adverse factors. Landsc Online 1:1–33

Holtmeier FK, Broll G (2010) Altitudinal and polar treelines in the Northern Hemisphere — causes and response to climate change. Polarforschung 79:139–153

Jensen JR (2004) Introductory digital image processing: a remote sensing perspective, 3rd edn. Prentice-Hall, Upper Saddle River, NJ

Jin S, Sader SA (2005) MODIS time-series imagery for forest disturbance detection and quantification of patch size effects. Remote Sens Environ 99:462–470

Kaplan JO, New M (2006) Arctic climate change with a 2°C global warming: timing, climate patterns and vegetation change. Clim Change 79:213–241

Kharuk VI, Ranson KJ, Im ST, Vdovin AS (2010) Spatial distribution and temporal dynamics of high-elevation forest stands in southern Siberia. Global Ecol Biogeog 19:822–830

Klinge M, Böhner J, Erasmi S (2015) Modeling forest lines and forest distribution patterns with remote-sensing data in a mountainous region of semiarid central Asia. Biogeosciences 12:2893–2905

Knorn J, Rabe A, Radeloff VC, Kuemmerle T, Kozak J, Hostert P (2009) Land cover mapping of large areas using chain classification of neighboring Landsat satellite images. Remote Sens Environ 113:957–964

Körner C (1998) A re-assessment of high elevation treeline positions and their explanation. Oecologia 115:445–459

Kozak J (2010) Forest cover changes and their drivers in the Polish Carpathian Mountains since 1800. In: Nagendra H, Southworth J (eds) Reforesting landscapes linking pattern and process. Springer, Dordrecht, p 253–273

Kuemmerle T, Hostert P, Radeloff VC, van der Linden S, Perzanowski K (2008) Cross-border comparison of post-socialist farmland abandonment in the Carpathians. Ecosystems 11:614–628

Kuemmerle T, Chaskovskyy O, Knorn J, Radeloff VC, Kruhlov I, Keeton WS, Hostert P (2009) Forest cover change and illegal logging in the Ukrainian Carpathians in the transition period from 1988 to 2007. Remote Sens Environ 113:1194–1207

Kullman L (1979) Change and stability in the altitude of the birch tree-limit in the southern Swedish Scandes 1915–1975. Acta Phytogeogr Suec 65:1–121

Kullman L (2007) Tree line population monitoring of *Pinus sylvestris* in the Swedish Scandes, 1973–2005: implications for tree line theory and climate change ecology. J Ecol 95:41–52

Lambin EF, Meyfroidt P (2010) Land use transitions: socio-ecological feedback versus socio-economic change. Land Use Policy 27:108–118

Leal S, Melvin TM, Grabner M, Wimmer R, Briffa KR (2007) Tree-ring growth variability in the Austrian Alps: the influence of site, altitude, tree species and climate. Boreas 36:426–440

Leonelli G, Pelfini M, Morra di Cella U, Valentina Garavaglia V (2011) Climate warming and the recent tree line shift in the European Alps: the role of geomorphological factors in high-altitude sites. Ambio 40:264–273

Lillesand TM, Kiefer RW, Chipman JW (2008) Remote sensing and image interpretation. John Wiley and Sons, New York, NY

Linderman M, Rowhani P, Benz D, Serneels S, Lambin E (2005) Land-cover change and vegetation dynamics across Africa. J Geophys Res 110:D12104

Lu D, Mausel P, Brondízio E, Moran E (2004) Change detection techniques. Int J Remote Sens 25:2365–2401

MacDonald GM, Szeicz JM, Claricoates J, Dale KA (1998) Response of the central Canadian tree line to recent climatic changes. Ann Assoc Am Geogr 88:183–208

Mandanici E, Bitelli G (2015) Multi-image and multi-sensor change detection for long-term monitoring of arid environments with Landsat series. Remote Sens 7: 14019–14038

Mas JF (1999) Monitoring land-cover changes: a comparison of change detection techniques. Int J Remote Sens 20: 139–152

Mathisen IE, Mikheeva A, Tutubalina OV, Aune S, Hofgaard A (2014) Fifty years of tree line change in the Khibiny Mountains, Russia: advantages of combined remote sensing and dendroeological approaches. Appl Veg Sci 17: 6–16

Moiseev PA, Bartysh AA, Nagimov ZY (2010) Climate changes and tree stand dynamics at the upper limit of their growth in the north Ural Mountains. Russ J Ecol 41: 486–497

Monserrat Martí JM (1992) Evolucion glaciar y post-glaciar del clima y la vegetacion en la vertiente sur del Pirineo: estudio palinologico. Consejo Superior de Investigaciones Cientificas. Monografias del Instituto Pirenaico de Ecologia, Vol. 6 Coop. de Artes Graficas, Zaragoza

Moran MS, Bryant R, Thome K, Ni W and others (2001) A refined empirical line approach for reflectance factor retrieval from Landsat-5 TM and Landsat-7 ETM+. Remote Sens Environ 78:71–82

Motta R, Morales M, Nola P (2006) Human land-use, forest dynamics and tree growth at the treeline in the Western Italian Alps. Annal For Sci 63:739–747

Müller D, Kuemmerle T, Rusu M, Griffiths P (2009) Lost in transition: determinants of post-socialist cropland abandonment in Romania. J Land Use Sci 4:109–129

Munteanu C, Nita MD, Abrudan IV, Radeloff VC (2016) Historical forest management in Romania is imposing strong legacies on contemporary forests and their management. For Ecol Manage 361:179–193

Oberman N, Mazhitova G (2003) Permafrost mapping of northeast European Russia based on the period of climatic warming 1970–1995. Nor Geogr Tidsskr 57:111–120

Pardo I, Camarero JJ, Gutiérrez E, García M (2013) Uncoupled changes in tree cover and field layer vegetation at two Pyrenean tree line ecotones over 11 years. Plant Ecol Divers 6:355–364

Payette S (2007) Contrasted dynamics of northern Labrador tree lines caused by climate change and migrational lag. Ecology 88:770–780

Pflugmacher D, Cohen WB, Kennedy ER (2012) Using Landsat-derived disturbance history (1972–2010) to predict current forest structure. Remote Sens Environ 122: 146–165

Potapov P, Turubanova S, Hansen MC (2011) Regional-scale boreal forest cover and change mapping using Landsat data composites for European Russia. Remote Sens Environ 115:548–561

Pretzsch H, Biber P, Schütze G, Uhl E, Rötzer T (2014) Forest stand growth dynamics in Central Europe have accelerated since 1870. Nat Comm 5:4967

Rammig A, Jonas T, Zimmermann NE, Rixen C (2010) Changes in alpine plant growth under future climate conditions. Biogeosciences 7:2013–2024

Schickhoff U, Bobrowski M, Böhner J, Bürzle B and others (2015) Do Himalayan tree lines respond to recent climate change? An evaluation of sensitivity indicators. Earth Syst Dynam 6:245–265

Seto KC, Woodcock CE, Song C, Huang X, Lu J, Kaufmann RK (2002) Monitoring land-use change in the Pearl River delta using Landsat TM. Int J Remote Sens 23: 1985–2004

Shiyatov S, Terentev M, Fomin V, Zimmermann N (2007) Elevational and horizontal shifts of the upper boundaries of open and closed forests in the Polar Urals in the 20th century. Russ J Ecol 38:223–227

Singh A (1989) Digital change detection techniques using remotely-sensed data. Int J Remote Sens 10:989–1003

Sveinbjörnsson B, Hofgaard A, Lloyd A (2002) Natural causes of the tundra–taiga boundary. Ambio 12:23–29

Theurillat JP, Guisan A (2001) Potential impact of climate change on vegetation in the European Alps: a review. Clim Change 50:77–109

Townshend JR, Masek JG, Huang C, Vermote EF and others (2012) Global characterization and monitoring of forest cover using Landsat data: opportunities and challenges. Int J Digit Earth 5:373–397

Vorovencii I (2014) A change vector analysis technique for monitoring land cover changes in Copsa Mica, Romania, in the period 1985–2011. Environ Monit Assess 186: 5951–5968

Wang T, Zhang QB, Ma KP (2006) Tree line dynamics in relation to climatic variability in the central Tianshan Mountains, northwestern China. Global Ecol Biogeogr 15:406–415

Woodcock CE, Macomber SA, Pax-Lenney M, Cohen WB (2001) Large area monitoring of temperate forest change using Landsat data: generalization across sensors, time and space. Remote Sens Environ 78:194–203

Wulder MA, Masek JG, Cohen WB, Loveland TR, Woodcock CE (2012) Opening the archive: how free data has enabled the science and monitoring promise of Landsat. Remote Sens Environ 122:2–10

13

Bioclimatic effects on different mountain birch populations in Fennoscandia

Oddvar Skre[1,*], Bogdan Wertz[2], Frans E. Wielgolaski[3], Paulina Szydlowska[2], Stein-Rune Karlsen[4]

[1]Skre Nature and Environment, Fanaflaten 4, 5244 Fana, Norway
[2]University of Agriculture in Krakow, Dept. of Biometry and Forest Productivity, Al. 29 Listopada 46, 31-425 Krakow, Poland
[3]Department of Bioscience, University of Oslo, PO Box 1066 Blindern, 0316 Oslo, Norway
[4]NORUT Northern Research Institute, PO Box 6434, 9294 Tromsø, Norway

ABSTRACT: Mountain birch *Betula pubescens* ssp. *tortuosa* is the main treeline species in northern Europe, and the recent increase in treeline elevation in Fennoscandia due to climate change and land use has made mountain birch an important bioindicator. Birch seedlings from 10 populations were therefore transplanted to 3 northern Fennoscandian sites (1 oceanic, 1 continental mountain and 1 Arctic coastal site). Annual measurements were carried out on growth parameters and phenology (date of budbreak) from 1992 onwards. At the coastal site, measurements covered the whole period 1993 to 2010, while at the 2 other sites measurements covered only the period 1993 to 1997 (mountain site) and 2010 to 2014 (both sites). During the last period (2010 to 2014), measurements were made on a new set of seedlings, transplanted in 2002. The plants suffered a temporary transplantation stress because of root damage. In general, budburst occurred earlier in populations from northern and continental sites than from southern and coastal sites. Survival rates were dependent on climate and were generally higher at the oceanic than at the mountain site, due to more autumn frost and insect damage at the latter. At the mountain site, survival rates were lowest in oceanic and southern populations; at the Arctic site, survival rates were lowest in continental populations; while at the oceanic site, only small differences were found between populations. The present results seem to indicate that climate is an important driver of change at the mountain birch treeline, although land-use changes may be more important in the short term. In the future, when a warmer climate is predicted, plants that are adapted to a coastal climate may expand their range, and increased winter temperatures are expected to favour insect-resistant polycormic birch inbreeding with dwarf birch *Betula nana*.

KEY WORDS: Treelines · Climate change · Survival · Height and diameter growth · Dieback

1. INTRODUCTION

The circumpolar treeline between the northern boreal forest and the treeless tundra is one of the main vegetation transition zones in the Northern Hemisphere. The high sensitivity of mountain forests to climate and land-use changes (e.g. Callaghan et al. 2013, Holtmeier 2007) makes the circumpolar treeline a good indicator of these changes. In contrast to most of the circumpolar area, mountain birch *Betula pubescens* ssp. *tortuosa* is the main treeline species in Fennoscandia. Although the high-elevation forest limits have been lowered by past intensive land use (Bryn & Daugstad 2001, Gehrig-Fasel et al. 2007), these forests are mainly influenced by climate (Körner & Paulsen 2004). Consequently, reduced land-use intensity combined with climatic warming would lead to increased forest cover and treeline elevation, while intensive or sustained land use would oppose the climatic effect (Bryn 2008, Aune et al. 2011). Mountain forests may be further influenced by seed production and quality, and soil conditions (Speed et al. 2011).

*Corresponding author: oddvar@nmvskre.no

§Advance View was available online June 28, 2017

At a large scale, the Arctic treeline seems to be mainly a result of the 'climate hazard' factor (e.g. Skre et al. 2002, Sveinbjörnsson et al. 2002), including seed reproduction and freezing stress. However, individual clones of trees may survive as vegetative individuals, and eventually produce viable seeds as soon as summer temperatures are high enough for seed reproduction (Kullman 1998, Payette et al. 2002). Other studies such as Karlsson & Nordell (1996) and Weih & Karlsson (1999) indicate that the uptake of nitrogen from soil is probably the most important limiting process for growth of *B. pubescens* at the treeline in northern Fennoscandia. In addition, the transition from open, treeless tundra to boreal forests implies increased carbon sequestration, but also increased CO_2 output because of decreased albedo, particularly in evergreen tree species like spruce and pine (e.g. de Wit et al. 2014).

Global warming may influence tree growth and treeline position through increased seed viability and production, as well as increased vegetative growth. The first indication of these changes is therefore often an increase in seedling density at or above the treeline, as observed by Juntunen & Neuvonen (2006) in northern Finland. The final outcome is dependent on these factors and on the degree of human interaction or other disturbance factors. Vitasse et al. (2014a) found that winter warming would increase the risk of premature dehardening and eventual spring frost damage. Plants and ecotypes of *B. pubescens* with high demands for dormancy breaking would have an advantage (e.g. Heide 1993). As a rule, *B. pubescens* needs a chilling period below 10°C for budbreak. This requirement is expected to decrease with increasing latitudes of their mother trees (Myking & Heide 1995) and to increase with increasing oceanity (Leinonen 1996, Myking 1999). Similar results were found earlier by Larsen (1976) in Douglas fir *Pseudotsuga menziesii* provenances from British Columbia. According to Myking & Heide (1995), there is little risk of chilling deficit in birch in Scandinavia, even with a climatic warming of 7 to 8°C in winter. Seedlings and young trees are generally more susceptible to spring frosts than adult trees because of earlier budbreak (Vitasse et al. 2014b)

Grazing by sheep and reindeer has formed the treeline as well as the forest line in northern Norway (Tømmervik et al. 2004, Dalen & Hofgaard 2005), and mammals or insects may have modifying effects on treelines in a variety of environments (e.g. Cairns et al. 2007). Wind may also be an important factor locally (Scott et al. 1993, Sveinbjörnsson et al. 2002). However, Speed et al. (2011) concluded that sheep grazing would usually dominate over climatic warming, dependent on the intensity of grazing. In connection, according to e.g. Anamthawat-Jonsson et al. (1993), dwarf birch *Betula nana* may be more resistant to sheep grazing than mountain birch *B. pubescens*, which may have some influence on the competition between these 2 species. Similar relationships may also be found regarding birch resistance against insect attacks (e.g. Haukioja et al. 1988, Neuvonen et al. 2005).

Insect herbivory that causes tree death is the most widely reported mode of animal activity, and the effect of the autumnal moth *Epirrita autumnata* on *B. pubescens* has been widely studied in northern Sweden and Finland. Tenow et al. (2005) found that *Epirrita* eggs would not survive winter temperatures below −35°C, hence global warming would increase their survival rates. They also found that polycormic birch would recover faster than monocormic birch after insect attacks, due to higher ability to form lateral shoots. According to Wielgolaski (2005), polycormic birch is partly a result of inbreeding with *B. nana* in the *B. pubescens* populations (cf. Nilsen & Wielgolaski 2001, Thórsson et al. 2001). As climatic warming is expected to favour more insect outbreaks, future warming would lead to more polycormic birch that will recover faster and restore the carbon reserves (Karlsson & Weih 2003). According to Jepsen et al. (2008) and Jepsen et al. (2013), who used MODIS-NDVI (moderate resolution imaging spectroradiometer normalized difference vegetation index) data to monitor the spatio-temporal dynamics of moth outbreaks in birch forests (Jepsen et al. 2009), the winter moth *Operophtera brumata* is more common along the coast of northern Norway than the autumnal moth, but has recently expanded into colder areas due to the change in climate. The total impact of defoliation on the birch populations is therefore expected to increase in the future, and with increasing frequency and more species involved, resistance to insect attacks is expected to be an increasingly important selective factor for survival and growth at the treeline.

Also, summer temperatures influence birch growth by improving foliage quality and the capacity of trees to recover from grazing and/or insect attacks (Neuvonen et al. 1999, Virtanen & Neuvonen 1999).

The aim of the present study was to determine the differences in survival rates and in phenological and growth responses in young plants from different provenances of mountain birch *B. pubescens* ssp. *tortuosa* transplanted to 3 sites with varying environmental conditions, and compare the results with earlier findings (e.g. Ovaska et al. 2005). The results from the present study are expected to give some indications about how future climatic changes may influence survival and growth of mountain birch, and will be used in an attempt to answer the following questions: (1) How do birch provenances respond to climatic stress? (2) How well adapted are birch provenances to the expected climate change? (3) What are the implications of (1) and (2) for the future?

The present study was designed to test abiotic stress, but as insect attacks are closely connected to climatic warming, these damages are included in the analysis, whereas land-use changes such as overgrazing or land abandonment (Callaghan et al. 2013) were prevented by fencing the birch gardens. On the other hand, reindeer overgrazing in Finland has been shown to prevent birch regeneration after insect attacks (Lehtonen & Heikkinen 1995) that may be favoured by global warming.

2. MATERIALS AND METHODS

2.1. Experimental design

Seedlings of mountain birch *Betula pubescens* var. *tortuosa* from southern and northern, as well as oceanic and more continental origin, and various elevations were transferred to 3 different transplant gardens north of the Arctic Circle in Fennoscandia. The transplant gardens were situated along an oceanity–continentality gradient and at different elevations (Fig. 1, Table 1). Of the established sites, 2 (Kilpisjärvi and Vardø) were situated close to the Arctic or alpine treelines, while the third (Melbu) was situated at an elevation about 50 m lower than the treeline.

The 10 seed populations from 5 different countries used in the present study (Table 1) were sown in 1991 in Pakatti nursery, Finland, and transplanted to the 3 gardens in 1992. In Table 2, the position and elevations of the seed populations are shown with the mean January (t_1) and July temperatures (t_7; °C). The continentality is determined mainly by the amplitude between t_1 and t_7 (Ovaska et al. 2005). As a result, populations from Utsjoki (FU), Kevo (FJ and FK), Kilpisjärvi (FKi) and Abisko (SAb) were classified as continental, and those from Narsassuaq (GNa), Hafnarskogur (IHa), Melbu (NMe) and Hammerfest (NHa) as oceanic, while the Blefjell population (NBl) was intermediate between continental and oceanic (cf. Fig. 1).

A total of 6 replicates, each with 25 plants per population, were established at each garden. The sites were fenced to prevent reindeer, sheep or cattle grazing, on 1 m^2 plots per plant. Survival rates during the period after transplantation (cf. Fig. 2), measurements of total and living height, stem base diameter,

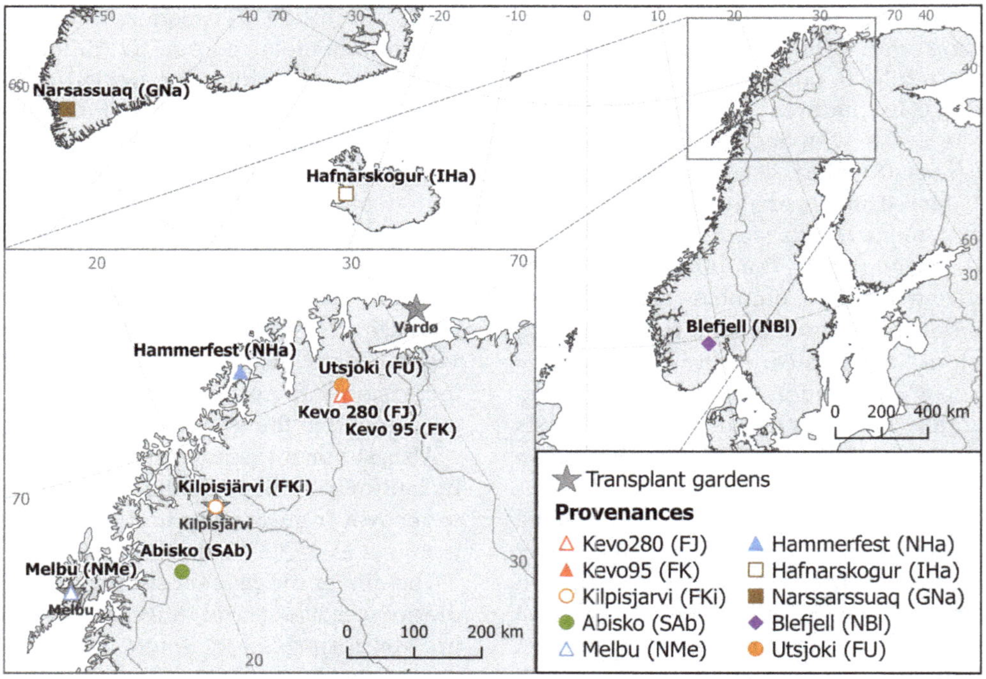

Fig. 1. Populations and transplant garden sites

Table 1. Origin of seed populations, with position, elevation (Elev.), mean temperatures 1961 to 1990 (°C) of January (t_1) and July (t_7), and dates of budbreak (1994). The localities are classified as oceanic when the difference between mean July and January temperatures (t_7 and t_1, respectively) is below 20°C

No.	Population	Continentality	Elev. (m a.s.l.)	°N	°E	t_1	t_7	Budbreak date Kilpisjärvi	Budbreak date Melbu
1	Kevo280 (FJ)	Continental	280	69.8	27.0	−12.1	11.8	23 June	20 May
2	Kevo95 (FK)	Continental	95	69.8	27.0	−13.6	13.3	23 June	20 May
3	Kilpisjärvi (FKi)	Continental	500	69.0	20.8	−10.6	11.2	19 June	19 May
4	Abisko (SAb)	Continental	360	68.3	18.8	−10.3	11.4	20 June	19 May
5	Melbu (NMe)	Oceanic	40	68.5	14.8	−1.0	13.1	13 July	21 May
6	Hammerfest (NHa)	Oceanic	70	70.7	23.7	−5.5	12.3	18 June	17 May
7	Hafnarskogur (IHa)	Oceanic	20	64.5	−21.9	−0.3	10.2	13 July	22 May
8	Narsassuaq (GNa)	Oceanic	70	61.2	−45.4	−3.3	7.0	17 July	20 May
9	Blefjell (NBl)	Intermediate	750	59.8	9.2	−5.6	13.4	20 July	23 May
10	Utsjoki (FU)	Continental	200	69.9	27.0	−12.8	12.5	22 June	19 May

Table 2. Site characteristics of the transplant gardens, showing elevation (Elev. m a.s.l.), position, normal mean temperatures 1961–1990 (°C) of January (t_1) and July (t_7), annual precipitation (P, mm), and number of days with snow cover (SD) and mean freezing temperatures (<0)

Site	Elev.	°N	°E	t_1^a	t_7^a	P	SD	<0
Vardø (Arctic oceanic)	13	70.3	31.1	−4.3	9.1	509	169	190
Kilpisjärvi (alpine continental)	500	69.0	20.8	−10.6	11.2	525	195	234
Melbu (oceanic)	40	68.5	14.8	−1.0	13.1	782	118	114

[a]Climate of seed populations and birch gardens is measured at nearby meteorological stations, except Kevo280, which is extrapolated from Kevo95 by elevation difference

and phenology (date of budbreak) were calculated annually during the period 1994 to 2010. However, many plants at the Vardø and Kilpisjärvi sites died during the first years after the transplantation due to frost and insect damage (R. Partanen pers. comm.). Owing to the strong decrease in survival rates at Kilpisjärvi, no observations were made after 1997 (Figs. 3, 4 & 5), except from the control measurement in 2010. At Vardø, the observations were also too few to arrive at any conclusions, while at Melbu, the observations continued annually until 2010.

Owing to low survival rates in the first set of plants (henceforth 'old plants') from 1992, a new set of plants from 8 of the same populations (cf. Table 1) were raised in the greenhouse at the University of Tromsø and transplanted to the Vardø and Kilpisjärvi sites, with 20 plants per population and replicate. Population #1 and 10 were omitted because of limited supply of viable seeds, and because preliminary measurements from Melbu and Kilpisjärvi indicated that they were not significantly different from population #2 (Kevo95) (cf. Figs. 3, 4 & 5). At the Vardø site, preliminary tests (O. Skre unpubl.) indicated that the results from 4 of the 5 replicates (I to IV) were significantly different from the fifth replicate (V). Replicates I to IV were all situated on a peat bog and strongly influenced by a local snowbed, while Replicate V was situated on a dry sandy terrace with early snowmelt and sheltered by a willow thicket (cf. Table A1). The measurements on the new set of plants were initiated in 2010, after the plants had recovered from transplantation, and continued until 2014. As in the first set of transplantations, the survival rates (%), and total and living height and diameter measurements were calculated annually. Owing to the strong correlation between height and diameter growth (Table 3), only the height is included (see the Appendix).

Table 3. Pearson's correlation coefficients between variables, where SR is survival rate (%) at plant level, D_{AV} is average stem base diameter (mm) and H_{AV} is average total height (cm). Significance — light grey: significant at α = 0.05; dark grey: significant at α = 0.01

Garden	Generation	Variable	SR	D_{AV}	H_{AV}
Melbu	Old	SR		[redacted]	[redacted]
Melbu	Old	D_{AV}	[redacted]		[redacted]
Melbu	Old	H_{AV}	[redacted]	[redacted]	
Kilpisjärvi	Old	SR		[redacted]	[redacted]
Kilpisjärvi	Old	D_{AV}	[redacted]		[redacted]
Kilpisjärvi	Old	H_{AV}	[redacted]	[redacted]	
Kilpisjärvi	New	SR		[redacted]	[redacted]
Kilpisjärvi	New	D_{AV}	[redacted]		[redacted]
Kilpisjärvi	New	H_{AV}	[redacted]	[redacted]	
Vardø	New	SR		0.20	0.24
Vardø	New	D_{AV}	0.20		0.64
Vardø	New	H_{AV}	0.24	0.64	

2.2. Statistical analysis

The datasets from old plants at Melbu 1994 to 2009, from Kilpisjärvi 1994 to 1997 and 2010, and from new plants at Kilpisjärvi and Vardø 2010 to 2014 were analysed. For evaluation of the bioecological success of each provenance and for their comparison, the following indicators were selected and computed from all years observed: (1) the survival rate (SR) at replicate level as a percentage of the total numbers of living plants during the observed years; (2) diameter (*D*) measured for each living plant at the stem base; and (3) living height (*H*) measured as the height of the longest living stem.

Preliminary measurements revealed quite frequent negative height increment (dieback) due to frost or insect damage, and sometimes also negative diameter growth (when a plant formed new stems after the death of old stems that had died or been destroyed). However, after careful initial data analysis (IDA), including detection of possible errors and influencing outliers, and inspection of variables' distributions and transformation attempts, we decided to provide comparison on the basis of quantitative analysis, using a linear-mixed model approach. Although some minor violations of its assumptions, such as deviations from normality (checked by Shapiro-Wilks test, at $\alpha = 0.05$) and heterogeneity of variance (checked by Levene's test, at $\alpha = 0.05$), were still present in the dataset, the applied method was found to be the best solution for processing the data, because it takes into account dependence of measurements (3 levels of nesting). Further, it can provide reliable results even with the missing data, and it provides quantitative estimators (parameters) that are easy to interpret, and their statistical significance can be assessed (B. Wertz unpubl.).

All provenances were characterised by the mean value of each indicator as well as 95% mean CI. For the analysis of temporal differences among tested provenances, the mean values in each year of observation in the old dataset were calculated and interconnected by lines, while the overall mean survival rates and total height were tested in the new dataset. For testing the relationships between selected indicators, the Pearson's correlation coefficient and its significance at $\alpha = 0.05$ was calculated. To estimate the differences between provenances, the linear mixed-model approach was applied separately to each garden. The indicators describing the biological success (SR, *D* and *H*) were treated as dependent variables, while time (Year of life—numerical variable) and origin (Provenance—categorical variable) were treated as fixed factors. The random factor consisted of particular plants, nested within replicates and provenances. The constructed model allowed a different slope for each plant, and included both independent variables (time and origin) because they had the lowest values of Akaike's information criterion (AIC). All statistical analysis was performed in the R environment (R Core Team 2015) with the help of the following packages: lme4 (Bates et al. 2015), enabling access to the creation of mixed models; lmerTest (Kuznetsova et al. 2016), providing p-value calculations that are helpful for an assessment of the statistical significance of differences; sjPlot (Lüdecke 2016) for visualization of results; and AICcmodavg (Marc & Mazerolle 2016), calculating AIC, suitable for optimal model selection.

3. RESULTS AND DISCUSSION

3.1. Phenology

The mean dates of budbreak in Table 1 varied from one year to another, according to winter temperatures and snow cover, but 1994 is chosen as representative because no temporal variations were found in the relationships between populations during the sample period (O. Skre unpubl.).

In 1994, budburst was almost 1 mo later in the southern and oceanic populations (NB, NMe, GNa and IHa) compared to the northern and continental populations (NHa, FKi, FK, FJ, SAb and FU), when transplanted to the continental Kilpisjärvi site (Table 1). According to Myking & Heide (1995), this means that the thermal requirements for dormancy breaking are higher in the former than in the latter group. This is most obvious when comparing the oceanic NMe plants with the continental SAb and FKi populations originating from approximately the same latitude. According to Myking (1999) and Larsen (1976), oceanic populations need to be more resistant against occasional spring frosts than continental relatives, because of the slower transition between winter and summer temperatures in oceanic areas. On the other hand, foliar senescence is almost solely dependent on day length and elevation (Ovaska et al. 2005).

3.2. Growth and survival rates

Many plants at the Vardø and Kilpisjärvi sites died during the first years after the transplantation, due to climatic stress and insect damage (R. Partanen pers.

comm.). Because of the strong decrease in survival rates at Kilpisjärvi, no observations were made after 1997, except from the control measurement in 2010. At Vardø, the observations were also too few to arrive at any conclusions, while at Melbu the observations continued annually until 2010. A much more rapid decrease in survival rates during the first years after transplantation (1994 to 1997) was detected at Kilpisjärvi than at Melbu (cf. Fig. 3). The main reason is probably that the Melbu site was located close to the sea and about 50 m below the treeline, with a much longer snow-free period (cf. Tables 1 & 2), and therefore was more protected from climatic damage than the site at Kilpisjärvi, which was located just below the treeline. The differences between sites (Fig. 2a,b) were strongest in the 4 southern and oceanic provenances (NMe, IHa, GNa and NBl). This result is in accordance with the basic assumption that the birch population has developed climatic adaptations to its different habitats that influence survival rates after transplantation (cf. Myking & Heide 1995, Gamache & Payette 2005). At the 3 most oceanic and southernmost sites (NB, GNa and IH), only a few plants had survived by the end of the period at Kilpisjärvi (Fig. 3), and consequently the estimates of growth parameters, particularly the total height, dropped to almost zero (Table 4, Fig. 2).

There was a strong correlation between the survival rates of individual plants and the corresponding growth parameters in the old datasets (Melbu and Kilpisjärvi), and in the new dataset from Kilpisjärvi (Table 3). The implication is that plants that are already damaged by frost and/or insect attacks have a lower chance of survival than healthy plants.

The plants growing at the Kilpisjärvi site varied substantially and were less successful in terms of survival rates than the Melbu plants (Fig. 2). This difference can possibly be explained by more severe winters (Tables 1 & 2) and more frequent insect attacks (R. Partanen pers. comm.) at the Kilpisjärvi site, leading to more frequent dieback (cf. Tenow et al. 2005) and strong variation in living height (Fig. 5). The 3 southern populations (NBl, GNa and IHa) and partly the northern oceanic NMe population seemed to be less successful than the northern continental populations, with Abisko (SAb) as an exception (cf. Fig. 2).

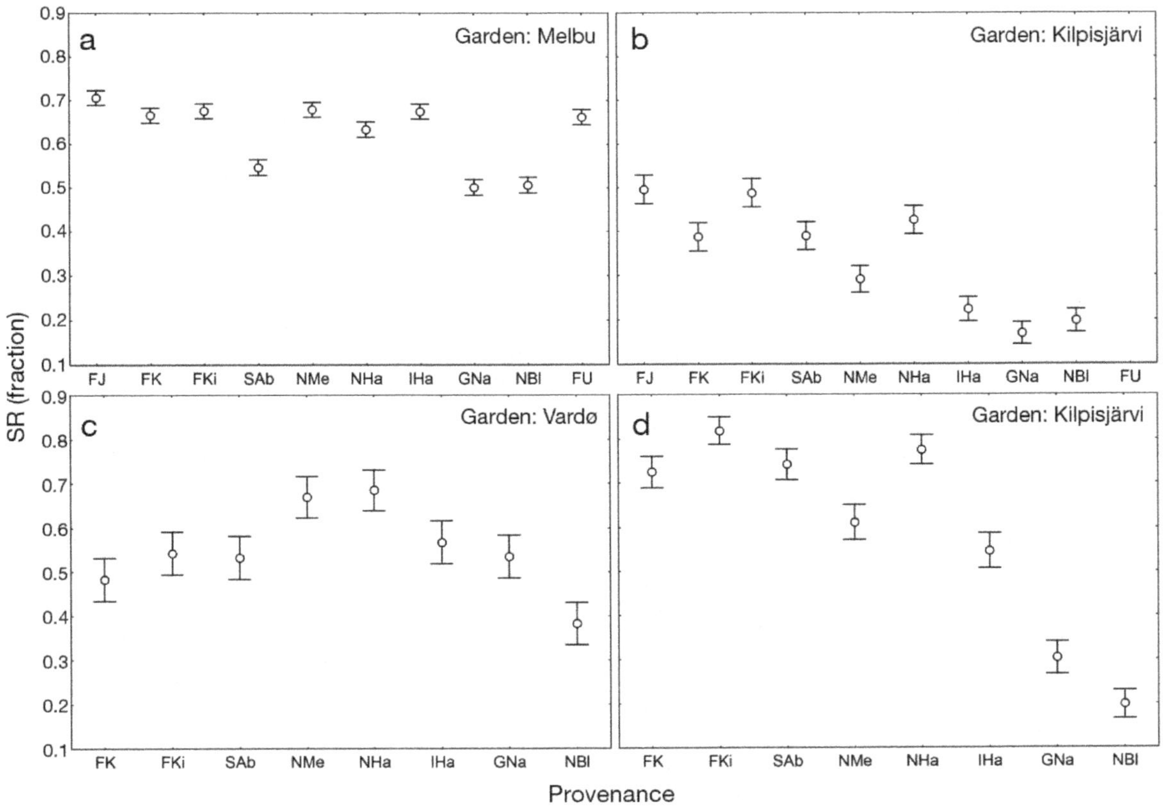

Fig. 2. Mean values of selected indicators, i.e. survival rates (SR, as fractions), with corresponding 95% CI for tested provenances of (a,b) old generation and (c,d) new generation plants at (a) Melbu, (b,d) Kilpisjärvi and (c) Vardø. See the Appendix for total height growth

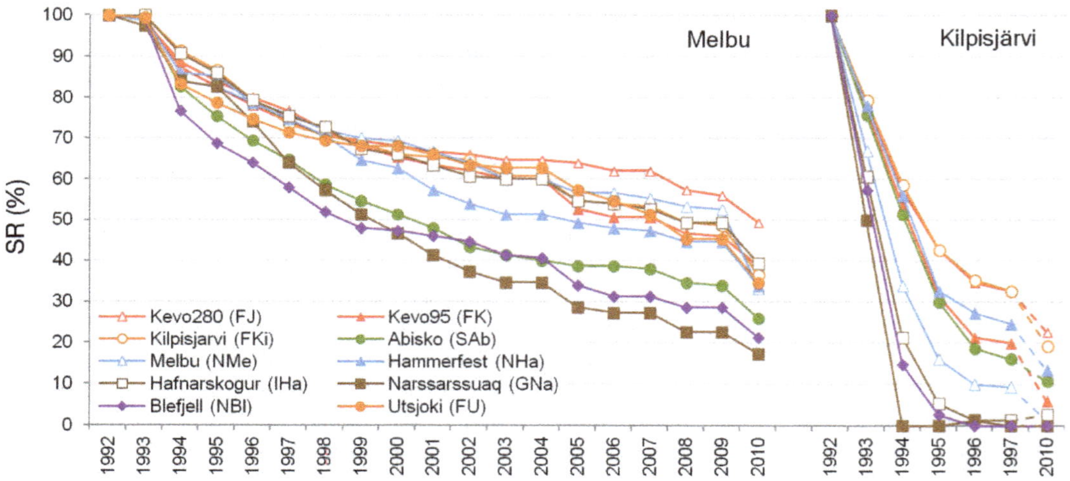

Fig. 3. Survival rates (SR) for tested provenances of old generations at Melbu (left) and Kilpisjärvi (right) as a function of time since 1992

The results of the linear mixed-model method (Table 4) agreed well with the conclusions from Fig. 2, i.e. there were significant differences between the survival rates in plants from Kevo and Kilpisjärvi (FJ, FK and FKi), and the western and southern plants (NBl and GNa) at both sites. However, at Kilpisjärvi the survival rates of northern coastal plants (IHa and NMe) were also different from those from northern Finland. Similar patterns were also found when comparing stem diameter (B. Wertz unpubl.) and total height (Fig. A1 in the Appendix). The implication is that when southern and oceanic ecotypes are moved northward and inland, they continue growing late into the autumn and are easily subjected to frost damages (cf. Heide 1993, Myking & Heide 1995). On the other hand, when northern ecotypes like NHa are moved southward, they start growing earlier in spring (cf. Table 1) and are protected from autumn frost by their early growth cessation (cf. Heide 1993, Taulavuori et al. 2004, Skre et al. 2008). They are also often characterized by so-called metabolic compensation as a response to low temperatures to compensate for a short growing season (Billings et al. 1971). This may be the main reason why the northern coastal Hammerfest population (NHa) showed the highest growth rates in terms of diameter and height growth (Figs. 4 & 5) among populations. Hence, the northern coastal/oceanic NHa population was among the most successful, presumably because it is protected from late autumn frosts by its northern origin and from spring frosts by adaptation to a coastal climate (cf. Larsen 1976, Myking & Heide 1995). The high stem growth in the NHa population may also partly be explained by the strong monocormic growth pattern in this population (H. Tømmervik pers. comm.).

Figs. 4 & 5 show a relatively long recovery time (7 to 8 yr) at the Melbu site after transplantation. At Kilpisjärvi, many plants died during the first years after transplantation, but the remaining plants recovered faster than at the oceanic Melbu site, probably because of better soil conditions (O. Skre unpubl.).

3.3. Temporal changes

The most obvious temporal change is the high mortality in all provenances at the Kilpisjärvi garden, particularly in the southern and coastal populations, where the survival rates (Fig. 3) dropped from 100% to 0–40% during the first 3 yr after transplantation (1992 to 1995). In contrast, survival rates at the Melbu site decreased only to 70–80% during the same period. After 1995, the mortality seemed to decrease, and at the time of the last control measurement in 2010, the survival rates at Kilpisjärvi had stabilized at 0 to 25%, with the southern populations (GNa and NBl) at the lower end of the range and the northern continental populations at the higher end. This was only 20 to 25% lower than the corresponding survival rates at Melbu in 2010, where there was an almost constant decrease during the whole period. The constructed model (Table 4) also supported those findings, and although the general trend in both gardens seems to be very similar — decrease of survival rate per year was −0.032 in Melbu and −0.031 at the Kilpisjärvi site — the general intercept in the latter is much lower (0.544 vs. 0.865).

Table 4. Parameters of the linear mixed-model, expressing changes of plant characteristics survival rate (SR), stem base diameter (*D*) and total height (*H*), due to time (Year of life) and provenance for the old (top) and new generation (bottom) of plants. The intercept estimates mean value for the Abisko (SAb) provenance. Light shading corresponds to p-value parameters significant at $\alpha = 0.05$ and dark shading at $\alpha = 0.01$. The predicted general trend value of dependent variable (*D*, *H*, SR) for a particular year of life can be obtained by adding the intercept, estimate for provenance, and mean change per year multiplied by number of years. Values of σ^2 express variation not explained by the model, while $\tau00$ values express variation explained by each level of nested group (Plants-Replicants-Provenances). N describes the number of groups distinguished at each level and pseudo-R^2 is an equivalent of ratio of variation explained by the model to unexplained variation left, and expresses overall goodness of model fit

Variable	SR	*D*	*H*	SR	*D*	*H*
Garden	**Melbu**			**Kilpisjärvi**		
Fixed parts						
Intercept-Abisko (SAb)	[redacted]			[redacted]		
Provenance						
Blefjell (NBl)	−0.035	[redacted]	−7.528	[redacted]		
Hafnarskogur (IHa)	0.087	0.43	5.366			
Hammerfest (NHa)	0.066	0.75	[redacted]			
Kevo280 (FJ)	0.074	−0.04	3.395			
Kevo95 (FK)	0.084	−0.60	−0.250			
Kilpisjärvi (FKi)	0.089	0.11	2.519			
Melbu (NMe)	0.079	−0.93	−3.637			
Narssarssuaq (GNa)	0.006	[redacted]	−6.762			
Utsjoki (FU)	0.067	0.32	3.044			
Year of life	[redacted]			[redacted]		
Random parts						
σ^2	0.051	1.239	60.703	0.14	1.37	156.54
$\tau_{00,\ Plant:(Replicant:Provenance)}$	0.069	2.639	177.108	0.03	1.46	309.42
$\tau_{00,\ Replicant:Provenance}$	0.002	0.459	20.285	0.01	0.32	46.13
$\tau_{00,\ Provenance}$	0	0	0	0.00	0.00	0.00
$N_{Plant:(Replicant:Provenance)}$	1500	1394	1283	1350	526	519
$N_{Replicant:Provenance}$	60	60	60	54	50	50
$N_{Provenance}$	10	10	10	9	9	9
Pseudo-R^2	0.802	0.901	0.895	0.433	0.951	0.914
Garden	**Vardø**			**Kilpisjärvi**		
Fixed Parts						
Intercept-Abisko (SAb)	[redacted]			[redacted]		
Provenance						
Blefjell (NBl)	−0.030	−0.05	−2.41	[redacted]	0.86	−4.34
Hafnarskogur (IHa)	0.000	1.40	1.23	−0.025	0.55	−2.90
Hammerfest (NHa)	0.001	1.25	1.61	−0.015	1.38	−2.40
Kevo95 (FK)	−0.005	−0.79	−9.89	−0.022	0.87	4.62
Kilpisjärvi (FKi)	−0.012	0.40	−3.11	0.012	0.76	3.62
Melbu (NMe)	0.026	1.38	−3.99	−0.028	0.67	1.70
Narssarssuaq (GNa)	0.000	0.30	0.52	[redacted]	0.22	−1.44
Year of life	[redacted]			[redacted]		0.68
Random parts						
σ^2	0.05	1.35	38.30	0.05	2.87	93.30
$\tau_{00,\ Plant:(Replicant:Provenance)}$	0.00	21.24	603.21	0.00	0.61	1127.04
$\tau_{00,\ Replicant:Provenance}$	0.00	0.85	290.49	0.00	4.01	309.77
$\tau_{00,\ Provenance}$	0.00	0.19	0.00	0.00	0.00	16.39
$N_{Plant:(Replicant:Provenance)}$	800	435	438	800	471	498
$N_{Replicant:Provenance}$	40	40	40	40	39	39
$N_{Provenance}$	8	8	8	8	8	8
Pseudo-R^2	0.846	0.946	0.951	0.820	0.810	0.861

Figs. 4 & 5 show that the lower growth rates and high mortality at the Kilpisjärvi site seem to have been stronger during the first years after transplantation than during the period from 1997 to 2010. This is partly due to spring frost damage, leading to dieback of annual shoots and whole stems, followed by formation of new stems. There seems to have been a strong selection caused by freezing stress, particularly in the non-adapted provenances (Taulavuori et al. 2004, Skre et al. 2008). This may be related to higher frost sensitivity in young plants and seedlings, due to earlier budbreak, than in adult trees (Vitasse et al. 2014b). The transplantation shock and strong selection pressure may also be seen when looking at the temporal changes in total height growth (Fig. 5) and, to a lesser extent, stem base diameter (Fig. 4) in all provenances. In fact, there is a negative growth in total height, or dieback of annual shoots, at Melbu during the first 3 yr after transplantation (Fig. 5), followed by a slow recovery as a result of new root formation and growth (cf. Table 4). A similar development was probably taking place at Kilpisjärvi, as the height increase from 1997 to 2010 is of the same order as at the Melbu site. There was also a slight negative growth in the stem diameter after transplantation that may be due to loss of primary stems and formation of new stems from the roots. The stem diameter measurements were then continued on these new stems, resulting in negative diameter growth caused by the discontinuity (J. Nilsen pers. comm.).

At the Melbu site, the plants from Kevo showed the highest survival rates, while plants from Hammerfest showed the strongest diameter growth (Figs. 3 & 4). On the other hand, the Kilpisjärvi plants were most successful at Kilpisjärvi (Fig. 3). In both gardens, however, the south-

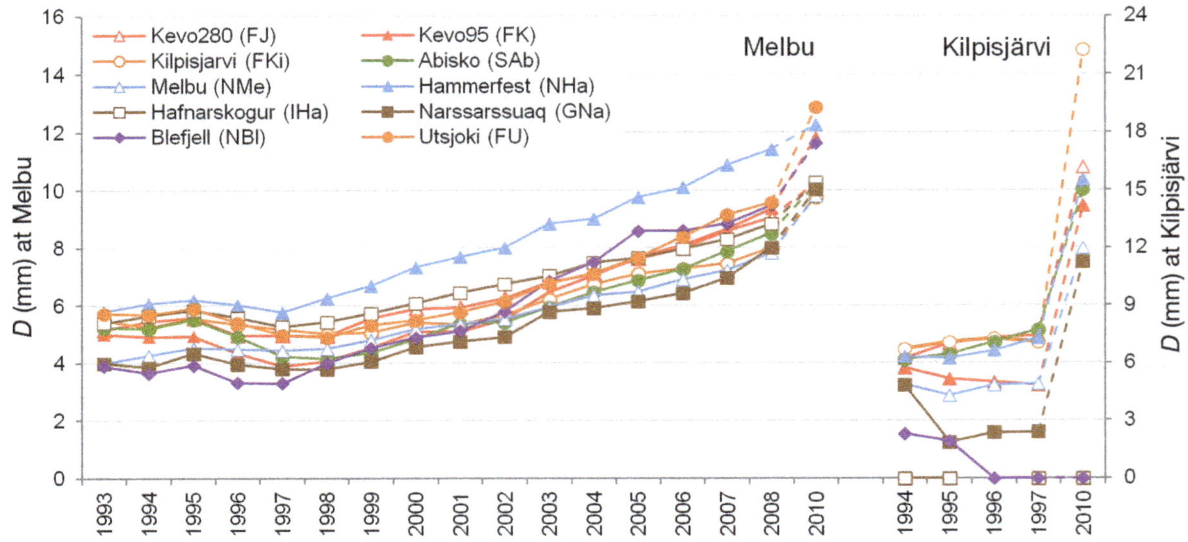

Fig. 4. Stem base diameter (D) for tested provenances of old generations at Melbu (left) and Kilpisjärvi (right) gardens as a function of time since 1992

ern populations (GNa and NBl) had the lowest survival rates.

Comparing the height and diameter growth (Figs. 4 & 5) with the survival rates in Fig. 3, it may be concluded that there also seems to have been a strong decline in survival rates after 1997 at Kilpisjärvi due to climatic stress (Tables 1 & 2) and insect attacks (R. Partanen pers. comm.). However, the surviving plants seem to have recovered and reached very similar stem diameters and total heights in 2010 to the corresponding plants at Melbu.

Studies on plant growth and survival after the combined climate and insect damage in the Kilpisjärvi garden during 1994 to 1997 (Figs. 3, 4 & 5) show that only a few plants from the 3 southernmost populations survived, presumably as a latitudinal effect where plants were subjected to late autumn frost. This latitudinal effect is seen more clearly in Fig. 2, where mean values with 95% CI are shown for the old set of plants at Melbu and Kilpisjärvi. The 2 southernmost populations (NB and GNa) were both less successful in terms of survival rates, height

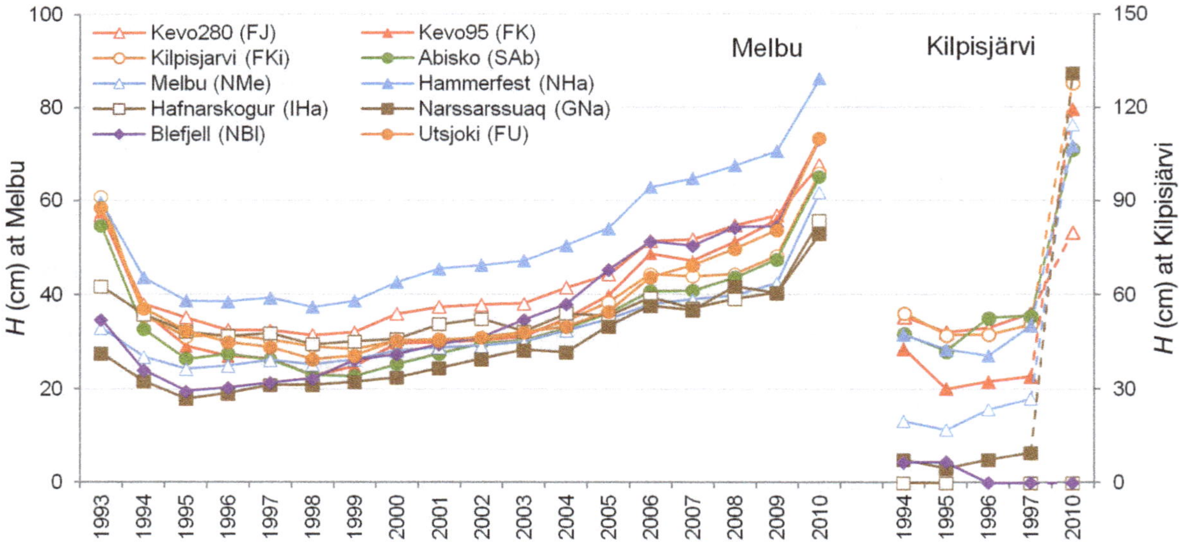

Fig. 5. Total height (H) for tested provenances of old generations at Melbu (left) and Kilpisjärvi (right) gardens as a function of time since 1992

growth and diameter growth than their northern and continental relatives, and in addition, the less southern but oceanic populations from Iceland (IHa) and Melbu (NMe) were also less successful.

3.4. Supplementary transplantation

The survival rates during 2010 to 2014 in the second set of transplanted plants at Vardø and Kilpisjärvi (2002) are shown in Figs. 2 & 6. At the Kilpisjärvi site, there was a strong latitudinal effect (Fig. 2), where the 3 southernmost populations (NBl, GNa and IHa) were less successful than their northern relatives, but also a certain effect of different oceanity, as the Melbu plants (NMe) were less successful than their continental relatives (FK, FKi and SAb). These results confirmed earlier results on the old dataset from 1992 (see Fig. 2). At the Vardø site, the oceanic provenances were relatively more successful than at Kilpisjärvi, compared with the 3 continental provenances. The northern oceanic plants from Hammerfest (NHa) were the most successful in terms of survival rates at both sites (Fig. 2). Measurements of total living height (Fig. A1) partly confirmed the results of the survival rates, but with generally much lower growth rates in the plants growing at the Vardø site than at the Kilpisjärvi site, in agreement with the differences in summer temperature (Table 1).

However, a clear difference was found in survival rates at Vardø between the 2 different parts of the site (Table A1), illustrating the effect of different snow cover and growing season (cf. Holtmeier & Broll 2005). In this respect, the 3 continental populations (FK, FKi and SAb) seemed to be most sensitive. Negative height increments owing to dieback were found in all populations at the 4 cold and wet plots (O. Skre unpubl.). In terms of survival rates and height increment, the warm and dry plot at the Vardø site was similar to the Kilpisjärvi site (Fig. 6). It is interesting to notice that plants from Iceland (IHa) showed relatively high survival rates in spite of its southern origin (cf. Table 1). This adaptation may be partly climatic (Myking 1999), but also genetic as a result of inbreeding with *Betula nana* (Anamthawat-Jonsson et al. 1993, Thórsson et al. 2001) followed by sheep grazing (cf. Bløndal 1993). Further, results from Swedish studies on insect outbreaks (Tenow et al. 2005) indicate that polycormic birch with a high degree of inbreeding with *B. nana* also is more resistant than monocormic birch, simply because of its ability to rejuvenate after damage (cf. Wielgolaski 2005).

The low number of measurements associated with huge within-site variations at the Vardø site was the main reason why no significant differences were found between provenances there. At the Kilpisjärvi site, the 2 southern and oceanic provenances (NB and GNa) were less successful than the others (Table 4), probably as a latitudinal effect (cf. Myking & Heide 1995, Skre et al. 2005). The most successful plants at the warm part of the Vardø site in terms of growth increment (Fig. 6) were those from the northern oceanic populations (e.g. NHa, NMe and IHa).

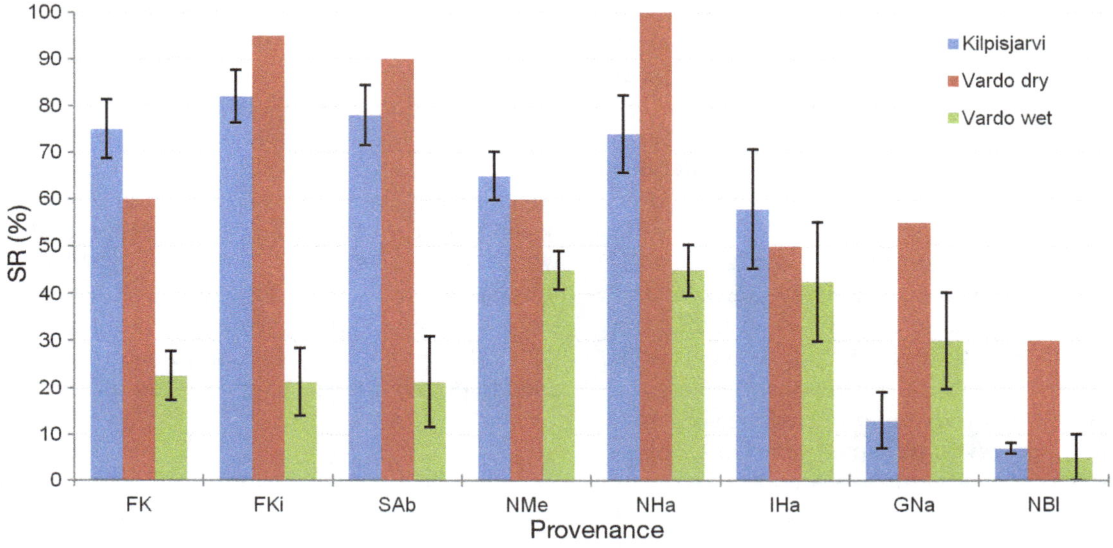

Fig. 6. Mean SR (%) with 95 % CI ± 2 SE in plants at Kilpisjärvi in 2002 and the wet–cold and dry–warm plot at the Vardø site in 2010 to 2014. The dry–warm plot at Vardø is only shown with its mean values

This is in accordance with e.g. Myking (1999), who found higher frost resistance in oceanic birch provenances than in continental relatives, probably as a result of adaptation to a longer spring transition period (cf. Taulavuori et al. 2004, Skre et al. 2008).

The present study showed that the Kilpisjärvi site was subjected to at least 2 severe insect attacks, probably by *Epirrita autumnata* (1995 and 2013), but there was no indication of similar damage at the 2 other sites (Melbu and Vardø), although Melbu is within the potential area of *Operophtera brumata* (Jepsen et al. 2008). Vardø is still well outside the area, but may be included if the present climate change continues. It may be noted that because Melbu is located 50 m below the actual treeline, the plants may be in better shape than if they were located at the treeline, and are therefore able to produce more defence substances against insect attacks than the climatically more stressed plants at Kilpisjärvi, close to the treeline (cf. Haukioja et al. 1988, Neuvonen et al. 2005).

4. CONCLUSIONS

Returning to the initial questions in the introduction, the present results indicate the following:

(1) Climate change with milder winters and a longer growing season, as well as increasing precipitation, may favour mountain birch plants adapted to an oceanic climate, which are less sensitive to frost and insect damage than continental populations. This strongly suggests that mountain birch would be a good indicator for such changes in the future. The survival rates and growth responses in birch seedlings transplanted to the oceanic site at Melbu and the continental site at Kilpisjärvi varied according to the origin of the seed populations. Consequently, northern continental populations were more successful than their southern oceanic relatives at the Kilpisjärvi site. Northern oceanic populations were superior to southern and continental populations at the oceanic Melbu site, but because this site is located 50 m below treeline, the conclusions are less significant than at Kilpisjärvi. As expected, the various birch provenances seem to be best adapted to survive and grow in the climate of their origin. However, transplantation to a different climate makes the plants less adapted, but continental provenances seem to suffer more when transplanted to a coastal site (e.g. Vardø) than oceanic provenances when transplanted to a continental site (Kilpisjärvi). In the future, when a warmer climate is expected, this would favour plants that are already adapted to a coastal climate, while continental plants would be more restricted to inland areas. Plants with high demands for dormancy breaking (cf. Myking & Heide 1995), i.e. southern and oceanic provenances, would have an advantage (Vitasse et al. 2014a) as spring frosts caused by premature dehardening would also be more common (cf. Myking 1999).

(2) The present results seem to indicate that climate is an important driver of change (cf. Körner & Paulsen 2004), although land-use change such as overgrazing by reindeer and abandonment of farmland may be more important in the short run (Bryn 2008, Aune et al. 2011, Callaghan et al. 2013). The frequent insect attacks by geometrid moths are also related to climate change. Increased reindeer grazing as well as insect attacks would be expected to slow down treeline advance as stated by Virtanen & Neuvonen (1999), Jepsen et al. (2008) and Callaghan et al. (2013), while reduced reindeer grazing would enhance treeline advance (Neuvonen et al. 1999). The present study also indicated that large differences in survival and growth responses may occur locally, as shown in the Arctic Vardø transplant garden. Different snow and temperature conditions are the main selective factors, and the northern Hammerfest population (NHa) seemed to be most successful at this site. Finally, birch plants growing close to the treeline seem to be subjected to frequent frost damage, causing dieback of annual shoots and negative height increment.

(3) Future predictions (ACIA 2004) for northern areas indicate that a continued temperature increase by up to 2°C in the next 50 years, and a more oceanic climate along the coast of northern Scandinavia, is associated with increased risk of insect attacks (Jepsen et al. 2008). Polycormic birch with a high degree of inbreeding with *B. nana* would then have an advantage (Nilsen & Wielgolaski 2001). Similarly, oceanic birch provenances (IH, NMe and NHa) would be expected to expand at the expense of continental relatives (FKi, FK, Sab) because they are more adapted to spring frosts (e.g. Taulavuori et al. 2004, Skre et al. 2008). Similar relationships may be used to predict future changes at the species level.

Acknowledgements. Jarle Nilsen is acknowledged for his annual observations in the transplant garden at Melbu, Norway. The study was carried out as part of COST Action ES1203 'Enhancing the resilience capacity of SENSitive mountain FORest ecosystems under environmental change' (SENSFOR).

LITERATURE CITED

ACIA (2004) Impacts of a warming arctic climate. Arctic Climate Impact Assessment (ACIA) Overview Report. Cambridge University Press, New York, NY

Anamthawat-Jonsson K, Heslop-Harrison JS, Tomasson T (1993) Genetics, cytogenetics and molecular genetics of Icelandic birch: implications for breeding and reforestation in cold climates. In: Alden J, Mastrantonio JL, Ødum S (eds) Forest development in cold climates. Plenum Press, New York, NY, p 357–368

Aune S, Hofgaard A, Söderström L (2011) Contrasting climate and land-use driven tree encroachment patterns of subarctic tundra in northern Norway and the Kola Peninsula. Can J Res 41:437–449

Bløndal S (1993) Socioeconomic importance of forests in Iceland. In: Alden J, Mastrantonio JL, Ødum S (eds) Forest Development in cold climates. Plenum Press, New York, NY, p 1–14

Bryn A (2008) Recent forest limit changes in southeast Norway: effects of climate change on regrowth after abandoned utilization? Nor Geogr Tidsskr 62:251–270

Bryn A, Daugstad K (2001) Summer farming in the mountain birch forest. In: Wielgolaski FE (ed) Nordic mountain birch ecosystems. Parthenon Publishing, New York, NY, p 307–315

Cairns DM, Lafon C, Moen J, Young A (2007) Influence of animal activity on treeline position and pattern: implications for treeline responses to climate change. Phys Geogr 28:419–433

Callaghan TV, Jonasson C, Thierfelder T, Yang Z and others (2013) Ecosystem change and stability over multiple decades in the Swedish subarctic: complex processes and multiple drivers. Philos Trans R Soc Lond B 368: 20120488

Dalen L, Hofgaard A (2005) Differential regional treeline dynamics in the Scandes mountains. Arct Antarct Alp Res 37:284–296

de Wit HA, Bryn A, Hofgaard A, Karstensen J, Kvalevåg MM, Peters G (2014) Climate warming feedback from mountain birch forest expansion: reduced albedo dominates carbon uptake. Glob Change Biol 20:2344–2355

Gamache I, Payette S (2005) Latitudinal response of subarctic tree lines to recent climate change in eastern Canada. J Biogeogr 32:849–862

Gehrig-Fasel J, Guisan A, Zimmermann NE (2007) Tree-line shifts in the Swiss Alps: climate change or land abandonment? J Veg Sci 18:571–582

Haukioja E, Neuvonen S, Hanhimäki S, Niemelä P (1988) The autumnal moth in Fennoscandia. In: Berryman AA (ed) Dynamics of forest insect populations, patterns, causes, implications. Plenum Press, New York, NY, p 163–178

Heide OM (1993) Daylength and thermal time responses of budburst during dormancy release in some northern deciduous trees. Physiol Plant 88:531–540

Holtmeier FK (2007) Mountain timberlines: ecology, patchiness and dynamics. Springer, Berlin

Holtmeier FK, Broll G (2005) Sensitivity and response of Northern Hemisphere altitudinal and polar treelines to environmental change at landscape and local scales. Glob Ecol Biogeogr 14:395–410

Jepsen JU, Hagen SB, Ims RA, Yoccoz NG (2008) Climate change and outbreaks of the geometrids *Operophtera brumata* and *Epirrita autumnata* in subarctic birch forests: evidence of a recent outbreak range expansion. J Anim Ecol 77:257–264

Jepsen JU, Hagen SB, Høgda KA, Ims RA, Karlsen SR, Tømmervik H, Yoccoz NG (2009) Monitoring the spatio-temporal dynamics of geometrid moth outbreaks in birch forests using MODIS-NDVI data. Remote Sens Environ 113:1939–1947

Jepsen JU, Biuw M, Ims RA, Kapari L, Schott T, Vindstad OPL, Hagen SB (2013) Ecosystem impacts of a range expanding forest defoliator at the forest-tundra ecotone. Ecosystems 16:561–575

Juntunen V, Neuvonen S (2006) Natural regeneration of Scots pine and Norway spruce close to the timberline in northern Finland. Silva Fenn 40:443–458

Karlsson PS, Nordell KO (1996) Effects of soil temperature on nitrogen economy and growth of mountain birch near its presumed low temperature distribution limit. Ecoscience 3:183–189

Karlsson PS, Weih M (2003) Long-term patterns of leaf, shoot and wood production after insect herbivory in the mountain birch. Funct Ecol 17:841–850

Körner C, Paulsen J (2004) A world-wide study of high-altitude treeline temperatures. J Biogeogr 31:713–732

Kullman L (1998) Tree-limits and montane forests in the Swedish Scandes: sensitive biomonitors of climate change and variability. Ambio 27:312–321

Kuznetsova A, Brockhoff KA, Christensen RHJ (2016) lmerTest: tests in linear mixed effects models. R package version 2.0-30. Available at https://CRAN.R-project.org/package=lmerTest

Larsen JB (1976) Untersuchungen über die Frostempfindlichkeit von Douglasienherkünften und über den Einfluss der Nährstoffversorgung auf die Frostresistenz der Douglasie. Forst Holz Wirtsch 31:299–302

Lehtonen J, Heikkinen RK (1995) On the recovery of mountain birch after *Epirrita* damage in Finnish Lapland, with particular emphasis on reindeer grazing. Ecoscience 2: 349–356

Leinonen I (1996) A simulation model for the annual frost hardiness and freeze damage of Scots pine. Ann Bot (Lond) 78:687–693

Lüdecke D (2016) sjPlot: data visualization for statistics in social science. R package version 1.9.4. Available at http://CRAN.R-project.org/package=sjPlot

Marc J, Mazerolle A (2016) AICcmodavg: model selection and multimodel inference based on (Q)AIC(c). R package version 2.0-4.0. Available at http://CRAN.R-project.org/package=AICcmodavg

Myking T (1999) Winter dormancy release and budburst in *Betula pendula* Roth. and *B. pubescens* Ehrh. ecotypes. Phyton (Horn) 39:139–146

Myking T, Heide OM (1995) Dormancy release and chilling requirements of buds of latitudinal ecotypes of *Betula pendula* and *B. pubescens*. Tree Physiol 15:697–704

Neuvonen S, Niemelä P, Virtanen T (1999) Climate change and insect outbreaks in boreal forests: the role of winter temperatures. Ecol Bull 47:63–67

Neuvonen S, Bylund H, Tømmervik H (2005) Forest defoliation risks in birch forest by insects under different climate and land use scenarios in northern Europe. In: Wielgolaski FE (ed) Plant ecology, herbivory and human impact in Nordic mountain birch forests. Springer-Verlag, Berlin, p 125–138

Nilsen J, Wielgolaski FE (2001) Effects of fertilization and watering on morphology in young mountain birch plants of different provenances — a pilot study. In: Wielgolaski FE (ed) Nordic mountain birch ecosystems. UNESCO, Paris and Parthenon, New York, NY, p 71–76

Ovaska JA, Nilsen J, Wielgolaski FE, Kauhanen H and others (2005) Phenology and performance of mountain birch provenances in transplant gardens: latitudinal, altitudinal and oceanity–continentality gradients. In: Wielgolaski FE (ed) Plant ecology, herbivory and human impact in Nordic mountain birch forests. Springer-Verlag, Berlin, p 99–115

Payette S, Eronen M, Jasinski JJP (2002) The circumboreal tundra–taiga interface: Late Pleistocene changes. Ambio 12:15–22

R Core Team (2015) R: a language and environment for statistical computing. R Foundation for Statistical Computing, Vienna. Available at https://www.R-project.org/

Scott PA, Hansell IRC, Ericsson WR (1993) Influence of wind and snow on northern treeline environments at Churchill, Manitoba, Canada. Arctic 46:316–323

Skre O, Baxter R, Crawford RMM, Callaghan TV, Fedorkov A (2002) How will the tundra–taiga interface respond to climate change? Ambio 12:37–46

Skre O, Nilsen J, Næss M, Igeland B, Taulavuori K, Taulavuori E, Laine K (2005) Effects of temperature changes on survival and growth in mountain birch populations. In: Wielgolaski FE (ed) Plant ecology, herbivory and human impact in Nordic mountain birch forests. Springer-Verlag, Berlin, p 87–98

Skre O, Taulavuori K, Taulavuori E, Nilsen J, Igeland B, Laine K (2008) The importance of hardening and winter temperature for growth in mountain birch populations. Environ Exp Bot 62:254–266

Speed JDM, Austrheim G, Hester AJ, Mysterud A (2011) Browsing interacts with climate to determine tree-ring increment. Funct Ecol 25:1018–1023

Sveinbjörnsson B, Hofgaard A, Lloyd A (2002) Natural causes of the tundra-taiga boundary. Ambio 12:23–29

Taulavuori K, Taulavuori E, Skre O, Nilsen J, Igeland B, Laine K (2004) Dehardening of mountain birch (*Betula pubescens* ssp. *czerepanovii*) ecotypes at elevated temperatures. New Phytol 162:427–436

Tenow O, Bylund H, Nilsen AC, Karlsson PS (2005) Long-term influence of herbivores on northern birch forests. In: Wielgolaski FE (ed) Plant ecology, herbivory and human impact in Nordic mountain birch forests. Springer-Verlag, Berlin, p 165–182

Thórsson AT, Salmela E, Anamthawat-Jonsson K (2001) Morphological, cytogenetic and molecular evidence for introgressive hybridization in birch. J Hered 92:404–408

Tømmervik H, Johansen B, Tombre I, Thannheiser D, Høgda KA, Gaare E, Wielgolaski FE (2004) Vegetation changes in the Nordic mountain birch forests: the influence of grazing and climate change. Arct Antarct Alp Res 36:323–332

Virtanen T, Neuvonen S (1999) Performance of moth larvae on birch in relation to altitude, climate, host quality and parasitoides. Oecologia 120:92–101

Vitasse Y, Lenz A, Körner C (2014a) The interaction between freezing tolerance and phenology in temperate, deciduous trees. Front Plant Sci 5:541

Vitasse Y, Lenz A, Hoch G, Körner C (2014b) Earlier leaf-out rather than difference in freezing resistance puts juvenile trees at greater risk of damage than adult trees. J Ecol 102:981–988

Weih M, Karlsson PS (1999) Growth responses of altitudinal ecotypes of mountain birch to temperature and fertilization. Oecologia 119:16–23

Wielgolaski FE (2005) History and environment of the Nordic mountain birch. In: Wielgolaski FE (ed) Plant ecology, herbivory and human impact in Nordic mountain birch forests. Springer-Verlag, Berlin, p 3–18

Appendix

Table A1. Mean dates of budbreak at Replicates I to IV and V in Vardø during the observation years 2010 to 2015, and estimated snow cover in days (SMD = date of snowmelt, SFD = date of first snowfall, SD = duration of snow-free period)

Year[a]	I to IV	V
2010	24.6	17.6
2012	23.6	14.6
2014	23.7	16.6
2015	15.6	9.6
Mean	21.6	14.6
SMD	17.5	10.5
SFD	17.10	30.10
SD	155	169

[a]Observations on budbreak were missing in 2011 and 2013, but were taken in 2015, after the end of the sample period (2014)

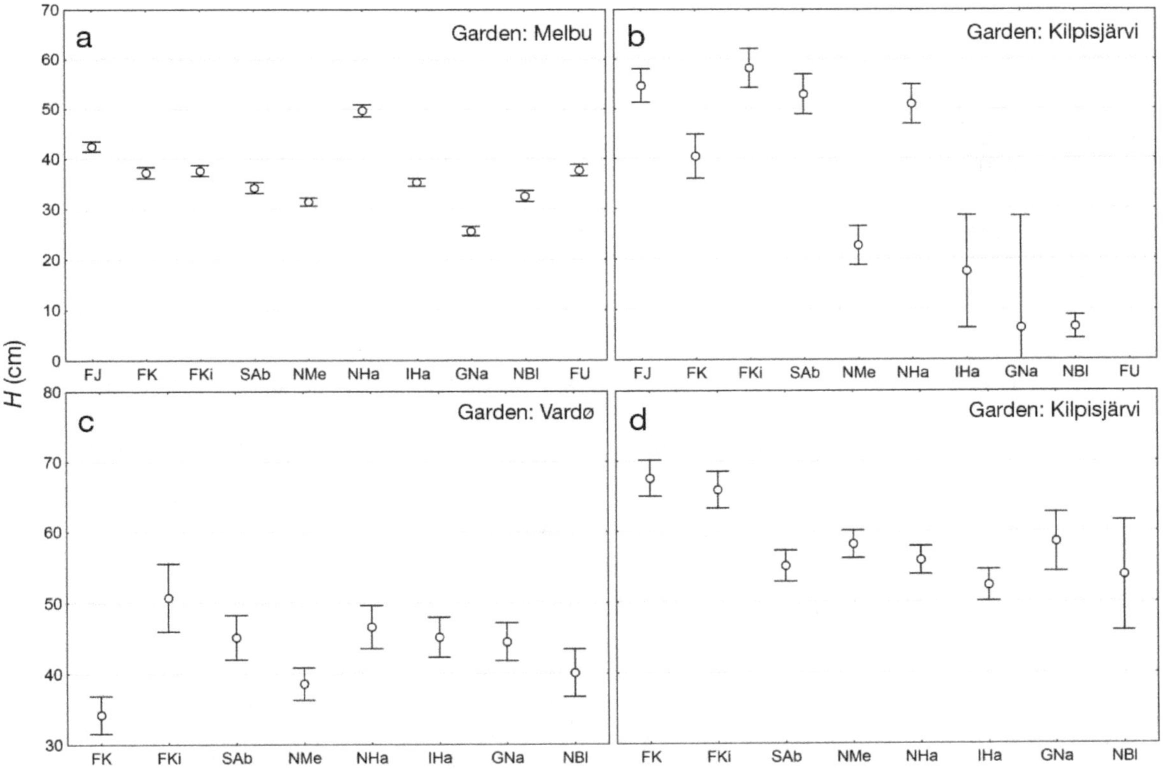

Fig. A1. Mean values of total height (H), with corresponding 95% CI for tested provenances of (a,b) old generation of plants at (a) Melbu and (b) Kilpisjärvi and (c,d) new generation of plants at (c) Vardø and (d) Kilpisjärvi

Future forest distribution on Finnmarksvidda, North Norway

Stein Rune Karlsen[1,*], Hans Tømmervik[2], Bernt Johansen[1], Jan Åge Riseth[1]

[1]Norut Northern Research Institute, PO Box 6434, 9294 Tromsø, Norway
[2]Norwegian Institute for Nature Research, FRAM – High North Centre for Climate and the Environment, PO Box 6606 Langnes, 9296 Tromsø, Norway

ABSTRACT: Finnmarksvidda is Norway's largest mountain plateau, located in the Arctic/alpine-boreal transition area. The area is also a central winter grazing area for the reindeer herds of the indigenous Sámi people. This study develops a bioclimatic-based model to simulate future potential vegetation, with a focus on forest types. The model utilizes a bioclimatic study in the region, where vegetation types have been grouped according to minimum summer temperature demands. This is then used as a base for modelling of future vegetation. Due to the flat landscape of Finnmarksvidda, the model shows that a 1°C increase in summer temperatures will potentially lead to an increase of forested areas by 4485 km^2, which is a 70% increase from the current 6900 km^2 to a simulated 11 706 km^2. This in turn will lead to a reduction of Arctic-alpine heaths from 4440 km^2 today to potentially only 670 km^2. Such changes will have consequences for the reindeer grazing system, as the predicted changes will lead to a decrease in the vegetation types that have high winter grazing accessibility for reindeer, from 2386 km^2 today to potentially only 377 km^2. On the other hand, vegetation types with medium accessibility will experience an increase, from 2857 to 3366 km^2.

KEY WORDS: Climate warming · Bioclimatic model · Temperature driver · Future forest · Treeline · Reindeer herding

1. INTRODUCTION

A predicted and to some extent realized consequence of global warming is the spread of shrub and birch forests to the Arctic tundra and its low-alpine fringes (e.g. Hofgaard et al. 2013). Recent research results show that shrub encroachment and tree invasion on the northern Fennoscandian tundra have been more rapid than previously expected (Kullman 2002, Tømmervik et al. 2009), and these changes are mainly human-driven and occurring concurrently with climatic change (Tømmervik et al. 2009, Hofgaard et al. 2013). Finnmarksvidda is Norway's largest mountain plateau and is located in the low alpine–northern boreal transition zone in northern Norway. This area therefore might potentially undergo considerable vegetation changes in terms of increased forest cover in a future warming climate with summer temperatures as the main driver. Finnmarksvidda is also the central area for the indigenous Sámi people's reindeer herds. The Sámi people with their reindeer herds utilize the differences in vegetation composition between the coast and Finnmarksvidda situated further inland when they move their herds according to the changing seasons. In winter, reindeer graze the alpine lichen heaths in the interior, and in spring, they move to the coast and take advantage of the early greening, grazing on herbs and grasses (Johansen & Karlsen 2005). Shifts in the vegetation composition affect this finely tuned reindeer grazing system (Tømmervik et al. 2012, Riseth et al. 2016). In particular, the spread of birch and pine forests to the Arctic and alpine parts will greatly influence the amount of lichen. During the last decades, there has been evidence of altitudinal

*Corresponding author: stein.rune.karlsen@norut.no

§Advance View was available online May 18, 2017

and latitudinal advances in forest limits in northern Fennoscandia (Tømmervik et al. 2009, Aune et al. 2011, Van Bogaert et al. 2011, Hofgaard et al. 2013, Franke et al. 2015). However, the pattern is not clear, and these studies also indicate that it is difficult to separate the climatic drivers that lie behind the forest advances from other drivers, and the advances are possibly more dependent on the grazing regime. Understanding the response of terrestrial ecosystems to climatic warming is a challenging task because of the complex interactions of drivers such as climate, disturbance and recruitment across the landscape. Hence, the transition from alpine heaths to boreal forests is potentially controlled by many drivers that include climate, soil substrate changes, grazing, topography and disturbance, and may be expressed as either gradual or abrupt spatial changes (see e.g. Sveinbjörnsson et al. 2002, Holtmeier & Broll 2007). An attempt using dynamic vegetation modelling (LPJ-GUESS model) for the Barents Region (Wolf et al. 2008) overestimated the forest abundance for the northern parts of the Kola Peninsula (Wolf et al. 2008). This modelling was carried out using a very coarse vegetation map (Olson Ecosystem map; spatial resolution of 0.5° × 0.5° grid). Hence, by studying the modelling maps from this study for Finnmark county, Norway, an overestimation of the forest abundance also seems to be the case, since this part of the Barents Region has more rugged and high mountainous terrain that includes steep barren mountains and glaciers, where no tree can sprout and survive. One of the most difficult challenges in projecting vegetation dynamics from climate change is the development of rules for vegetation responses to climate.

This study aims to create a bioclimatic-based model to simulate potential future vegetation and forest distribution on Finnmarksvidda. This is done by utilizing a bioclimatic study of the region (Karlsen et al. 2005), where the vegetation types have been grouped according to minimum summer temperature demands and then used as a base for modelling future vegetation, where a method for this has recently been developed (Sjögren et al. 2015). Next, we also discuss some possible impacts on the reindeer winter grazing system.

2. MATERIALS AND METHODS

2.1. Study area and current treeline

Finnmarksvidda is situated in the low-alpine–northern boreal transition ecotone between 68–70° N and 22–26° E (Fig. 1). The municipalities Karasjok (5453 km^2) and Kautokeino (9708 km^2) constitute most of this plateau. The Precambrian bedrock of Finnmarksvidda was reshaped by glacier activity during the Pleistocene era and is covered with a moraine layer, which forms a flat and gentle landscape with nutrient-poor soils such as podzols and nanopodzol (Lindström 1987). The current treeline is made up of birch *Betula pubescens* and varies a lot across Finnmarksvidda. It lies as low as 360 m altitude in the central north and can reach up to 520 m altitude in the southwestern parts, where it is common to find individual trees even at 600 m altitude. However, most often, the treeline lies between 420 and 480 m altitude. In most cases, on dry and nutrient-poor soil, the treeline gradually changes from open multi-stemmed birch forest to dwarf birch heaths (Tømmervik et al. 2009). In Karasjok and Kautokeino, as much as 22 and 63 % of the municipalities are between 400 and 600 m altitude, respectively. Hence, large parts of Finnmarksvidda are situated between the birch treeline and the birch forest line.

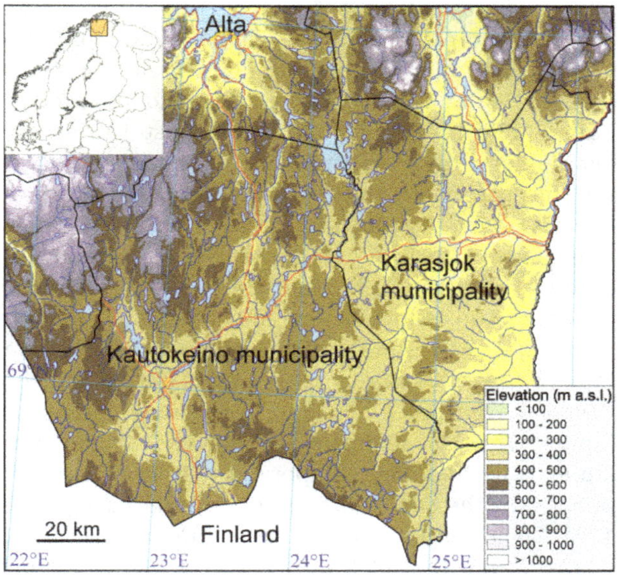

Fig. 1. An elevation model of the study area, Finnmarksvidda in northern Norway (inset: Fennoscandia). Current tree line is formed by birch and is between 420 and 480 m altitude

2.2. Previous, present and future climate

Finnmarksvidda has a continental climate where the meteorological stations in the villages Karasjok and Kautokeino show a mean July temperature (standard normal 1961–1990) of 13.1 and 12.4°C, respectively (Aune 1993). The mean annual precipitation is 366 mm in Karasjok and 325 mm in Kautokeino (Førland 1993), with about half of the precipitation falling between June and August. During the last century, the Finnmarksvidda summer temperature has shown a trend of about 0.07°C increase per decade (Vikhamar-Schuler et al. 2010, Hanssen-Bauer et al. 2015). Precipitation has been variable, with wet summers in the mid-1960s and in the last 2 decades. The length of the phenologically defined growing season is in the range of 90–130 d (Karlsen et al. 2008), without any clear trends during the last decades (Høgda et al. 2013). The future scenarios for Finnmarksvidda indicate an increase in summer (June, July, August) temperature in the range of 2–5.5°C by the end of this century compared with the 1971–2000 period, depending on the scenario used (Hanssen-Bauer et al. 2015). The prediction for summer precipitation (Hanssen-Bauer et al. 2015) is in the range of 15–24% increase, while it is 13–14% increase for winter precipitation (December, January, February).

2.3. Simulation of potential future forest distribution

Recently, a bioclimatic-based method was developed to model vegetation backwards in time (Sjögren et al. 2015). In this study, we apply the method to model potential future vegetation on Finnmarksvidda, with a focus on forest types. The starting point of the modelling is the result from an extensive bioclimatic study on Varangerhalvøya, northeast Norway. On the peninsula, methods for using plant species and vegetation types have been developed and applied (Karlsen & Elvebakk 2003, Karlsen et al. 2005). The bioclimatic methods for climatic mapping are based on the fact that most plant species and plant communities both in the Arctic and adjacent areas have to some extent a distribution pattern limited by growing season temperature. Varangerhalvøya is a peninsula located approximately 150 km east-northeast of Finnmarksvidda, and most of the results from the peninsula can be transferred to the study area. The vegetation on the peninsula was first mapped using Landsat TM satellite data, and then the mapped vegetation units were defined as temperature indicators based on their total distribution patterns and the temperature indicator value of their high frequency and dominant species, as found by relevé analysis. Vegetation units were grouped according to their minimum temperature requirements in terms of mean July temperature. Vegetation types were also grouped according to their habitat preferences. In this study, we started with a Landsat TM-based vegetation map from Finnmarksvidda (Johansen & Karlsen 2007). This map was then compared with vegetation units that had been previously grouped according to their minimum temperature requirements in terms of mean July temperature mapped on Varangerhalvøya (Karlsen et al. 2005). All mapped vegetation types on Finnmarksvidda, except pine (*Pinus sylvestris*) forests and lichen (*Cladonia* spp.) heaths, have related types on Varangerhalvøya. Mires were excluded from the study due to their large variation in species composition and fine mosaic structures that are not easily detected by Landsat TM satellite data. The mapped vegetation types on Finnmarksvidda were then grouped according to their minimum temperature preferences, with increments of 1°C difference in mean July temperature. Also, the pine forest types and lichen heaths were forced into temperature and habitat groups, by judgement of species composition found by literature studies (Wehberg 2007).

After the grouping of all vegetation types (except for mires) had been carried out according to minimum temperature and habitat preferences, we proceeded to the next step, which involved analysing the altitude distribution of all vegetation types in relation to the closest treeline today. The current treeline is formed by birch and varies across the study area. Based on the forest data supplied by the Norwegian Mapping Authority (N50 data) and modified by hyperclustering-based classification (Hofgaard et al. 2013), polylines of the upper treeline were extracted with the use of ArcGIS software. We then compared the upper treeline with a terrain model and created a gridline raster where each pixel was assigned altitude information. Since these gridlines only covers small parts of the study area, we interpolated the data (kriging) to create a raster that covers the whole study area, where each pixel was assigned a value corresponding to the altitude of the closest upper treeline. To avoid outliers when determining the current upper limit of the different vegetation types, we used the height limit where 95% of the area covered by each vegetation types occurred. The final step was to simulate future vegetation expected from increased July temperature. The simulation was performed using Python scripts in the ArcGIS software. We assumed that a 1°C change in July temperature would correspond to a 171 m change in altitude of the treeline, based on a temperature change of 0.585°C per 100 m. We also assumed that current temperature-dependent (thermophilous) vegetation types would expand into colder areas and replace less temperature-dependent vegetation types with equal habitat preferences. For instance, a blueberry-birch forest type (which requires approximately 10°C mean July temperature) will replace a blueberry-crowberry alpine heath (which requires approximately 9°C mean July temperature). Both vegetation types occur on mesic habitats. In cases with no climatic/habitat counterparts, the vegetation types were reclassified to coarser units. The simulation is regulated such that the replacement from one group to another must not exceed a 171 m altitude change per 1°C change in mean July temperature. For instance, if a forest type at present reaches 100 m below the current treeline, it will potentially reach 71 m above the current treeline with a 1°C increase in mean July temperature. Finally, all mapped and modelled units were judged according to winter reindeer grazing accessibility (cf. Johansen & Karlsen 2005, 2007).

3. RESULTS

3.1. Present vegetation cover

The present and potential future vegetation cover on Finnmarksvidda is illustrated in Fig. 2. Table 1 shows the present and future cover of the modelled vegetation types for Karasjok and Kautokeino municipalities respectively. Fig. 3 summarizes the changes in forest distribution, and at present, forest covers 62% of the total area in Karasjok and 36% in Kautokeino. In Karasjok, birch is the most common forest type, but at lower altitudes, pine and mixed birch-pine forests are also present. In Kautokeino, the forests consist almost completely of birch. Alpine heaths and meadows cover 25% in Karasjok and 43% in Kautokeino, while mires cover 10 and 16% respectively. Vegetation types with high or medium winter grazing accessibility covers 25% of the total area in Karasjok and 40% in Kautokeino (Table 1).

3.2. Potential future vegetation cover

The results of the simulation indicate potentially dramatic changes in vegetation cover due to temperature increase (Table 1, Figs. 2 & 3). In Karasjok, forest-covered areas will potentially increase from 62 to 84%. Such changes will cause a reduction in heaths and meadows from 25 to only 3%. In particular, the pine forest and mixed pine-birch forests types will increase in Karasjok. In Kautokeino, forested areas will double from 36 to 74%, and correspondingly, alpine heaths and meadows will be reduced from 43% cover to 5%. The mixed pine-birch forest will increase from 1.5 to 18% cover. Lichen-rich

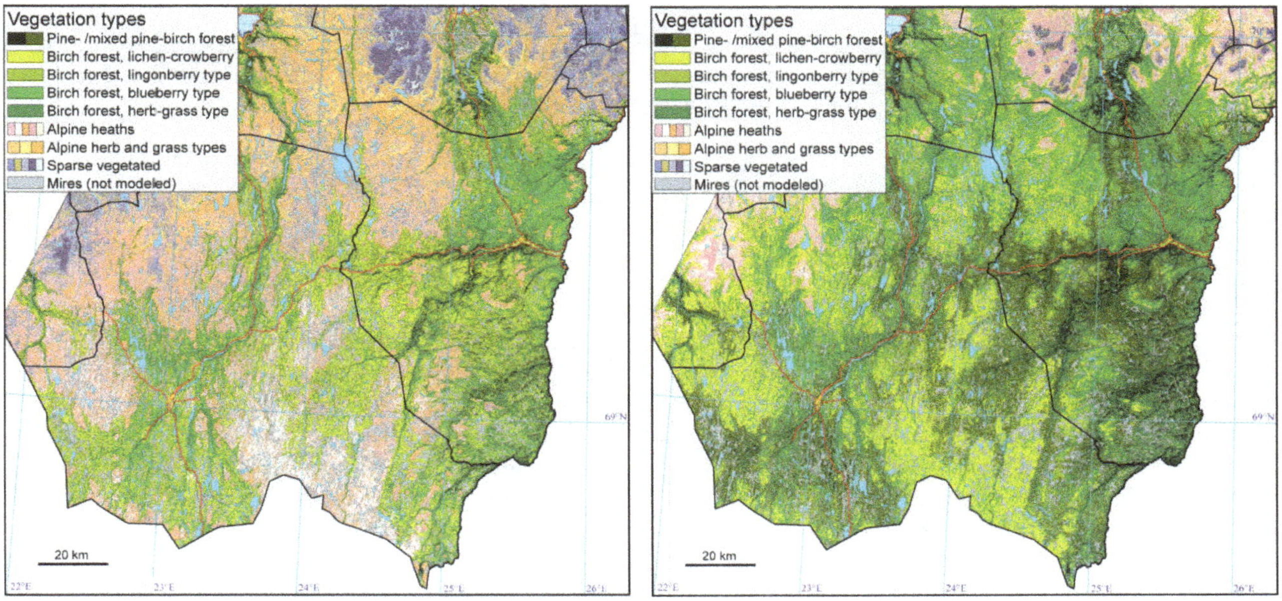

Fig. 2. Left: current vegetation, redrawn from Johansen & Karlsen (2007). Right: scenario map showing simulated potential forest distribution at 1°C increase of July temperature

Table 1. Current and simulated potential future cover of main vegetation units in Karasjok and Kautokeino municipalities. The cover of current (year 2006) main vegetation units is extracted from Johansen & Karlsen (2007). The area of simulated potential future cover is based on 1°C increase of mean July temperature. Mire types are not included in the simulation

	Winter grazing accessibility for reindeer	Karasjok Present km²	%	Karasjok +1°C km²	%	Kautokeino Present km²	%	Kautokeino +1°C km²	%
Pine forests	Medium	205	3.8	697	12.8	184	1.9	321	3.3
Mixed pine-birch forest	Low	549	10.1	1432	26.3	145	1.5	1768	18.2
Birch forest, open lichen-crowberry type	Medium	484	8.9	353	6.5	1002	10.3	1947	20.1
Birch forest, dense lingonberry type	Low	1070	19.6	474	8.7	1146	11.8	1060	10.9
Birch forest, mesic type, blueberry and grass-dominated	Low	524	9.6	661	12.1	343	3.5	620	6.4
Birch forest, herb and grass-dominated	Low	562	10.3	979	18.0	686	7.1	1394	14.4
Exposed heaths	High	336	6.2	30	0.6	1462	15.1	218	2.2
Lichen heaths	High	24	0.4	26	0.5	564	5.8	103	1.1
Crowberry-dwarf birch heaths	Medium	301	5.5	9	0.2	681	7.0	39	0.4
Fresh heaths and meadows	Low	576	10.6	56	1.0	1037	10.7	167	1.7
Unvegetated/sparsely vegetated		105	1.9	17	0.3	391	4.0	5	0.1
Mires (not modelled)		556	10.2	556	10.2	1527	15.7	1527	15.7
Water		154	2.8	154	2.8	533	5.5	533	5.5
Total		5447	100.0	5447	100.0	9702	100.0	9702	100.0

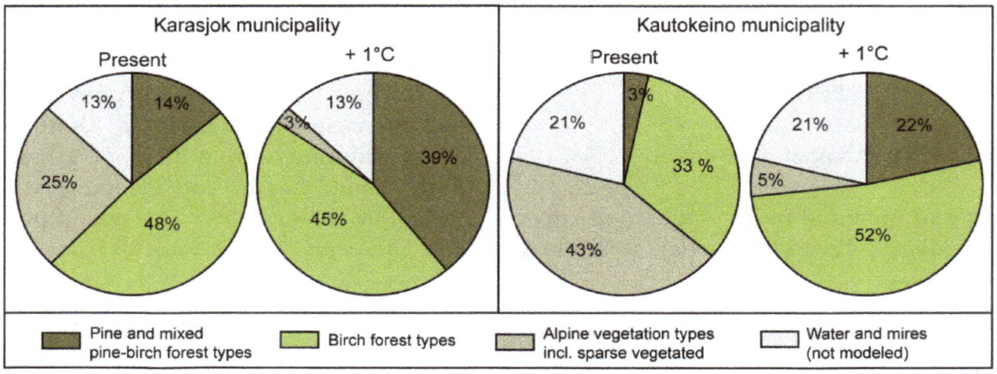

Fig. 3. Present and potential future (1°C increase of July temperature) forest distribution in Karasjok (left) and Kautokeino (right) municipalities, extracted from Table 1

crowberry (*Empetrum hermaphroditum*)-birch forest and grass and herb-rich birch forest will also approximately double their cover (Table 1). In total for Karasjok and Kautokeino, forested areas will increase by 4485 km², which corresponds to a 70% increase from 6900 km² at present to a simulated 11706 km². This will also lead to a reduction of mountain heaths from 4440 km² at present to potentially only 670 km². These changes will result in a dramatic decrease in vegetation types with high winter grazing accessibility from 2386 km² today to potentially only 377 km². Vegetation types with low winter grazing accessibility will increase in area by 3010 km² (from 5601 to 8611 km²). The area covered by vegetation types with medium winter grazing accessibility will also increase from 2857 to 3366 km².

4. DISCUSSION

4.1. Reliability of simulation of future forest distribution

The modelling in this study indicates that a dramatic increase in the distribution of the forests on Finnmarksvidda may take place with only a 1°C increase in mean July temperature. This modelled change will completely change all aspects of the ecosystems on Finnmarksvidda, by having effects on biodiversity, energy exchange, reindeer husbandry and other land use. However, can this be true? It is challenging to develop rules for vegetation response to climate change. To simulate changes in tree cover and vegetation, advanced models like the ALFRESCO model in Alaska (Rupp et al. 2000a,b) have been developed, which take into account several drivers like temperature, precipitation, seed dispersal and fire. Or for example the dynamic vegetation modelling framework LPJ-GUESS, which is designed to simulate global and regional dynamics and composition of vegetation in response to changes in climate, atmospheric CO_2 concentration and nitrogen deposition (Smith et al. 2001, 2014). On the other hand, an example of a very simple model uses only a terrain model and altitudinal replacement, assuming that the current forest distribution is in equilibrium with temperature and would change with increased summer temperature. Such a model is used for simulation of future forest distribution in current alpine areas of Sweden (Moen et al. 2004). A similar reasoning is also common when inferring past temperatures from pollen records in Fennoscandia (Seppä & Hicks 2006). Temperature as a determinant of tree distributions is a question of scale (e.g. Holtmeier & Broll 2005). The importance of temperature on a global scale is clearly reflected in the gradual descent of the treeline from its maximum altitudinal position in the subtropics towards the pole. Grace et al. (2002) show that temperature seems to limit forest growth more significantly than it limits photosynthesis over the temperature range 5–20°C. If we assume that growth and reproduction of birch are controlled by temperature, an advance of the birch treeline would be predicted. Grace et al. (2002), Lucht et al. (2002) and Hofgaard et al. (2013) have provided evidence using remotely sensed data that suggest that this is occurring. Our results presented in Figs. 2 & 3, showing the simulated potential increase of forested areas in the future, are in agreement with these authors. On a local scale, hydrology/snow or rather, topography and substrate texture governing hydrology, are probably the most decisive drivers determining the distribution of plants. This study works on a regional scale with pixel size on 30 m and hence drivers other than temperature influence the forest distribution. In addition to temperature, our modelling also takes into account available habitats. The model is thereby able to simulate the future distribution of 5 main forest types. This also ensures for instance that it is not possible for forest to grow into an area that is currently without soil, since soil formation is a slow process in the mountains. However, the role that trees themselves have on the environment is not taken into account. Trees affect the physical and edaphic conditions. New trees will change soil and hydrological properties and thereby change the habitats. Trees may also increase snow distribution and accumulation, reduce wind, nutrient conditions and energy exchange, since trees and shrubs reduce surface albedo (e.g. Sveinbjörnsson et al. 2002, Holtmeier & Broll 2007, de Wit et al. 2014). Still, it is reasonable to believe that summer temperature is the main limiting factor. Evidence that Finnmarksvidda was most likely covered by forests in warmer periods during the Holocene (Høeg 2000, Seppä & Hicks 2006, Huntley et al. 2013) support this view.

There is already a general trend towards more forest cover on Finnmarksvidda. During the last decades, the vegetation cover has changed dramatically on Finnmarksvidda; lichen heaths have been reduced and birch forest has increased (Tømmervik et al. 2004, 2009, 2012, Johansen & Karlsen 2005, 2007, Hofgaard et al. 2013). This is mainly due to changes in grazing pressure by reindeer. The removal of 'the barrier effect' of the thick lichen coverage (Sedia & Ehrenfeld 2003) by heavy reindeer grazing and trampling provides open sites that makes it easier for birch seeds to germinate and sprout (Houle & Filion 2003, Tømmervik et al. 2004). Eskelinen & Virtanen (2005) stated that natural grazing by reindeer favours species colonization and seedling emergence. This has in turn led to the establishment of clusters of trees and subsequent forests (Houle & Filion 2003, Tømmervik et al. 2004). Furthermore, this has led to an elevation of the forest upper boundary by the filling in of the gap between the 'old' forest line and the treeline with forest (Sveinbjörnsson et al. 2002, Tømmervik et al. 2004). On the other hand, intense reindeer grazing can have negative impacts and cause erosion, and seedling establishment can be hampered by drought (e.g. Holtmeier 2012). However, factors other than temperature and grazing may have influenced these changes. Increased precipitation, which has occurred on Finnmarksvidda during the last few decades, is also important (Tømmervik et al. 2009). Another important disturbance factor influencing growth at the treeline in the region is insect defoliation resulting from outbreaks of geometrid moths (Tenow et al. 2007, Jepsen et al. 2009). Such outbreaks may cause treeline retreat. Finnmarksvidda experienced large outbreaks in both 2004 and 2005 (Bjerke et al. 2014), which may have influenced the treeline locally (authors' pers. obs., Holtmeier 2012). However, disturbances such as moth outbreaks (Karlsen et al. 2013) and winter warming (Bokhorst et al. 2009, Bjerke et al. 2014) might also lead to more rapid birch forest growth as it may damage crowberry *Empetrum hermaphroditum*, which possibly prevent birch seeds germinating and sprouting (Nilsson & Zackrisson

1992, Bråthen et al. 2010). In contrast to the described effects of reindeer grazing on lichen pastures in the winter grazing areas, the effects on herbs, willows and trees in summer pasture areas are quite different, as reindeer can stabilize and move the treeline downwards (Cairns & Moen 2004, Cairns et al. 2007, Tømmervik et al. 2009), and also promote the growth of rare mountain plants (Olofsson & Oksanen 2005). Prevailing disturbance regimes in the coastal parts of Finnmark are to a large extent species-specific, but directly or indirectly related to climate (Hofgaard et al. 2013). Also, grazing by reindeer, moose, cows and sheep might prevent shrubification and forest advance (Tømmervik et al. 2009, Hofgaard et al. 2013). In particular, this is evident for birch forests on the coast and the near-coast inland. However, at the coast, the reduced number of cattle and sheep during the last decades have increased the forest cover in these areas (Fylkesmannen i Finnmark 2012, Hofgaard et al. 2013).

4.2. Lagged response to climate change

We are then left with the question: when will the modelled increase in forest distribution happen? How long is the lagged response? Our modelling indicates the direct magnitude of the change, but reveals little about the rate and pattern of change. Due to the flat landscape of Finnmarksvidda and the large variation in the altitude of the treeline, the main increase in forest distribution in our modelling largely results from the filling in of gaps between the existing trees and tree groups, and to a lesser extent from the establishment of trees at sites above the present tree limit. It is reasonable to believe that this will happen gradually and that the altitudinal shift of the treeline may lag behind climate change on a timescale of decades. There will likely be local differences whereby changes will take place according to variations in soil conditions, insect outbreaks, winter warming effects, precipitation and grazing pressure.

Pine has a slower growth rate and dispersal rate than birch and therefore responds on an even longer timescale (Kellomäki & Kolström 1994). Spruce (*Pica abies*) forests occur at nearby locations in northern Finland. Possible future spruce forests were not included in the modelling, since a northward advance could be inhibited by factors other than summer temperature, such as soil properties or other climatic factors (Siren 1955, Oksanen 1995, Sutinen et al. 2005).

In a synthesis study on tundra and treeline ecosystems, Epstein et al. (2004) found that in the Seward Peninsula in Alaska, most of the species present in the shrub tundra communities above the treeline are also present in the understory of forested areas, and that the transition from tundra to forested vegetation therefore typically involves gradual changes in tree density and morphology. It is reasonable to believe that a similar development occurs on Finnmarksvidda. Also from the Seward Peninsula study, it was shown by simulation using the AFRESCO model that under a 2°C warming scenario, there would be a time lag of 70–290 yr for the doubling of the initial total forest cover to take place (Rupp et al. 2000b). However, the simulation was performed for spruce forest and a 2°C increase, while the development of birch forest in Finnmarksvidda might occur faster, perhaps on a time frame of half a century. The relatively rapid and ongoing increase of birch forest cover has also been shown by Hofgaard et al. (2013). These analyses revealed an average northward advance of birch and pine latitudinal forest lines in northernmost Norway since the early 20th century of 156 and 71 m yr^{-1}, respectively. This shows that birch has a more rapid northbound advance on the tundra than pine. However, the more rapid advance for birch than pine might partly be a result of extensive illegal logging, especially at the latitudinal pine treeline, but also at the latitudinal treeline, as discussed by Hofgaard et al. (2013). A focus on factors limiting the advance of certain species as well as the response of the forest–tundra ecotone to climate change are needed in order to refine the output from more dynamic regional and global vegetation models (e.g. Wolf et al. 2008).

4.3. Impacts of future forests on reindeer grazing system

The modelling of future vegetation cover indicates a dramatic change in forest distribution on Finnmarksvidda, and major parts of the open lichen heaths will change to woodland with less accessibility for winter reindeer grazing. This will impact the reindeer grazing system in several ways. Reindeer are adapted to utilize lichens as energy sources, which is sufficient to ensure survival if enough lichens are available. The establishment of forest on previous lichen and dwarf shrub-dominated heaths may increase snow accumulation and reduce wind (Holtmeier & Broll 2007). This in turn typically tends to create more compact types of snow in forests, making penetrability of the snow pack, and hence grazing due to lower accessibility, gradually more difficult as the winter proceeds (Collins & Smith 1991, Heggberget et al. 2002, Riseth et al. 2011). Therefore, snow conditions are crucial in determining the availability of the lichens, and in particular, late-season ice-crust formations may have strong effects on the condition and survival of the animals (Moen 2008). Snow hardness, snow depth and animal mobility are also factors affecting reindeer selection of feeding areas (Collins & Smith 1991). Greater snow depths in forests makes digging of grazing hollows more demanding and energy-consuming, and thus requires relocation to a more open landscape (Sara 1999). Accordingly, birch expansion increases the grazing pressure on the remaining open tundra and sparsely populated pine and birch forest area, since the relocation will need to be conducted at an earlier stage of the winter. On the other hand, greater snow depths might hinder grazing and promote lichen growth, especially on heavily grazed areas in the forests (Tømmervik et al. 2009).

Acknowledgements. This study is mainly financed by The Norwegian Reindeer Administration Authorities, 'Development Fund of Reindeer Husbandry', and in part by the Nordic Centre of Excellence (NCoE) project 'Climate-change effects on the epidemiology of infectious diseases and the impacts on Northern Societies (CLINF)'. This article is based upon work from COST Action ES1203 SENSFOR, supported by COST (European Cooperation in Science and Technology). We thank Dr. Hannah Vickers for the English editing.

LITERATURE CITED

Aune B (1993) Temperaturnormaler, normalperiode 1961–1990. Klima rapp 02/93. Det Norske Meteorologiske Institutt, Oslo

Aune S, Hofgaard A, Söderström L (2011) Contrasting climate- and land-use-driven tree encroachment patterns of subarctic tundra in northern Norway and the Kola Peninsula. Can J For Res 41:437–449

Bjerke JW, Karlsen SR, Høgda KA, Malnes E and others (2014) Record-low primary productivity and high plant damage in the Nordic Arctic Region in 2012 caused by multiple weather events and pest outbreaks. Environ Res Lett 9:084006

Bokhorst SF, Bjerke JW, Tømmervik H, Callaghan TV, Phoenix GK (2009) Winter warming events damage sub-Arctic vegetation: consistent evidence from an experimental manipulation and a natural event. J Ecol 97: 1408–1415

Bråthen KA, Fodstad CH, Gallet C (2010) Ecosystem disturbance reduces the allelopathic effects of *Empetrum hermaphroditum* humus on tundra plants. J Veg Sci 21: 786–795

Cairns D, Moen J (2004) Herbivory influences tree lines. J Ecol 92:1019–1024

Cairns D, Lafon C, Moen J, Young A (2007) Influences of animal activity on treeline position and pattern: implications for treeline responses to climate change. Phys Geogr 28:419–433

Collins WB, Smith TS (1991) Effects of wind-hardened snow on foraging by reindeer (*Rangifer tarandus*). Arctic 44: 217–222

de Wit HA, Bryn A, Hofgaard A, Karstensen J, Kvalevåg M, Peters GP (2014) Climate warming feedback from mountain birch forest expansion: reduced albedo dominates carbon uptake. Glob Change Biol 20:2344–2355

Epstein HE, Beringer J, Gould WA, Lloyd AH and others (2004) The nature of spatial transitions in the Arctic. J Biogeogr 31:1917–1933

Eskelinen A, Virtanen R (2005) Local and regional processes in low-productive mountain plant communities: the roles of seed and microsite limitation in relation to grazing. Oikos 110:360–368

Førland JE (1993) Nedbørsnormaler, normalperiode 1961–1990. Klima rapp 39/93. Det Norske Meteorologiske Institutt, Oslo

Franke AK, Aatsinki P, Hallikainen V, Huhta E and others (2015) Quantifying changes of the coniferous forest line in Finnish Lapland during 1983–2009. Silva Fennica 49: 1408

Fylkesmannen i Finnmark (2012) Skog- og klimaprogram for Finnmark 2013-2016. Fylkesmannen i Finnmark, Vadsø

Grace J, Berninger F, Nagay L (2002) Impacts of climate change on the tree line. Ann Bot 90:537–544

Hanssen-Bauer I, Førland EJ, Haddeland I, Hisdal H and others (2015) Klima i Norge 2100. NCCS rep no. 2/2015. Norsk klimaservicesenter, Oslo

Heggberget TM, Gaare E, Ball JP (2002) Reindeer (*Rangifer tarandus*) and climate change: importance of winter forage. Rangifer 22:13–32

Høeg HI (2000) Pollenanalytiske undersøkelser i Finnmark, Nord Norge. AmS-Varia 37:53–97

Hofgaard A, Tømmervik H, Rees G, Hanssen F (2013) Latitudinal forest advance in northernmost Norway since the early 20th century. J Biogeogr 40:938–949

Høgda K, Tømmervik H, Karlsen SR (2013) Trends in start of the growing season in Fennoscandia 1982-2011. Remote Sens 5:4304–4318

Holtmeier FK (2012) Impact of wild herbivorous mammals and birds on the altitudinal and northern treeline ecotones. Landsc Online 30:1–28

Holtmeier FK, Broll G (2005) Sensitivity and response or northern hemisphere altitudinal and polar treelines to global environmental change at landscape and local scales. Glob Ecol Biogeogr 14:395–410

Holtmeier FK, Broll G (2007) Treeline advance—driving processes and adverse factors. Landsc Online 1:1–33

Houle G, Filion L (2003) The effects of lichens on white spruce seedling establishment and juvenile growth in a spruce-lichen woodland of subarctic Quebec. Ecoscience 10:80–84

Huntley B, Long AJ, Allen JRM (2013) Spatio-temporal patterns in Lateglacial and Holocene vegetation and climate of Finnmark, northernmost Europe. Quat Sci Rev 70: 158–175

Jepsen JU, Hagen SB, Høgda KA, Ims RA, Karlsen SR, Tømmervik HA, Yoccoz NG (2009) Monitoring the spatio-temporal dynamics of geometrid moth outbreaks in birch forest using MODIS NDVI. Remote Sens Environ 113: 1939–1947

Johansen B, Karlsen SR (2005) Monitoring vegetation changes on Finnmarksvidda, Northern Norway, using Landsat

MSS and Landsat TM/ETM+ satellite images. Phytocoenologia 35:969–984

Johansen B, Karlsen SR (2007) Finnmarksvidda – kartlegging og overvåkning av reinbeiter – status 2006. Norut IT Rapp IT394/1-2007. Norut, Tromsø

Karlsen SR, Elvebakk A (2003) A method using indicator plants to map local climatic variation in the Kangerlussuaq/Scoresby Sund area, East Greenland. J Biogeogr 30:1469–1491

Karlsen SR, Elvebakk A, Johansen B (2005) A vegetation-based method to map climatic variation in the arctic-boreal transition area of Finnmark, north-easternmost Norway. J Biogeogr 32:1161–1186

Karlsen SR, Tolvanen A, Kubin E, Poikolainen J and others (2008) MODIS-NDVI based mapping of the length of the growing season in northern Fennoscandia. Int J Appl Earth Obs Geoinf 10:253–266

Karlsen SR, Jepsen JU, Odland A, Ims RA, Elvebakk A (2013) Outbreaks by canopy-feeding geometrid moth cause state-dependent shifts in understorey plant communities. Oecologia 173:859–870

Kellomäki S, Kolström M (1994) The influence of climate change on the productivity of Scots pine, Norway spruce, pendula birch and Pubescent birch in southern and northern Finland. Forest Ecol Manag 5:201–217

Kullman L (2002) Rapid recent range-margin rise of tree and shrub species in the Swedish Scandes. J Ecol 90:68–77

Lindström M (1987) Northernmost Scandinavia in the geological perspective. Ecol Bull 38:17–37

Lucht W, Prentice IC, Myneni RB, Sitch S and others (2002) Climatic control of the high-latitude vegetation greening trend and Pinatubo effect. Science 296:1687–1689

Moen J (2008) Climate change: effects on the ecological basis for reindeer husbandry in Sweden. Ambio 37:304–311

Moen J, Aune K, Edenius L, Angerbjörn A (2004) Potential effects of climate change on treeline position in the Swedish mountains. Ecol Soc 9:16

Nilsson MC, Zackrisson O (1992) Inhibition of Scots pine seedling establishment by *Empetrum hermaphroditum*. J Chem Ecol 18:1857–1870

Oksanen L (1995) Isolated occurrences of spruce, *Picea abies*, in northernmost Fennoscandia in relation to the enigma of continental mountain birch forests. Acta Bot Fennica 153:81–92

Olofsson J, Oksanen L (2005) Effects of reindeer density on vascular plant diversity on North Scandinavian mountains. Rangifer 25:5–18

Riseth JÅ, Tømmervik H, Helander Renvall E, Labba N and others (2011) Sámi TEK as a guide to science: snow, ice and reindeer pasture facing climate change. Polar Rec 47:202–217

Riseth JÅ, Tømmervik H, Bjerke JW (2016) 175 years of adaptation: North Scandinavian Sámi reindeer herding between government policies and winter climate variability (1835–2010). J For Econ 24:186–204

Rupp TS, Starfield AM, Chapin FS III (2000a) A frame-based spatially explicit model of subarctic vegetation response to climate change: comparison with a point model. Landsc Ecol 15:383–400

Rupp TS, Chapin FS III, Starfield AM (2000b) Responses of subarctic vegetation to transient climatic change on the Seward Peninsula in north-west Alaska. Glob Change Biol 6:541–555

Sara MN (1999) Praktisk reinbeitebruk—tradisjonelle kunnskaper. Rangifer Rep 3:93–102

Sedia EG, Ehrenfeld JG (2003) Lichens and mosses promote alternate stable plant communities in the New Jersey Pinelands. Oikos 100:447–458

Seppä H, Hicks S (2006) Integration of modern and past pollen accumulation rate (PAR) records across the arctic tree-line: a method for more precise vegetation reconstructions. Quat Sci Rev 25:1501–1516

Siren G (1955) The development of spruce forest on raw humus sites in Northern Finland and its ecology. Acta Forest Fennica 62, No. 4, article ID7453

Sjögren P, Karlsen SR, Jensen C (2015) The use of quantitative models to assess long-term climate–vegetation dynamics—a case study from the northern Scandinavian Mountains. Holocene 25:1124–1133

Smith B, Prentice IC, Sykes MT (2001) Representation of vegetation dynamics in the modelling of terrestrial ecosystems: comparing two contrasting approaches within European climate space. Glob Ecol Biogeogr 10:621–637

Smith B, Warlind D, Arneth A, Hickler T, Leadley P, Siltberg J, Zaehle S (2014) Implications of incorporating N cycling and N limitations on primary production in an individual-based dynamic vegetation model. Biogeosciences 11: 2027–2054

Sutinen R, Hyvönen E, Ruther A, Ahl A, Marja-Liisa Sutinen ML (2005) Soil-driven timberline of spruce (*Picea abies*) in Tanaelv belt–Lapland granulite transition, Finland. Arct Antarct Alp Res 37:611–619

Sveinbjörnsson B, Hofgaard A, Lloyd A (2002) Natural causes of the tundra-taiga boundary. Ambio Spec Rep 12:23–29

Tenow O, Nilssen AC, Bylund H, Hogstad O (2007) Waves and synchrony in *Epirrita autumnata/Operopthera brumata* outbreaks. I. Lagged synchrony: regionally, locally and among species. J Anim Ecol 76:258–268

Tømmervik H, Johansen B, Tombre I, Thannheiser D, Høgda KA, Gaare E, Wielgolaski FE (2004) Vegetation changes in the Nordic mountain birch forest: the influence of grazing and climate change. Arct Antarct Alp Res 36:323–332

Tømmervik H, Johansen B, Riseth JÅ, Karlsen SR, Solberg B, Høgda KA (2009) Above ground biomass changes in the mountain birch forests and mountain heaths of Finnmarksvidda, Northern Norway, in the period 1957-2006. For Ecol Manage 257:244–257

Tømmervik H, Bjerke JW, Gaare E, Johansen B, Thannheiser D (2012) Rapid recovery of recently overexploited winter pastures for reindeer in northern Norway. Fungal Ecol 5:3–15

Van Bogaert R, Haneca K, Hoogesteger J, Jonasson C, Dapper MD, Callagan TV (2011) A century of tree line changes in sub-Arctic Sweden shows local and regional variability and only a minor influence of 20th century climate warming. J Biogeogr 38:907–921

Vikhamar-Schuler D, Hanssen-Bauer I, Førland E (2010) Long-term climate trends of Finnmarksvidda, Northern-Norway. Met.no rep no. 6/2010. Norwegian Meteorological Institute, Oslo

Wehberg J (2007) Der Fjellbirkenwald in Lappland. Eine vegetationsökologische Studie. Mitt Geogr Ges Hamburg 99:1–215

Wolf A, Callaghan TV, Larson K (2008) Future changes in vegetation and ecosystem function of the Barents Region. Clim Change 87: 51

15

Drivers of treeline shift in different European mountains

Pavel Cudlín[1,*], Matija Klopčič[2], Roberto Tognetti[3,4], Frantisek Máliš[5,6], Concepción L. Alados[7], Peter Bebi[8], Karsten Grunewald[9], Miglena Zhiyanski[10], Vlatko Andonowski[11], Nicola La Porta[12], Svetla Bratanova-Doncheva[13], Eli Kachaunova[13], Magda Edwards-Jonášová[1], Josep Maria Ninot[14], Andreas Rigling[15], Annika Hofgaard[16], Tomáš Hlásny[17], Petr Skalák[1,18], Frans Emil Wielgolaski[19]

[1]Global Change Research Institute CAS, Academy of Sciences of the Czech Republic, České Budějovice 370 05, Czech Republic
[2]University of Ljubljana, Biotechnical Faculty, Department of Forestry and Renewable Forest Resources, Slovenia
[3]Dipartimento di Bioscienze e Territorio, Università degli Studio del Molise, Contrada Fonte Lappone, 86090 Pesche, Italy
[4]MOUNTFOR Project Centre, European Forest Institute, 38010 San Michele all Adige (Trento), Italy
[5]Technical University Zvolen, Faculty of Forestry, 960 53 Zvolen, Slovakia
[6]National Forest Centre, Forest Research Institute Zvolen, 960 92 Zvolen, Slovakia
[7]Pyrenean Institute of Ecology (CSIC), Apdo. 13034, 50080 Zaragoza, Spain
[8]WSL Institute for Snow and Avalanche Research SLF, 7260 Davos Dorf, Switzerland
[9]Leibniz Institute of Ecological Urban and Regional Development, 01217 Dresden, Germany
[10]Forest Research Institute, BAS 132, Kl. Ohridski Blvd. 1756 Sofia, Bulgaria
[11]Faculty of Forestry, University Ss. Cyril and Methodius, Skopje, Macedonia
[12]Research and Innovation Centre, Fondazione Edmund Mach (FEM) and MOUNTFOR Project Centre, European Forest Institute, 38010 San Michele all Adige (Trento) Italy
[13]Division of Ecosystem Research, IBER-Bulgarian Academy of Sciences, 1113 Sofia, Bulgaria
[14]University of Barcelona, Department of Plant Biology, 08028 Barcelona, Spain
[15]WSL Swiss Federal Institute for Forest, Snow and Landscape Research, Zürcherstrasse 111, 8903 Birmensdorf, Switzerland
[16]Norwegian Institute of Nature Research, 7485 Trondheim, Norway
[17]Faculty of Forestry and Wood Sciences, Czech University of Life Sciences, 165000 Prague, Czech Republic
[18]Czech Hydrometeorological Institute, 143 06 Prague, Czech Republic
[19]University of Oslo, 0316 Oslo, Norway

ABSTRACT: A growing body of evidence suggests that processes of upward treeline expansion and shifts in vegetation zones may occur in response to climate change. However, such shifts can be limited by a variety of non-climatic factors, such as nutrient availability, soil conditions, landscape fragmentation and some species-specific traits. Many changes in species distributions have been observed, although no evidence of complete community replacement has been registered yet. Climatic signals are often confounded with the effects of human activity, for example, forest encroachment at the treeline owing to the coupled effect of climate change and highland pasture abandonment. Data on the treeline ecotone, barriers to the expected treeline or dominant tree species shifts due to climate and land use change, and their possible impacts on biodiversity in 11 mountain areas of interest, from Italy to Norway and from Spain to Bulgaria, are reported. We investigated the role of environmental conditions on treeline ecotone features with a focus on treeline shift. The results showed that treeline altitude and the altitudinal width of the treeline ecotone, as well as the significance of climatic and soil parameters as barriers against tree species shift, significantly decreased with increasing latitude. However, the largest part of the commonly observed variability in mountain vegetation near the treeline in Europe seems to be caused by geomorphological, geological, pedological and microclimatic variability in combination with different land use history and present socio-economic relations.

KEY WORDS: Vegetation zone shift · Climate change · Climate models · Treeline ecotone · European mountains · Ecosystem services

*Corresponding author: cudlin.p@czechglobe.cz

1. INTRODUCTION

Mountain regions are crucial areas for studying the impact of climate change on vegetation communities; steep climatic gradients enable testing of ecological, ecophysiological and evolutionary responses of flora to changing geophysical influences related to climate change. In addition, most of the species living there grow in conditions classified as their ecological limits (Körner 2012). Vegetation in European mountainous regions has been subjected to severe changes during the remote (e.g. Kullman 1988) and recent pasts (e.g. Vrška et al. 2009, Boncina 2011, Bodin et al. 2013, Elkin et al. 2013, Schwörer et al. 2014). Changes in species distribution and plant community composition have been the most often observed, although evidence of complete community exchange has not yet been registered.

A strong elevational zonation is typical of montane vegetation, caused mainly by steep climatic gradients. Vegetation zones are characterized by a specific structural-functional type of phytocoenosis constructed by particular main plant species (edificators) and roughly following latitude (latitudinal vegetation zones) or altitude (altitudinal vegetation zones or vegetation belts). Climatic zones along altitudinal gradients are compressed, with large habitat and species diversity in successive altitude vegetation zones. In these steep gradients, species richness decreases with increasing altitude, although topographic isolation results in increased levels of endemism (Pedrotti 2013).

The most obvious vegetation boundary at high elevations is that of the upper forest limit (Harsch & Bader 2011). Owing to its diffuse character, it might be better to refer to the treeline ecotone, which consists of the belt between the boundary of the closed forest stand–timberline and the uppermost or northernmost scattered and stunted individuals of the forest-forming tree species regardless of their growth form and height (Holtmeier & Broll 2005). Within this ecotone, the treeline (i.e. the line connecting the tallest patches of forest composed of trees of ≥3 m height) is often delimited; at this elevation, the mean temperatures for the growing season are ca. 6.4°C (Körner 2012). Climate is one of the most important limiting factors defining the spatial distribution of any species (Pearson & Dawson 2003, Wieser et al. 2014). Temperature, especially summer mean temperature and temperature sums, is the primary factor causing the formation of treelines at the global scale (Grace et al. 2002, Moen et al. 2004), although it may be substantially affected by several other non-climatic factors, e.g. mass elevation effect (Ellenberg 1988), past land use or forest management (Gehrig-Fasel et al. 2007). The main factors affecting the expansion and/or retreat of tree stands at their upper limits are scale-dependent (Holtmeier 2009). At the global scale, treeline position is determined by growing-season temperatures (Körner & Paulsen 2004), whereas at the landscape scale, second-order factors (i.e. climatic stress caused by wind or precipitation, natural disturbances and geomorphological factors), in addition to temperature, significantly affect treeline elevation and dynamics (Holtmeier & Broll 2005, Hagedorn et al. 2014, Treml & Chuman 2015). A recent review on the causes producing the upper limits of tree occurrence, introduced 6 current concepts: climatic stress (e.g. frost damage, winter desiccation), disturbances (e.g. wind, ice blasting, avalanches), insufficient carbon supply, limitation to cell growth and tissue formation, nutritional limitation and limited regeneration (Wieser & Tausz 2007). Air temperature, as the strongest factor, influences the treeline in 2 different ways: temperatures during the warm part of the year are the main control of treeline elevation while temperatures during the cold part of the year can damage the living tissue of the trees (evergreen or deciduous species) (Jobbágy & Jackson 2000). Climate change is likely to trigger latitudinal and elevational vegetation zone shifts, mainly by altering species mortality and recruitment, by exceeding physiological thresholds and changing natural disturbance regimes (Gonzalez et al. 2010).

Vegetation shifts need to be considered as an inherent adaptation mechanism allowing populations to track climatically suitable sites. Such shifts, however, can be limited by a variety of non-climatic factors, such as nutrient availability, soil conditions, landscape fragmentation or species-specific traits, including dispersal capacity, competition with ground vegetation, presence of mycorrhizal fungal symbionts or increasing virulence of pests and pathogens (Camarero et al. 2015b). In particular, species geographic ranges are expected to shift depending on their habitat preferences and their ability to adapt to new conditions. A growing body of evidence suggests that processes of treeline upward expansion, drought-induced retraction of species distributions and shifts of some dominant tree species may occur in response to global climate change (Kullman 1999, Kittel et al. 2000, Hansen et al. 2001, Payette et al. 2001, Theurillat & Guisan 2001). Using a meta-analysis, treeline advance was recorded in 52% of 166 sites around the world (Harsch et al. 2009). On the other hand, there is

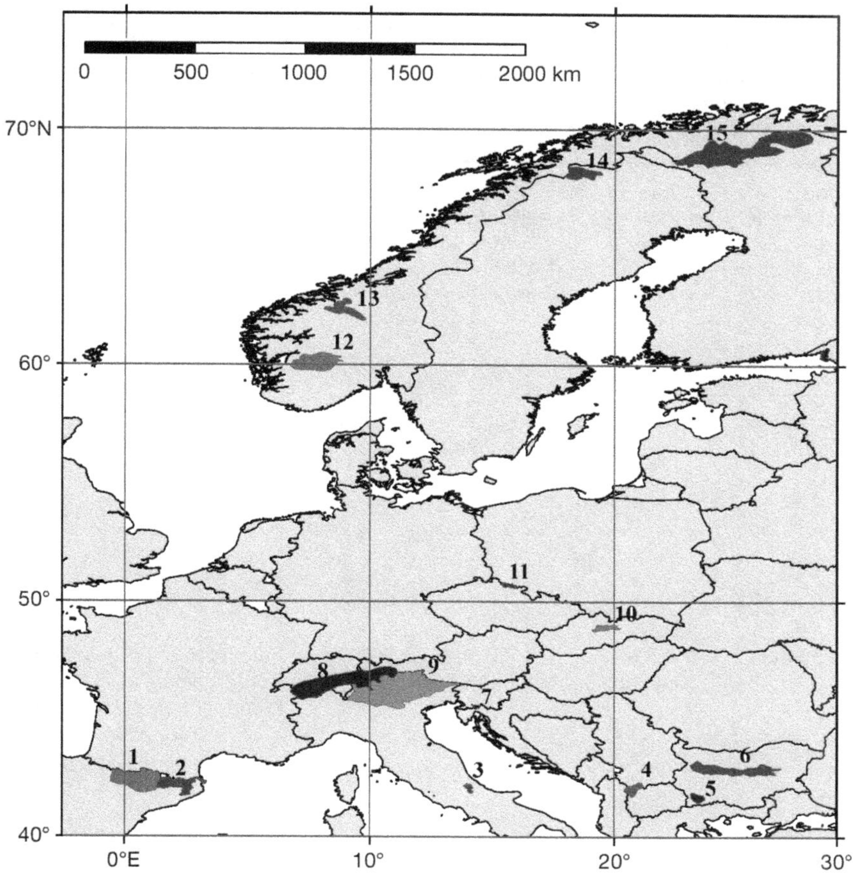

Fig. 1. Selected European mountains. Grey shading differentiates mountain areas. 1 Central Pyrenees (CP), 2 Eastern Pyrenees (EP), 3 Apennines (AP), 4 Shara Mts. (SH), 5 Pirin (PI), 6 Central Balkan Mts. (CB), 7 Northern Dinaric Mts. (DI), 8 Central Alps (CA), 9 Eastern Alps (EA), 10 Low Tatra Mts. (LT), 11 Giant Mts. (GM), 12 Hardangervidda (HG), 13 Dovre (DO), 14 Northern Swedish Lapland (NS), 15 Inner Finnmark/northernmost Finnish Lapland (FL)

evidence that treelines have always been dynamic and influenced by climate change and forest development cycles in the past (Kullman 1988, 2007, Gehrig-Fasel et al. 2007, Vladovi et al. 2014, Treml & Chuman 2015).

There are 3 main aspects of environmental change to which trees are likely to respond: increasing temperature, rising concentrations of CO_2 and increasing deposition of nitrogen (Grace et al. 2002, Lindner et al. 2014). The trees at the treeline accumulate carbohydrates, because autotrophic respiration is more limited by low temperature than photosynthesis. This means that factors such as temperature and nitrogen abundance, both of which affect the capacity of a tree to use the products of photosynthesis, will probably be more important than factors directly affecting photosynthesis, such as elevated CO_2 (Fajardo et al. 2012).

However, climate signals can be confounded with the effects of human activity. Stronger treeline dynamics due to a coupled effect of climate change and highland pasture abandonment frequently occur in European mountain ranges (Ellenberg 1988). Other human-related influential factors, such as (overly) intensive forest management, pasture and grazing, may be regionally important as well. Direct human-related factors may be crucial in changing the species composition and structure of mountain forests, especially in treeline ecotones where climate change does not represent the key factor (Alados et al. 2014).

The aims of this study were to evaluate the effects of geographical position and different regional climatic and other site conditions on treeline ecotone characteristics of selected European mountain areas, to identify the natural and anthropogenic drivers of treeline shift. We further discuss the effects of climate and land use changes on biodiversity in forests below the treeline.

2. METHODS

To demonstrate and discuss the possibilities and limits of treeline species shift, including latitudinal and longitudinal heterogeneity, the current situation in 11 mountain areas from Italy to Norway and from Spain to Bulgaria were analysed (Fig. 1). To determine the influence of mountain massif size on the mesoclimatic and vegetation conditions, including treeline formation, we included large mountain complexes (Pyrenees, Alps, Scandes) as well as relatively small, separated mountains (Pirin and Giant Mountains) in this study. The characteristics of the treeline ecotone, including historical and recent human impacts on mountain ecosystems for the 11 mountain areas of interest, are summarized in Table 1. Extensive mountain massifs (Pyrenees, Alps, Scandes) were further divided into more homogeneous units (Central Pyrenees, Eastern Pyrenees, Central Alps, Eastern Alps, Hardangervidda, Dovre, Northern Swedish Lapland, Inner Finnmark/northernmost Finnish Lapland). To esti-

Table 1. Timberline and treeline parameters and characteristics of human activities, changes according to climatic models and limits to climate change induced species shifts in selected European mountains. A1B, A2: emissions scenarios (IPPC); A.a.: *Abies alba*; A.p.: *Acer pseudoplatanus*; F.s.: *Fagus sylvatica*; L.d.: *Larix decidua*; P.a.: *Picea abies*; P.c.: *Pinus cembra*; P.h.: *Pinus heldreichii*; P.m.: *Pinus mugo*; P.p.: *Pinus peuce*; P.s.: *Pinus sylvestris*; P.u.: *Pinus uncinata*; T: timberline; Tr: treeline; TSL: tree species limit; VZ: vegetation zone; C: century; mng: management. Elevation range of T and Tr: mean (min–max)

Country; part of mountains	Coordinates of central site	Elevation range of T (m a.s.l.)	Elevation range of Tr (m a.s.l.)	Range of tree species limit (m a.s.l.)	Annual altitudinal shift of T or Tr (m yr^{-1})	Area of study (km^2)	VZs near Tr (m a.s.l.)
Spain; Spanish Pyrenees	42°37′N; 0°27′E	1930 (1800–2050)	No data	P.u. 2700	0.49 (T; 1956–2006)	15 800	Pine with A.a., P.s. and F.s. (1400–2050); subalpine with P.u. (1800–2300)
Italy; Majella Massif, part of Apennines	42°05′N; 14°05′E	1800 (1700–1850)	No data	F.s. 2100, P.m. 2300	1 (Tr [P.m.]; 1954–2007)	741	Beech forest (800–1850); mixed beech–dwarf pine (1800–2000); dwarf pine (2100–2300)
Macedonia; Shara Mountain (Balkan Peninsula)	41°47″N; 20°33′E	1850 (1800–1900)	1900 (1850–1950)	P.a. 2100, P.p. 2200	~1 (T;) 1934–2010	829	Mixed fir–beech–spruce (800–1900); beech (1600–1900); dwarf pine (1600–2200)
Bulgaria; Pirin	41°42′N; 23°31′E	1900 (1800–2100)	2100 (1900–2300)	P.p. 2500	No data	384	Beech (1000–1500); coniferous forest (P.s., P.p., P.h., P.a., 1300–2300); dwarf pine (2000–2500)
Bulgaria; Central Balkan Mts.	42°47′N 24°36′E	Beech T 1600 (1500–1700)	1600 (1500–1850)	F.s. 1700, P.a., A.a. 1850	No data	717	Beech (800–1700); coniferous (P.a., A.a., 1500–1850); juniper (1500–1850)
Slovenia; Northern Dinaric Mts.	45°36′N; 14°28′E	1540 (1455–1600)	1600 (1550–1650)	F.s. 1650, P.a. 1700, A.a. 1650	No data	~200	Mixed fir–beech–spruce (600–1400); beech (1300–1550); dwarf pine with small groups or individual trees of F.s., P.a., A.a., A.p. (1500–1700)
Switzerland, Italy; Central and Eastern Alps	46°46′N; 9°52′E; 46°17′N; 11°45′E	1900 (1700–2150)	2100 (1850–2300)	P.a. 2400, L.d. 2450	ca.1.3 (T; 1954–present)	~250 + ~750	Spruce with L.d. (1000–1900); forest with P.c., L.d. and P.a. (1700–2200); dwarf shrub vegetation, partly pastures (2200–2500)

Human influence in VZs in past and recently	Annual temperature (T) and precipitation (P) differences between 1961–1990 and 2021–2050; mean of CORDEX models	Limits to climate change induced species shift	References
Since 11th C: clear-cutting and grazing, moderate on irregular slopes and ridges; since 1930: significant release of influence but recovery mainly at moderate elevations	HIRHAM5 model: period ΔT2021–2050 = 1.5–2°C; ΔT2051–2080 = 4°C; ΔP2021–2050 = +5%; ΔP2051–2080 = +30% compared to 1960–1990	Inadequacy of substrata (rocky); climatic constraints (including growing-season extreme events, such as late-spring freezing, summer drought, etc.)	Batllori et al. (2009), Alados et al. (2014), Gartzia et a. (2014), Camarero et al. (2015a)
Since 1000 BC: cutting, burning and grazing mng; since the mid-20th C: grazing intensity decreased; now occasionally managed and regeneration of P.m.	HadCM3_A2, period 2020–2080: ΔT_{min}_Jan2050 = 1.7°C; ΔT_{min}_Jan2080 = 3.1°C; ΔP2050 = –7%; ΔP2080 = –23%	Low density of F.s. in T, dispersal distances of F.s. seeds upslope; poor soils; excessive exposure to solar radiation in open areas	van Gils et al. (2008), Palombo et al. (2013, 2014)
Since 14th C: clear-cutting and grazing to enlarge alpine pastures and to produce charcoal; since 1970: abandonment of traditional agricultural practices	Mean of 4 GCMs (CSIRO/Mk2, HadCM3, ECHAM4/OPYC3, NCAR – PCM); ΔT2050 = 2.6°C; ΔT2100 = 5.3°C; ΔP2050 = –3%; ΔP2100 = –8%	Differences in soil types between subalpine beech and mixed forest (poor calcareous soils or rankers on silicate bedrock); cliffs and rocks; cattle grazing	Em (1986), Strid et al. (2003), Bergant (2006) Amidžic et al. (2012)
Since 680 AD: intensive fires, deforestation and grazing; since 1962: traditional mng has changed; last decades tourism impact has increased	HADCM3_A2, CGGM2, ΔT2050 = 1.7°C, ΔT2080 = 2.8°C; ΔP2080 = –15%	Rocky sites; extreme hydrothermal conditions; competition of shrubs and tree seedlings; locally intense pasturing	Velchev (1997), Grunewald & Scheithauer (2011), Raev et al. (2011)
Human impact (mostly burning) started in 17th C; excessive and improper forest mng; release of influence since 1980	HADCM3_A2, CGCM2, ΔT2050 = 1.7°C, ΔT2080 = 2.8°C; ΔP2050 = –15%; ΔP2080 = –15%	Low density of F.s. in T, dispersal distances of F.s. seeds upslope; pasture intensity, new expansion of juniper stands	Raev et al. (2011), Gikov et al. (2016), Zhiyanski et al. (2008, 2016)
15–18th C: slash, burn, grazing mng; 18–19th C: uncontrolled cutting; 20th C: irregular shelterwood systems and over-exploitation mng	DMI-HIRHAM5_A1B, period 2001–2100: ΔT2050 = 1.2°C; ΔT2100 = 2.5°C; ΔP2050 = +15 mm; ΔP2100 = –18 mm	Impact of wild large ungulates on tree seedlings (especially A.a.); shallow, nutrient-poor soils (i.e. ranker and rendzina)	Klopčič et al. (2010, 2015), Mina et al. (2017)
Since 12th C: slash, burn, grazing mng; 13–19th C: intensive forest mng; at the end of the 19th C: decrease in grazing	ENSEMBLES models_A1B, period 1980–2100: ΔT2050 = 1.2–1.8°C; ΔT2100 = 2.7–4.1°C; ΔP2050 = 0 mm; ΔP2100 = –5 mm	Frequent disturbances by snow avalanches and snow gliding; ongoing cattle grazing; game grazing in the coldest years	Tattoni et al. (2010), Kulakowski et al. (2011), Barbeito et al. (2012), Bebi et al. (2012), Dawes et al. (2015)

Continued on next page

Table 1 (continued)

Country; part of mountains	Coordinates of central site	Elevation range of T (m a.s.l.)	Elevation range of Tr (m a.s.l.)	Range of tree species limit (m a.s.l.)	Annual altitudinal shift of T or Tr (m yr^{-1})	Area of study (km^2)	VZs near Tr (m a.s.l.)
Slovakia; Dumbier Tatra (part of Low Tatra Mts)	48°55'N; 19°31'E	1400 (1350–1450)	1530 (1400–1630)	P.a. 1730	~0.3 (Tr; 1950–present)	~400	Spruce–fir–beech (1200–1350); spruce (1300–1550); P.m. with individual P.a. trees (1400–1700)
Czech Republic; Giant Mts	50°42'N; 15°38'E	1250 (1130–1460)	1320 (1250–1460)	P.a. 1500	0.43 (Tr; 1936–2005)	550	Beech–spruce (700–1050), P.a. dominates with increasing altitude; spruce (1000–1400); P.m. with P.a. trees (1400–1560)
Norway Southern Scandes	60°10'N; 7°40'E; 62°20'N; 10°05'E	850	600–1050	1500	0.8 (Tr; 1915–2007)	40 600	Birch forest dominance at high altitudes, but with some scattered P.s. or P.a.; mixed birch–conifer below 800
Norway, Sweden, Finland Northern Scandes	68°20'N; 18°20'E; 69°50'N; 27°00'E	West: 650 East: 100	West: 750 East: 290	West: 1000 East: 0	0.6 (Tr ; 1958–2008)	35 000	Birch forest dominance at all altitudes, but with some scattered pines or pine stands on sandy soils

mate the geographical position of these mountain units, 1 value was used for the geographical latitude and longitude of the locus of the studied mountain units. The analysed parameters of these 15 mountain units are shown in Table 2.

In addition, we calculated the following climate characteristics for all mountain units: annual mean air temperature, annual sum of precipitation, the growing season length and the date of the onset of the growing season. These characteristics were then compared for 2 distinct periods, 1961–1990 and 1991–2015 (Table 3). The climate characteristics were derived from the E-OBS gridded dataset of station observations version 13.1 (Haylock et al. 2008). We used the E-OBS version on the regular longitude–latitude grid with a horizontal resolution of 0.25 degrees. The climate data from all grid points within the mountain units or near their geographical borders were considered.

Data processing for 15 mountain units was made on the basis of data obtained from the literature and personal studies by the co-authors; these are summarized in Tables 1 & 2 and in Tables S1 & S2 in the Supplement at www.int-res.com/articles/suppl/c073p135_supp.pdf. In addition, a semi-quantitative valuation of abiotic, biotic and anthropogenic factors limiting an upward treeline shift was performed (Fig. 2). To answer the question how natural and anthropogenic factors have influenced treeline ecotone characteristics with a focus on treeline shift, redundancy analysis (RDA) was used to describe and test the effect of the explanatory variables (geographical position, size of the whole mountain massif, start of human influence and start of the decrease in human influence) on treeline ecotone characteristics (timberline elevation, width of the treeline ecotone, tree species forming the treeline and treeline shift per year) and identify groups of mountain units with similar variability of the dependent data (ter Braak & Šmilauer 2012). The whole data set for this analysis is shown in Table S2. We tested both their simple effects, which show how much variation every explan-

Human influence in VZs in past and recently	Annual temperature (T) and precipitation (P) differences between 1961–1990 and 2021–2050; mean of CORDEX models	Limits to climate change induced species shift	References
From 14th C to 1922: mining, forest change to spruce monocultures; from 13th C to 1978: pastures in upper parts of mountain	Average of 10 RCM_B1: $\Delta T2050$ = 1.8°C; $\Delta T2100$ = 3.7°C; $\Delta P2050$ = +24 mm; $\Delta P2100$ = −35 mm	More rocky and nutrient-poor soils; late frosts; steep slopes with avalanches; absence of mature trees	Körner (2003), Fridley et al. (2011), Hlásny et al. (2011, 2016)
9–11th C: forest clear cutting, grazing in dwarf pine VZ; since 18th C: forest change to spruce monocultures, since 19th C: artificially planted	ALADIN_A1B, period 1961–2100: $\Delta T2050$ = 1.3°C; $\Delta T2100$ = 3.4°C; $\Delta P2050$ = +25 mm; $\Delta P2100$ = −14 mm	Different soil types for F.s. and P.a. (cambisol versus podzol); only TSL of P.a. could elevate on ranker soils in P.m. VZ	Treml and Banaš (2000), Cudlín et al. (2013), Treml & Chuman (2015), Treml & Migo (2015)
Grazed and browsed by reindeer and livestock (sheep and cattle) over thousands of years	RegClim/MPI/Hadley: $\Delta T2050$ = 1.2°C; $\Delta T2100$ = 2.2°C; $\Delta P2050$ = +23 mm; $\Delta P2100$ = −12 mm	Sheep and reindeer grazing and episodic insect outbreaks (*Epirrita*)	Dalen & Hofgaard (2005), Wielgolaski (2005), Hofgaard et al. (2009), Kullman, Öberg (2009)
Grazed and browsed by semi-domestic reindeer for hundreds of years; increased grazing from year 1960	RegClim/MPI/Hadley: $\Delta T2050$ = 1.6°C; $\Delta T2100$ = 2.9°C; $\Delta P2050$ = +18 mm; $\Delta P2100$ = −10 mm	Continued grazing regime and frequent episodic insect outbreaks (*Epirrita*)	Dalen & Hofgaard (2005), Wielgolaski (2005), Tømmervik et al. (2005), Hofgaard et al. (2009), Aune et al. (2011), Mathisen et al. (2014)

atory variable can explain separately, without using the other variables, and their conditional effects, which depend on the variables already selected in the model. The statistical significance of the explanatory variables was tested using a Monte Carlo permutation test, and only predictors with $p \leq 0.05$ were included in the subsequent RDA (Šmilauer & Lepš 2014).

To find the drivers of biodiversity loss in forests, meadows and animal communities (dependent variables), the explanatory variables (geographical position, timberline elevation, tree species forming the treeline, start of human influence, start of the decrease in human influence and temperature increase between the 2 periods 1961–1990 and 2021–2050) were tested again by RDA in Canoco 5 as mentioned above (Figs. 3 & 4). The whole data set for this analysis — including additional information about treelines in the mountain areas of interest, with a focus on treeline shift (its rate, drivers, limits and problems with its assessment) — is presented in Tables S1 & S2.

Selected univariate graphs were constructed to visualize the data of relationships between the explanatory and dependent variables of all analyses, which were not seen distinctly from the results of the multivariate statistics or linear regression (Fig. 5).

3. RESULTS

3.1. Effect of environmental variables on treeline ecotone characteristics with a focus on treeline shift

The variation in the dependent variables (timberline elevation, width of treeline ecotone, tree species forming the treeline and treeline shift per year) was significantly affected only by latitude and size of the whole mountain massif (Fig. 3). The adjusted explained variability by all explanatory variables was 34.2% (F = 4.6, p = 0.01). The RDA diagram shows that in the mountains located more to the north, the

Table 2. Factors influencing treeline ecotone characteristics in selected European mountains. The mountain size is the size of the whole mountain massive, functioning as a mezoscale climate unit, not only the studied mountain area. Width of the treeline ecotone was determined by subtracting the value of the mean timberline elevation from the value of the upper tree species limit. C: conifers, B: broadleaved trees; HI: human influence

Mountain unit	North latitude	East longitude	Mountain size (km^2)	Timberline (m a.s.l.)	Width of treeline ecotone (m)	Tree species forming the treeline	Temperature increase between 1961–1990 and 1991–2015 (°C)	Treeline shift (m yr^{-1})	Start of HI (yr)	Start of HI decrease (yr)
Central Pyrenees (CP)	42°21'N	0°30'W	55 000	2250	300	C	0.8	1.92 (1956–2006)	1000 BC	1930
Eastern Pyrenees (EP)	42°28'N	1°30'E	55 000	2400	300	C	0.8	0.71 (1956–2006)	1000	1930
Apennines (AP)	42°05'N	14°05'E	50 000	1800	300	B	1.4	1 (1954–2007)	1000 BC	1950
Shara Mts. (SH)	41°47'N	20°33'E	1600	1850	350	C, B	0.7	1 (1934–2010)	1300	1970
Pirin (PI)	41°42'N	23°31'E	2585	1900	400	C	0.5	no data	680	1962
Central Balkan Mts. (CB)	42°47'N	24°36'E	11 600	1600	250	C, B	0.7	no data	1700	1980
Northern Dinaric Mts. (DI)	45°36'N	14°28'E	1000	1540	160	B	1,3	no data	1400	1900
Central Alps (CA)	46°46'N	9°52'E	180 000	2100	550	C	0.7	1.3 (1954–2015)	1300	1880
Eastern Alps (EA)	46°17'N	11°45'E	180 000	1800	400	C	0.9	1.3 (1954–2015)	1150	1950
Low Tatra Mts. (LT)	48°55'N	19°31'E	1240	1400	330	C	0.9	0.57 (1975–2016)	1400	1922
Giant Mts. (GM)	50°42'N	15°38'E	630	1250	250	C	0.7	0.43 (1936–2005)	1000	1963
Hardangervidda (HG)	60°03'N	7°47'E	9000	900	150	C, B	0.5	0.8 (1915–2007)	800	1970
Dovre (DO)	62°21'N	8°52'E	6000	1000	400	B	0,7	0.8 (1915–2007)	1000	no decrease
Northern Swedish Lapland (NS)	68°12'N	18°33'E	4000	700	300	B	1.1	0.6 (1958–2008)	1000	no decrease
Inner Finnmark/northernmost Finnish Lapland (FL)	69°10'N	25°12'E	60 000	290	90	B	1.0	0.6 (1958–2008)	1600	no decrease

treeline ecotone is narrower, treeline shift is smaller, the timberline occurs at lower elevations, and the treeline is formed more by broadleaved species. Higher values of treeline ecotone width and treeline shift were found in the mountains with a greater size of the whole mountain massif.

When every explanatory variable was tested separately, values of the start of the decrease of human influence in the mountains also had a significant effect on the variation of the tree line ecotone data (explained variation = 20.6%, pseudo-F = 3.4, p = 0.024). Some selected relationships of the explanatory and dependent variables, even those not showing significant effects on the variation in the tree line ecotone parameters, are depicted in Fig. 5.

3.2. Effect of recent changes in climate and land use on the biodiversity

Biodiversity loss in forests, meadows and animal communities, analysed by RDA, was explained by geographical position, treeline species composition and temperature increase between the 2 periods 1961–1990 and 2021–2050 (Fig. 4). The adjusted explained variation by all explanatory variables was 63.2% (F = 5.8, p = 0.001). The RDA results indicated that biodiversity loss in forest communities increased with increasing latitude and longitude as well as in the case where broadleaved species formed the treeline. In contrast, the highest biodiversity loss in meadows was found in the mountains positioned more to the south and with higher temperature increase. The biodiversity loss in animal communities was negatively correlated with longitude and temperature increase.

4. DISCUSSION

The comparison of several mountain areas situated across Europe shows a variation in understanding of what constitutes the treeline across countries. The primary issue is the different approach to the definition of forest when applying a tree height threshold. This threshold decreases from

5 m (Jeník & Lokvenc 1962) to 3 m (Körner 2012) in Northern and Central Europe to only 2 m (Holtmeier 2009) in Central and Southern Europe, where even shrubby stands (e.g. *Pinus mugo*) are considered as a forest in some countries, e.g. in Spain, Italy and Macedonia (Batllori et al. 2009). Another problem is a different designation of forest stands below the treeline: 'upper montane forests' in Central Europe and 'subalpine forests' in Southern Europe (Ellenberg 1988). Further differences are related to the tradition of different branches of science (e.g. a more traditional 'geobotanical' approach in Central Europe versus a more experimental approach in Western Europe), especially in the rate of applying new progressive methods (e.g. climatic modelling or molecular biological methods).

Our comparison of selected regions (Tables 1 & 2, Figs. 1 & 2) showed that southern mountains, compared to those located more centrally and in the north had (1) a longer and more profound exploitation by humans in the past; (2) greater differences in climatic parameters (temperature, length of growing period) between the 2 periods 1961–1990 and 1991–2015 (Table 3) and (3) more dramatic climate change scenarios, especially concerning temperature increases. Longitude also had some influence on the start and intensity of human influence, and on treeline elevation and rate of treeline shift (Table 2, Fig. 5). The occurrence of broadleaf tree species in the treeline ecotone in the northern (and to some extent the southern) countries distinguishes these from the Central European countries, where conifers prevail. It is interesting that grazing, an important factor shaping treeline ecosystems (Dirnböck et al. 2003,

Table 3. Differences of selected climatic parameters between the 2 study periods (1961–1990 and 1991–2015) in selected European mountains

	Length of growing period (d)[a]	Beginning of growing season (d)[b]	Mean annual temperature (°C)	Mean annual precipitation (%)
Central Pyrenees	13.5	−8.9	0.8	−10.7
Eastern Pyrenees	11.7	−9.1	0.8	−9.2
Apennines	28.4	−23.2	1.4	5.7
Shara Mts.	1.6	0.1	0.7	−10.0
Pirin	2.6	−1.4	0.5	2.7
Central Balkan Mts.	16.9	−13.6	0.7	7.6
Northern Dinaric Mts.	16.0	−9.5	1.3	−5.6
Central Alps	14.7	−8.6	0.7	6.2
Eastern Alps	12.5	−9.6	0.9	−3.3
Low Tatra Mts.	2.4	−3.4	0.9	−9.1
Giant Mts.	−1.0	1.3	0.7	5.5
Hardangervidda Blefjell	6.3	3.4	0.5	5.6
Dovre	13.8	3.2	0.7	6.4
Northern Swedish Lapland	8.1	0.5	1.1	−2.7
Inner Finnmark/northernmost Finnish Lapland	1.7	0.8	1.0	7.7

[a]Positive numbers indicate a prolongation
[b]Negative numbers indicate an earlier beginning

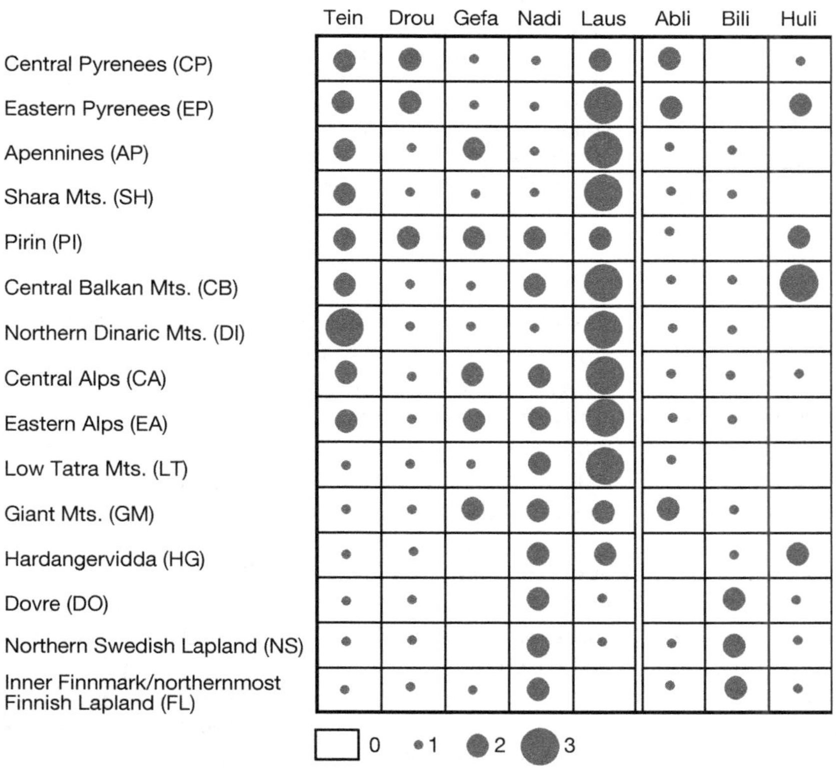

Fig. 2. Rates of treeline drivers (temperature increase [Tein], land use change [Laus]) and treeline shift limits (drought [Drou], geomorphological factors [Gefa], natural disturbances [Nadi], other abiotic, biotic and human factors) in selected European mountains. Abli, Bili and Huli: abiotic, biotic and human treeline shift limits, respectively). Rate of treeline driver influence — 0: no influence; 1: weak influence; 2: middle influence; 3: strong influence

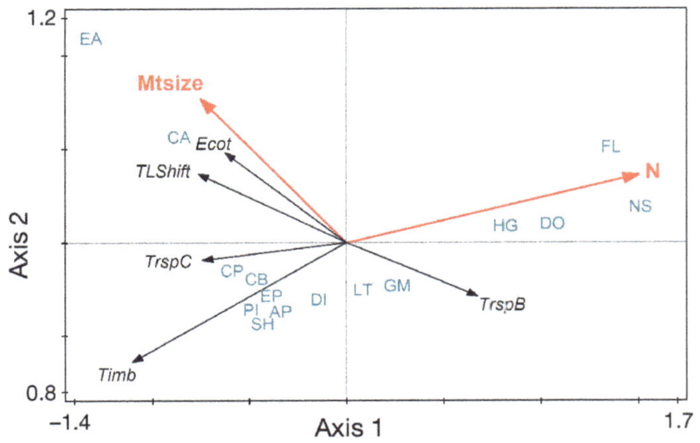

Fig. 3. Redundancy analysis diagram, with variation in timberline elevation, width of treeline ecotone, the tree species forming the treeline, and treeline shift used as dependent variables, explained by explanatory variables (latitude, mountain size). Mountain units (blue) are projected as the centres of abbreviations (see Table 2). Explanatory variables account for 43.6% of the total variation in the dependent data. The first canonical axis explained 46.6% of variation, the second axis explained 3.0% of variation. Dependent variables (black) are *Ecot*: altitudinal width of treeline ecotone, *TLShift*: treeline shift, *TrspC* (*TrspB*): treeline formed by conifers (broadleaved trees), *Timb*: timberline elevation. Explanatory variables (red) are N: north latitude, Mtsize: size of the mountain massif

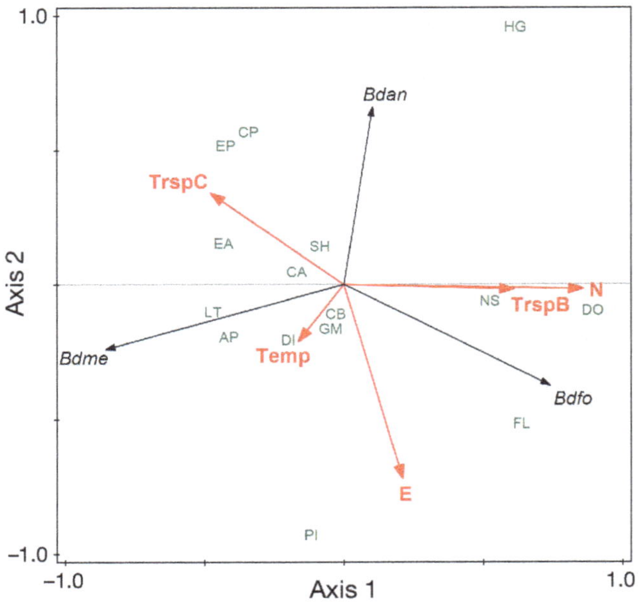

Fig. 4. Redundancy analysis diagram, with variation in biodiversity loss in forests, meadows and animal communities, used as dependent variables, explained by explanatory variables (geographical position, tree species forming the treeline and temperature increase between the 2 study periods [1961–1990 and 2021–2050]). Mountain units (green) are projected as centres of abbreviations (see Table 2). Explanatory variables account for 76.3% of total variation in the dependent data. The first canonical axis explained 48.4% of variation, the second axis explained 18.3% of variation. Dependent variables (black) are *Bdfo*/*Bdme*/*Bdan*: biodiversity loss in forests/meadows/animal communities. Explanatory variables (red) are N: north latitude, E: east longitude, TrspC (TrspB): treeline formed by conifers (broadleaved trees), Temp: temperature increase between the 2 study periods

Gehrig-Fasel et al. 2007), limits treeline shift in half of the countries of interest, regardless of latitude, longitude or recent political developments (Shara, Pirin, Central Balkans, Alps, Scandes; Fig. 2).

Despite the stated differences between studies looking at climate change impacts on the treeline, it is clear that current and future changes in temperature will seriously affect treeline ecotones in all mountain ranges of Europe. Winter is a period when treeline stands and individual trees have to survive severe, life-limiting conditions (Wieser & Tausz 2007). Therefore, winter warming might increase the chance of young trees surviving and passing the most critical period from seedling to sapling and further to the mature stage (Körner 2003). There is already much evidence worldwide that winter warming is one of the significant drivers of treeline advance (Harsch et al. 2009). Although it is expected that tree growth at the upper distributional margins (and in the northern European countries) will increase due to ongoing climate change (Peñuelas & Boada 2003, Chen et al. 2011, Hlásny et al. 2011, Lindner et al. 2014), extremes in temperature (e.g. black frosts, temperature reversals in the spring) or winter precipitation (e.g. lack of snow, drought) might also limit this process in the future in some regions and local situations (Holtmeier & Broll 2005, Hagedorn et al. 2014). An earlier start to the growing period (especially in the Apennines, Central Balkan and Spanish Pyrenees; see Table 2) can play a negative role in the resistance of trees and seedlings to early spring frosts. On the other hand, at lower distributional margins (and in Southern European countries), it is expected that tree growth will decrease or forests will experience some level of dieback due to drought (Breshears et al. 2005, Piovesan et al. 2008, Allen et al. 2010, Huber et al. 2013). A serious implication is the positive effect that warming can have on root rot fungi and populations of bark beetles resulting in large-scale disturbances to temperate and also high-altitude forests (Jankovský et al. 2004, Elkin et al. 2013, Millar & Stephenson 2015).

Although recent upward advances of treeline ecotones are widespread in mountain regions (Harsch et

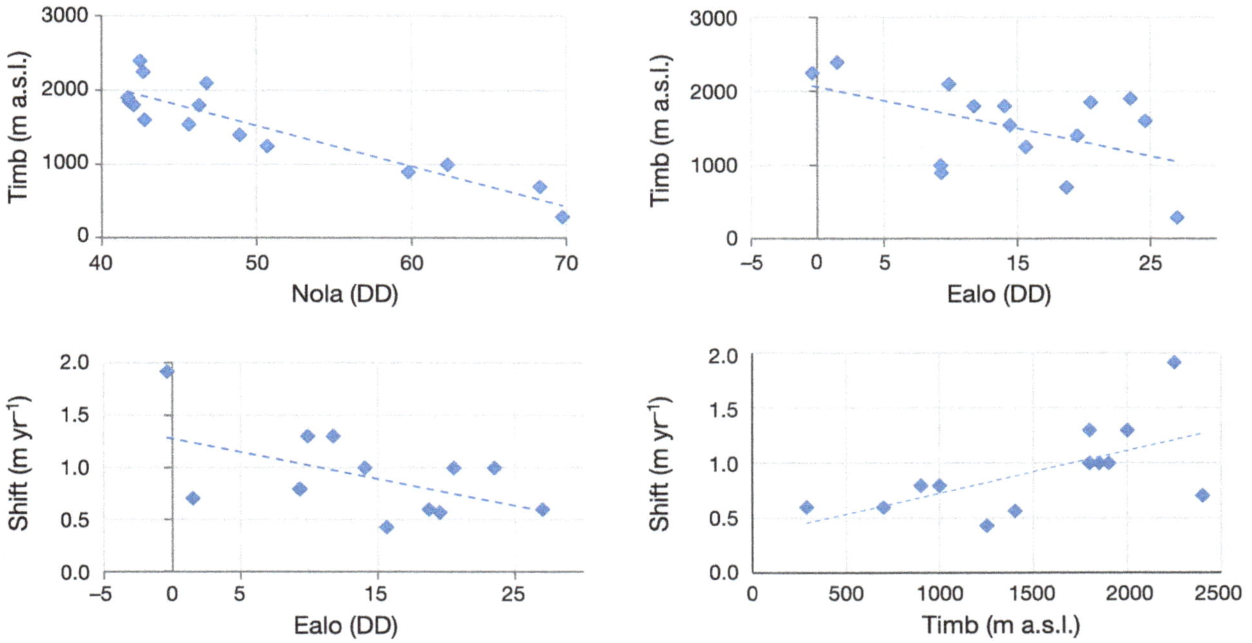

Fig. 5. Univariate graphs of the relationships of dependent variables, viz. timberline (Timb) and treeline shift per year (Shift), and their explanatory variables. Nola: north latitude; Ealo: east longitude; DD: decimal degrees

al. 2009), only a few published studies have included quantitative data about treeline shifts (Devi et al. 2008, Kullman & Öberg 2009, Diaz-Varela et al. 2010, Van Bogaert et al. 2011). Additionally, our knowledge of the spatial patterns in treeline ecotone shifts at the landscape scale is still surprisingly poor, except for some studies from subarctic areas, the Ural Mountains and the Alps (Lloyd et al. 2002, Diaz-Varela et al. 2010, Hagedorn et al. 2014). In the 15 studied mountain units, the values of treeline shift ranged from 0.43 to 1.9 m yr^{-1}, showing a rather distinct spatial pattern of the dynamics within European mountains. The observed treeline shift had a significant positive relationship only with northern latitude and a weak positive relationship with mountain-range size (i.e. the size of the whole mountain massif; Fig. 3). Nevertheless, this illustrates the importance of the mass elevation effect (Körner 2012), and partially explains why treelines could be at different altitudes at the same latitude. Small negative regressions with east longitude and timberline elevation are shown in Fig. 5. There was no apparent dependence of the treeline shift rate on climatic parameter changes between the periods 1961–1990 and 1991–2015.

A whole forest vegetation zone shift is a complex process, as not only the life strategies of an individual tree need to be considered, but plant, animal and microorganism species and their interrelationships, as well as the relationships to their specific microhabitats (including soil conditions), are also involved (Urban et al. 2012). Across all regions, we identified several important obstacles to treeline shifts, often related to site properties, such as significant rockiness, having shallow or low-nutrient soils, extreme relief causing disturbances by avalanches, snow gliding or wind damage (data from the Pyrenees, Apennines, Shara, Northern Dinaric, Low Tatra and Giant Mountains; see Fig. 2). Soil heterogeneity certainly has an important role in plant responses to climate change, and could also maintain the resilience of the community assemblages (Fridley et al. 2011). Another group of obstacles includes extreme climatic parameters, especially in winter (reported e.g. from the Pyrenees, Apennines, Pirin, Central Balkan, Northern Dinaric and Giant Mountains; Fig. 2). Their combination can result in edaphic and/or climatic unsuitability of habitats where species could potentially migrate. Climatically and edaphically suitable sites will likely decline over the next century, particularly in mountain landscapes (Bell et al. 2014). Therefore, trees in the treeline ecotone will colonize previously forested habitats. The colonization success of individual forest communities is affected by differences in species dispersal and recruitment behaviour (Dullinger et al. 2004, Jonášová et al. 2010). For these reasons, colonization of new forest habitats by the next tree generation may not be successful and may result in loss of species diversity (Honnay et al. 2002, Ibáñez et al. 2009, Dobrowski et al. 2015), as reported from Macedonia, Slovenia, the Czech Republic, Slovakia and Norway (Fig. 4).

Another limiting factor is the existence of strong adaptation mechanisms of the dominant tree species which allow them to survive in their current distribution areas. Species migration is likely to be slow due to the limited quantity of climatically and edaphically suitable sites and the slow velocity of seed dispersal and tree regeneration, and will be hampered even more by fragmentation of high mountain landscapes caused by human activities, including brush invasion in abandoned meadows (Gartzia et al. 2014). A vertical shift in a vegetation zone involves not only single species of plants, animals and microorganisms, but also their interrelationships and links to soil conditions. The speed at which climatic conditions change will be different from changes in the soil conditions due to the slowness of soil-forming processes. Soil types and conditions (cambisol versus podzol) are crucial obstacles for the shift from a beech forest zone into the spruce forest zone in the Czech Republic (Vacek & Mat jka 2010). According to other sources, the dominant tree species can influence soil formation processes (especially humus forms). Under favourable climatic and orographic conditions, beech is able to change its soil conditions over decades to centuries (J. Macku unpubl. data).

Significant changes in biodiversity must be expected in all of the mountain areas of interest. We found that biodiversity declined particularly in the southern regions of Europe, where the timberline is situated at higher elevations and human impact is mostly longer-lasting (Fig. 4). Similarly, Pauli et al. (2012) reported an increase in species richness of mountain grasslands across Europe except for Mediterranean regions, and assigned this different response of the southern regions to decreased water availability. On the other hand, ecosystems which exhibited biodiversity increases were not only enriched by migrant species, but the assemblages underwent thermophilization, i.e. cold-adapted species declined and more warm-adapted species spread (Gottfried et al. 2012). However, not all species are able to track climate changes. It is expected that around 40% of habitats will become climatically unsuitable for many mountain species, particularly endemic ones, increasing their extinction probability during this century (Dullinger et al. 2012). The abandonment of traditional land use forms is another source of this loss, which may be a more important driver than temperature increase (Fig. 4). Serious biodiversity losses in mountain meadows were reported from Spain, Italy, Macedonia, Bulgaria, Slovenia and Slovakia. Therefore, controlled grazing must occur in order to maintain alpine grasslands (Dirnböck et al. 2003). Unfortunately, grazing is still active only in smaller parts of European mountains (e.g. in the Central Alps and the Central Balkan Mountains, but pasturing can also negatively affect vegetation diversity) and sometimes does not serve to maintain alpine grassland. The same is true for grassland management, which can help to maintain mountain species and decrease habitat loss. A serious biodiversity loss in mountain meadows was reported from the Pyrenees, the Apennines, and the Shara, Pirin, Central Balkan, Northern Dinaric and Low Tatra Mountains (Fig. 4).

In forests, species responses to climate change may be equivocal; some recent studies indicated contradictory shifts in species distributions (e.g. Lenoir et al. 2008, Zhu et al. 2012, Rabasa et al. 2013). This disparity is widely discussed and assigned to the great variety of non-climatic factors or even tree ontogeny (Grytnes et al. 2014, Lenoir & Svenning 2015, Máliš et al. 2016). Disturbances leading to tree mortality may also play an important role (Cudlín et al. 2013); changes in tree canopy cover modify light availability and microclimate and can induce the thermophilization of forest vegetation (De Frenne et al. 2015, Stevens et al. 2015). The loss of this microclimate buffering effect of forests may induce a biotic homogenization of forests (Savage & Vellend 2015) or the creation of novel non-analogical communities, such as oak–pine forests (Urban et al. 2012), which may be a new threat to forest biodiversity.

The observed changes in treeline forest ecosystems are often related to changes in land-use intensity (Theurillat & Guisan 2001, Alados et al. 2014). According to climate change predictions and the recent and future exploitation intensity of European mountains, trees and forest communities will shift upward due to land use change or climate change or both. Reduced land-use intensity certainly will interact with climate change by facilitating or inhibiting species occurrence, and will accelerate forest expansion above the present treeline, particularly to previously forested habitats (Theurillat & Guisan 2001). The simultaneous action of both main treeline shift drivers, viz. temperature increase and decrease in land use intensity, was recorded from all of our studied mountain areas. The extensive differences in timberline and treeline elevations in almost all studied mountains (Table 1) indicate the anthropo-zoogenic treeline type (according to Ellenberg 1998). In most mountains (e.g. in the Apennines, Shara Mountains, Central Balkans and Alps), land use is the prevailing factor influencing vegetation drift (Fig. 5). Previous research showed that the upward shift of the treeline in the Swiss Alps was predominantly attributable to land abandonment, and only in some situations to climate change (Gehrig-Fasel et al. 2007). In the Apennines in the last few decades, tree establishment has been mainly controlled by land use, while tree growth has been controlled by climate, pointing to a minor role played by climate in shaping the current treeline (Palombo et al. 2013).

The impact of climate change and the connected land use change on biodiversity and ecosystem services provision in several European countries is summarized by Wielgolaski et al. (2017), and Kyriazopoulos et al. (2017) (both this Special), and Sarkki et al. (2016). One of the possible adaptive management options in response to climate change, assisted migration as human-assisted movement of species, has been frequently debated in the last few years (Ste-Marie et al. 2011). It is possible to apply it as a type of assisted colonization, i.e. intentional movement and release of an organism outside its indigenous range to avoid extinction of populations of the focal species (e.g. Macedonian pine species), or as an ecological replacement, i.e. the intentional movement and release of an organism outside its indigenous range to perform a specific ecological function (e.g. planting of *Pinus mugo* in the Alps; IUCN/SSC 2013).

5. CONCLUSIONS

The analysis of 11 mountain areas across Europe showed that with increasing latitude, the treeline and altitudinal width of the treeline ecotone significantly decreases, as do the significance of climatic and soil parameters as barriers against tree species shift. Although temperature is the overwhelming controlling factor of tree growth and establishment in temperate and boreal treeline ecotones, late-seasonal drought might also play a driving role in Mediterranean treeline ecotones. Longitude was less influential, mostly affecting climate and land use change effects on increased biodiversity loss, as well as the size of the area that forms one mesoclimate unit affecting altitudinal ecotone width. The biggest part of the commonly observed remaining variability in mountain vegetation near the treeline in Europe seems to be caused by geomorphological, geological, pedological and microclimatic variability in combination with different land use history and the present socio-economic relations. The observed differences in climatic parameters between the mountain areas of interest in comparison with the reference period 1960–1990 (0.9°C) have not explained the relatively high differences in the rate of treeline shift per year (1.49 m). The predicted variability in temperature increase due to global warming (1.2–2.6°C in 2050 and 2.2–5.3°C in 2100) could lead to a much bigger differentiation in treeline ecotone biodiversity and ecosystem processes between southern and northern European mountains in the future. Therefore, these differences must be taken into account by scientists and EU policy makers when formulating efficient adaptive forest management strategies for treeline ecosystems at the European level (e.g. assisted migration of adapted genotypes).

Acknowledgements. This paper is based firstly upon work from the COST Action ES 1203 SENSFOR, supported by COST (European Cooperation in Science and Technology; www.cost.eu). This international work was further supported by projects granted by the Ministry of Education Youth and Sports of the Czech Republic, grant NPU I LO1415 and LD 14039, by the agency APVV SR under projects APVV-14-0086 and APVV-15-0270. We acknowledge the E-OBS dataset from the EU-FP6 project ENSEMBLES (http://ensembles-eu.metoffice.com) and the data providers in the ECA&D project (www.ecad.eu)

LITERATURE CITED

Alados CL, Errea P, Gartzia M, Saiz H, Escós J (2014) Positive and negative feedbacks and free-scale pattern distribution in rural-population dynamics. PLOS ONE 9: e114561

Allen CD, Macalady AK, Chenchouni H, Bachelet D and others (2010) A global overview of drought and heat-induced tree mortality reveals emerging climate change risks for forests. For Ecol Manag 259:660–684

Amidžić L, Bartula M, Jarić S (2012) Syntaxonomy overview of vegetation of Šar Planina in the Balkan Peninsula. Vegetos 25:348–360

Aune S, Hofgaard A, Söderström L (2011) Contrasting climate and land use driven tree encroachment pattern of sub-arctic tundra in Northern Norway and Kola Peninsula. Can J For Res 41:437–449

Barbeito I, Dawes M, Rixen C, Senn J, Bebi P (2012) Factors driving mortality and growth at treeline: a 30-year experiment of 92000 conifers. Ecology 93:389–401

Batllori E, Camarero JJ, Ninot JM, Gutiérrez E (2009) Seedling recruitment, survival and facilitation in alpine *Pinus uncinata* treeline ecotones. Implications and potential responses to climate warming. Glob Ecol Biogeogr 18:460–472

Bebi P, Teich M, Hagedorn F, Zurbriggen N, Brunner SH, Grét-Regamey A (2012) Veränderung von Wald und Waldleistungen in der Landschaft Davos im Zuge des Klimawandels. Schweiz Z Forstwes 163:493–501

Bell DM, Bradford JB, Lauenroth WK (2014) Mountain landscapes offer few opportunities for high elevation tree species migration. Glob Change Biol 20:1441–1451

Bergant K (2006) Climate change scenarios for Macedonia. University of Nova Gorica, Center for Atmospheric

Research, Nova Gorica

Bodin J, Badeau V, Bruno E, Cluzeau C, Moisselin JM, Walther GR, Dupouey JL (2013) Shifts of forest species along an elevational gradient in Southeast France: climate change or stand maturation? J Veg Sci 24:269–283

Boncina A (2011) History, current status and future prospects of uneven-aged forest management in the Dinaric region: an overview. Forestry 84:467–478

Breshears DD, Cobb NS, Rich PM, Price KP and others (2005) Regional vegetation die-off in response to global-change-type drought. Proc Natl Acad Sci USA 102: 15144–15148

Camarero JJ, García-Ruiz JM, Sangüesa-Barreda G, Galván JD and others (2015a) Recent and intense dynamics in a formerly static Pyrenean treeline. Arct Antarct Alp Res 47:773–783

Camarero JJ, Gazol A, Galván JD, Sangüesa-Barreda G, Gutiérrez E (2015b) Disparate effects of global-change drivers on mountain conifer forests: warming-induced growth enhancement in young trees vs. CO_2 fertilization in old trees from wet sites. Glob Change Biol 21:738–749

Chen IC, Hill JK, Ohlemüller R, Roy DB, Thomas CD (2011) Rapid range shifts of species associated with high levels of climate warming. Science 333:1024–1026

Cudlín P, Seják J, Pokorný J, Albrechtová J, Bastian O, Marek M (2013) Forest ecosystem services under climate change and air pollution. In: Matyssek R, Clarke N, Cudlín P, Mikkelsen TN, Tuovinen JP, Wiesner G, Paoletti E (eds) Climate change, air pollution and global challenges: understanding and perspectives from forest research. Developments in Environmental Science Vol 13. Elsevier, Oxford, p 521–546

Dalen L, Hofgaard A (2005) Differential regional treeline dynamics in the Scandes mountains. Arct Antarct Alp Res 37:284–296

Dawes MA, Philipson CD, Fonti P, Bebi P, Hättenschwiler S, Hagedorn F, Rixen C (2015) Soil warming and CO_2 enrichment induce biomass shifts in alpine tree line vegetation 2014. Glob Change Biol 21:2005–2021

De Frenne P, Rodríguez-Sánchez F, De Schrijver A, Coomes DA, Hermy M, Vangansbeke P, Verheyen K (2015) Light accelerates plant responses to warming. Nat Plants 1: 15110

Devi N, Hagedorn F, Moiseev P, Bugmann H, Shiyatov SG, Mazepa V, Rigling A (2008) Expanding forests and changing growth forms of Siberian larch at the treeline of the Polar Urals during the 20th century. Glob Change Biol 14:1581–1591

Diaz-Varela R, Colombo R, Meroni M, Calvo-Iglesias MS, Buffoni A, Tagliaferri A (2010) Spatio-temporal analysis of alpine ecotones. A spatial explicit model targeting altitudinal vegetation shifts. Ecol Model 221:621–633

Dirnböck T, Dullinger S, Grabherr G (2003) A regional impact assessment of climate and land-use change on alpine vegetation. J Biogeogr 30:401–417

Dobrowski SZ, Swanson AK, Abatzoglou JT, Holden ZA, Safford HD, Schwartz MK, Gavin DG (2015) Forest structure and species traits mediate projected recruitment declines in western US tree species. Glob Ecol Biogeogr 24:917–927

Dullinger S, Dirnböck S, Grabherr G (2004) Modelling climate change-driven treeline shifts: relative effects of temperature increase, dispersal and invisibility. J Ecol 92:241–252

Dullinger S, Gattringer A, Thuiller W, Moser D and others (2012) Extinction debt of high-mountain plants under twenty-first-century climate change. Nat Clim Chang 2: 619–622

Elkin C, Gutierrez AG, Leuzinger S, Manusch C, Temperli C, Rasche L, Bugmann H (2013) A 2°C warmer world is not safe for ecosystem services in the European Alps. Glob Change Biol 19:1827–1840

Ellenberg H (1988) Vegetation ecology of Central Europe, 4th edn. Cambridge University Press, Edinburgh

Em H (1986) Spruce southern distribution range. Spruce forest on Shara Mountain in Macedonia. Contrib Macedonian Acad Sci Arts Skopje 5:11–28

Fajardo A, Piper FI, Pfund L, Körner CH, Hoch G (2012) Variation of mobile carbon reserves in trees at the alpine treeline ecotone is under environmental control. New Phytol 195:794–802

Fridley JD, Grime JP, Askew AP, Moser B, Stevens CJ (2011) Soil heterogeneity buffers community response to climate change in species-rich grassland. Glob Change Biol 17:2002–2011

Gartzia M, Alados CL, Pérez-Cabello F (2014) Assessment of the effects of biophysical and anthropogenic factors on woody plant encroachment in dense and sparse mountain grasslands based on remote sensing data. Prog Phys Geogr 38:201–217

Gehrig-Fasel J, Guisan A, Zimmermann NE (2007) Tree line shifts in the Swiss Alps: climate change or land abandonment? J Veg Sci 18:571–582

Gikov A, Dimitrov P, Zhiyanski M (2016) Land cover dynamics on the northern slope of the Troyan Passage during 30 years period. Probl Geogr 1-2:78–92

Gonzalez P, Neilson RP, Lenihan JM, Drapek RJ (2010) Global patterns in the vulnerability of ecosystems to vegetation shifts due to climate change. Glob Ecol Biogeogr 19:755–768

Gottfried M, Pauli H, Futschik A, Akhalkatsi M and others (2012) Continent-wide response of mountain vegetation to climate change. Nat Clim Chang 2:111–115

Grace J, Berninger F, Nagy L (2002) Impacts of climate change on the tree line. Ann Bot 90:537–544

Grunewald K, Scheithauer J (2011) Landscape development and climate change in Southwest Bulgaria (Pirin Mts.). Springer, Heidelberg

Grytnes JA, Kapfer J, Jurasinski G, Birks HH and others (2014) Identifying the driving factors behind observed elevational range shifts on European mountains. Glob Ecol Biogeogr 23:876–884

Hagedorn F, Shiyatov FG, Mazepa VS, Devi NM and others (2014) Treeline advances along the Urals mountain range — driven by improved winter conditions? Glob Change Biol 20:3530–3543

Hansen AJ, Neilson RP, Dale VH, Flather CH and others (2001) Global change in forests: responses of species, communities, and biomes. BioScience 51:765–779

Harsch MA, Bader MY (2011) Treeline form–a potential key to understanding treeline dynamics. Glob Ecol Biogeogr 20:582–596

Harsch MA, Hulme PE, McGlone MS, Duncan RP (2009) Are treelines advancing? A global meta-analysis of treeline response to climate warning. Ecol Lett 12:1040–1049

Haylock MR, Hofstra N, Klein Tank AMG, Klok EJ, Jones PD, New M (2008) A European daily high-resolution gridded dataset of surface temperature and precipitation. J Geophys Res Atmos 113:D20119

Hlásny T, Barcza Z, Fabrika M, Balázs B and others (2011)

Climate change impacts on growth and carbon balance of forests in Central Europe. Clim Res 47:219–236
Hlásny T, Trombik J, Dobor L, Barcza Z, Barka I (2016) Future climate of the Carpathians: climate change hotspots and implications for ecosystems. Reg Environ Change 16:1495–1506
Hofgaard A, Dalen L, Hytteborn H (2009) Tree recruitment above the treeline and potential for climate driven treeline change. J Veg Sci 20:1133–1144
Holtmeier FK (2009) Mountain timberlines. Ecology, patchiness, and dynamics, 2nd edn. Advances in Global Change Research 36. Springer Science & ScienceMedia B.V., Dordrecht
Holtmeier FK, Broll G (2005) Sensitivity and response of northern hemisphere altitudinal and polar treelines to environmental change at landscape and local scales. Glob Ecol Biogeogr 14:395–410
Honnay O, Verheyen K, Butaye J, Jacquemyn H, Bossuyt B, Hermy M (2002) Possible effects of habitat fragmentation and climate change on the range of forest plant species. Ecol Lett 5:525–530
Huber R, Rigling A, Bebi P, Brand F and others (2013) Sustainable land use in mountain regions under global change: synthesis across scales and disciplines. Ecol Soc 18:36
Ibáñez I, Clark JS, Dietze MC (2009) Estimating colonization potential of migrant tree species. Glob Change Biol 15: 1173–1188
IUCN/SSC (International Union for Conservation of Nature/Species Survival Commission) (2013) Guidelines for reintroductions and other conservation translocations. Version 1.0. IUCN/Species Survival Commission, Gland
Jankovský L, Cudlín P, Čermák P, Moravec I (2004) The prediction of development of secondary Norway spruce stands under the impact of climatic change in the Drahany highlands (The Czech Republic). Ekologia (Bratisl) 23(Suppl 2):101–112
Jeník J, Lokvenc T (1962) Die alpine Waldgrenze im Krkonoše Gebirge. Rozpravy SAV. Řada matematických a přírodních věd 72. Československá akademie věd, Praha
Jobbágy EG, Jackson RB (2000) Global controls of forest line elevation in the northern and southern hemispheres. Glob Ecol Biogeogr 9:253–268
Jonášová M, Vávrová E, Cudlín P (2010) Western Carpathian mountain spruce forest after a windthrow: natural regeneration in cleared and uncleared areas. For Ecol Manag 259:1127–1134
Kittel TGF, Steffen WL, Chapin FS (2000) Global and regional modelling of Arctic–boreal vegetation distribution and its sensitivity to altered forcing. Glob Change Biol 6:1–18
Klopcic M, Jerina K, Bončina A (2010) Long-term changes of structure and tree species composition in Dinaric uneven-aged forests: Are red deer an important factor? Eur J For Res 129:277–288
Klopcic M, Simončič T, Bončina A (2015) Comparison of regeneration and recruitment of shade-tolerant and light-demanding tree species in mixed uneven-aged forests: experiences from the Dinaric region. Forestry 88:552–563
Körner C (2003) Alpine plant life: functional plant ecology of high mountain ecosystems; with 47 tables. Springer Science & Business Media B.V., Dordrecht
Körner C (2012) Alpine treelines. Functional ecology of the global high elevation tree limits. Springer, Basel
Körner C, Paulsen J (2004) A world-wide study of high altitude treeline temperatures. J Biogeogr 31:713–732
Kulakowski D, Bebi P, Rixen C (2011) The interacting effects of land use change, climate change and suppression of natural disturbances on landscape forest structure in the Swiss Alps. Oikos 120:216–225
Kullman L (1988) Holocene history of the forest–alpine tundra ecotone in the Scandes Mountains (central Sweden). New Phytol 108:101–110
Kullman L (1999) Early holocene tree growth at a high elevation site in the northernmost Scandes of Sweden (Lapland): a palaeobiogeographical case study based on megafossil evidence. Geogr Ann 81:63–74
Kullman L (2007) Tree line population monitoring of *Pinus sylvestris* in the Swedish Scandes, 1973–2005: implications for tree line theory and climate change ecology. J Ecol 95:41–52
Kullman L, Öberg L (2009) Post-Little Ice Age treeline rise and climate warming in the Swedish Scandes: a landscape ecological perspective. J Ecol 97:415–429
Kyriazopoulos AP, Skre O, Sarkki S, Wielgolaski FE, Abraham EM, Ficko A (2017) Human–environment dynamics in European treeline ecosystems: a synthesis based on the DPSIR framework. Clim Res 73:17–29
Lenoir J, Svenning JC (2015) Climate-related range shifts — a global multidimensional synthesis and new research directions. Ecography 38:15–38
Lenoir J, Gégout JC, Marquet PA, de Ruffray P, Brisse H (2008) A significant upward shift in plant species optimum elevation during the 20th century. Science 320: 1768–1771
Lindner M, Fitzgerald JB, Zimmermann NE, Reyer C and others (2014) Climate change and European forests: What do we know, what are the uncertainties, and what are the implications for forest management? J Environ Manag 146:69–83
Lloyd AH, Rupp TS, Fastie CL, Srafield AM (2002) Patterns and dynamic of treeline advance on the Seward Penisula, Alaska. J Geophys Res 107:8161
Máliš F, Kopecký M, Petřík P, Vladovič J, Merganič J, Vida T (2016) Life-stage, not climate change, explains observed tree range shifts. Glob Change Biol 22:1904–1914
Mathisen IE, Mikheeva A, Tutubalina OV, Aune S and Hofgaard A (2014) Fifty years of tree line change in the Khibiny Mountains, Russia: advantages of combined remote sensing and dendroecological approaches. Appl Veg Sci 17: 6–16
Millar CI, Stephenson NL (2015) Temperate forest health in an era of emerging megadisturbance. Science 349: 823–826
Mina M, Bugmann H, Klopcic M, Cailleret M (2017) Accurate modeling of harvesting is key for projecting future forest dynamics: a case study in the Slovenian mountains. Reg Environ Change 17:49–64
Moen J, Aune K, Edenius L, Angerbjörn A (2004) Potential effects of climate change on treeline position in the Swedish mountains. Ecol Soc 9:16, www.ecologyandsociety.org/vol9/iss1/art16
Palombo C, Chirici G, Marchetti M, Tognetti R (2013) Is land abandonment affecting forest dynamics at high elevation in Mediterranean mountains more than climate change? Plant Biosyst 147:1–11
Palombo C, Marchetti M, Tognetti R (2014) Mountain vegetation at risk: current perspectives and research reeds. Plant Biosyst 148:35–41
Pauli H, Gottfried M, Dullinger S, Abdaladze O, Akhalkatsi

M, Alonso JLB, Grabherr G (2012) Recent plant diversity changes on Europe's mountain summits. Science 336: 353–355

Payette S, Fortin MJ, Gamachet I (2001) The subarctic forest–tundra: the structure of a biome in a changing climate. BioScience 51:709–718

Pearson GR, Dawson TP (2003) Predicting the impacts of climate change on the distribution of species: Are bioclimate envelope models useful? Glob Ecol Biogeogr 12: 361–371

Pedrotti F (2013) Plant and vegetation mapping. Geobotany Studies. Springer-Verlag, Berlin

Peñuelas J, Boada M (2003) A global change-induced biome shift in the Montseny mountains (NE Spain). Glob Change Biol 9:131–140

Piovesan G, Biondi F, Filippo AD, Alessandrini A, Maugeri M (2008) Drought driven growth reduction in old beech (*Fagus sylvatica* L.) forests of the central Apennines, Italy. Glob Change Biol 14:1265–1281

Rabasa SG, Granda E, Benavides R, Kunstler G and others (2013) Disparity in elevational shifts of European trees in response to recent climate warming. Glob Change Biol 19:2490–2499

Raev I, Zhelev P, Grozeva M, Markov I and others (2011) Programme of measures for adaptation of forests in Bulgaria and mitigate the negative impact of climate change on them. Project FUTURE forest helping Europe tackle climate change. INTERREG IVC, ERDF, Sofia

Sarkki S, Ficko A, Grunewald K, Nijnik M (2016) Benefits from and threats to European treeline ecosystem services: an exploratory study of stakeholders and governance. Reg Environ Change 16:2019–2032

Savage J, Vellend M (2015) Elevational shifts, biotic homogenization and time lags in vegetation change during 40 years of climate warming. Ecography 38:546–555

Schwörer C, Henne PD, Tinner W (2014) A model-data comparison of Holocene timberline changes in the Swiss Alps reveals past and future drivers of mountain forest dynamics. Glob Change Biol 20:1512–1526

Šmilauer P, Lepš J (2014) Multivariate analysis of ecological data using Canoco 5. Cambridge University Press, Cambridge

Ste-Marie C, Nelson EA, Dabros A, Bonneau ME (2011) Assisted migration: introduction to a multifaceted concept. For Chron 87:724–730

Stevens JT, Stafford HD, Harrison S, Latimer AM (2015) Forest disturbance accelerates thermophilization of understory plant communities. J Ecol 103:1253–1263

Strid A, Andonoski A, Andonovski V (2003) Alpine biodiversity in Europe. Ecological Studies 167. Springer-Verlag, Berlin

Tattoni C, Ciolli M, Ferretti F, Cantiani MG (2010) Monitoring spatial and temporal pattern of Paneveggio forest (Northern Italy) from 1859 to 2006. iForest Biogeosci For 3:72–80

Ter Braak CJF, Šmilauer P (2012) Canoco reference manual and user's guide: software for ordination (version 5.0). Microcomputer Power, Ithaca, NY

Theurillat JP, Guisan A (2001) Potential impact of climate change on vegetation in the European Alps: a review. Clim Change 50:77–109

Tømmervik H, Wielgolaski FE, Neuvonen S, Solberg B, Høgda KA (2005) Biomass and production on a landscape level in the mountain birch forests. In: Wielgolaski FE (ed) Plant ecology, herbivory and human impact in Nordic mountain birch forests. Ecos Studies 180. Springer, Berlin, p 53–70

Treml V, Banaš M (2000) Alpine timberline in the High Sudetes. Acta Univ Carol Geogr 15:83–99

Treml V, Chuman T (2015) Ecotonal dynamics of the altitudinal forest limit are affected by terrain and vegetation structure variables: an example from the Sudetes Mountains in Central Europe. Arct Antarct Alp Res 47:133–146

Treml V, Migo P (2015) controlling factors limiting timberline position and shifts in the Sudetes: a review. Geogr Pol 88:55–70

Urban MC, Tewksbury JJ, Sheldon KS (2012) On a collision course: competition and dispersal differences create no-analogue communities and cause extinctions during climate change. Proc R Soc Lond B Biol Sci 279:2072–2080

Vacek S, Matějka K (2010) State and development of phytocenoses on research plots in the Krkonoše Mts. forest stands. J For Sci 56:505–517

Van Bogaert R, Haneca K, Hoogesteger J, Jonasson C, de Dapper M, Callaghan TV (2011) A century of tree line changes in sub-Arctic Sweden shows local and regional variability and only a minor influence of 20th century climate warming. J Biogeogr 38:907–921

Van Gils H, Batsukh O, Rossiter D, Munthali W, Liberatoscioli E (2008) Forecasting the pattern and pace of *Fagus* forest expansion in Majella National Park, Italy. Appl Veg Sci 11:539–546

Velchev V (1997) Types of vegetation. In: Yordanova M, Donchev D (eds) Geography of Bulgaria. BAN, Sofia, p 269–283 (in Bulgarian)

Vladovič J, Merganič J, Máliš F, Križová E and others (2014) The response of forest vegetation diversity to changes in edaphic-climatic conditions in Slovakia. Technická univerzita vo Zvolene, Zvolene (in Slovak)

Vrška T, Adam D, Hort L, Kolář T, Janík F (2009) European beech (*Fagus sylvatica* L.) and silver fir (*Abies alba* Mill.) rotation in the Carpathians — a developmental cycle or a linear trend induced by man? For Ecol Manag 258: 347–356

Wielgolaski FE (ed) (2005) Plant ecology, herbivory, and human impact in Nordic mountain birch forests. Ecological Studies 180. Springer Verlag, Berlin

Wielgolaski FE, Hofgaard A, Holtmeier FK (2017) Sensitivity to environmental change of the treeline ecotone and its associated biodiversity in European mountains. Clim Res 73:151–166

Wieser G, Tausz M (2007) Trees at their upper limit. Treelife limitation at the alpine timberline. Springer, Dordrecht

Wieser G, Holtmeier FK, Smith WK (2014) Treelines in a changing global environment. In: Tausz M, Grulke N (eds) Trees in a changing environment. Plant Ecophys 9. Springer, Dordrecht, p 221–263

Zhiyanski M, Kolev K, Sokolovska M, Hursthouse A (2008) Tree species effect on soils in Central Stara Planina Mountains. Nauka Gorata 4:65–82

Zhiyanski M, Gikov A, Nedkov S, Dimitrov P, Naydenova L (2016) Mapping carbon storage using land cover/land use data in the area of Beklemeto, Central Balkan. In: Koulov B, Zhelezov G (eds) Sustainable mountain regions: challenges and perspectives in southeastern Europe. Springer, Basel, p 53–65

Zhu K, Woodall CW, Clark JS (2012) Failure to migrate: lack of tree range expansion in response to climate change. Glob Change Biol 18:1042–1052

16

Sensitivity to environmental change of the treeline ecotone and its associated biodiversity in European mountains

F. E. Wielgolaski[1,*], A. Hofgaard[2], F. K. Holtmeier[3]

[1]Department of Bioscience, University of Oslo, PO Box 1066 Blindern, 0316 Oslo, Norway
[2]Norwegian Institute for Nature Research, PO Box 5685 Sluppen, 7485 Trondheim, Norway
[3]Institute of Landscape Ecology, Heisenbergstrasse 2, Westfälische Wilhelms-Universität, 48149 Münster, Germany

ABSTRACT: Transition zones between mountain forests and treeless tundra, i.e. treeline ecotones, are characterized by great regional variety. In this paper, we discuss the biodiversity in various trophic levels in treeline ecotones throughout Europe, with particular focus on recent changes in land use and climate in northern and central mountains. In northernmost Europe, mountain birch prevails, while conifers (spruce, pine, larch) are the dominating species further south. While at continent-wide to global scales, the ecotone position is largely controlled by heat deficiency, it depends on a multitude of partly interacting abiotic and biotic factors other than climate at smaller scales. Climate change is a driving factor in treeline ecotone change, including physiognomic structure and biodiversity, although the effects of climate and other factors often overlap. Historical legacy plays an important role in this respect, and human impacts are particularly important. The recent decline in pastoral use of many European treeline areas often strongly influences plant diversity and re-growth of trees and other woody species. Climate change together with changing tree cover may influence snow cover, moisture regime, and nutrient conditions. Subsequently changed site conditions influence plant–plant interactions, favoring some species and disfavoring others, and plant–animal interactions. Native animals may cause widespread or local disturbances in treeline ecotone areas. Mass outbreaks of leaf-eating insects, for example, usually affect comparatively large forested areas whereas mammalian herbivores and birds have more local impact. However, high numbers of wild or domestic mammalian herbivores may challenge the carrying capacity of treeline ecotone areas at the same time as they preserve an open pasture character. This calls for cross-disciplinary study approaches, addressing the complexity of the ecotone regarding both causal background and biogeographic diversity.

KEY WORDS: Treeline ecotone · Ecotone change · Land use change · Climate change · Animal impact

1. INTRODUCTION

Transition zones between mountain forests and treeless tundra, i.e. treeline ecotones, are characterized by great regional variety regarding abiotic and biotic components structuring the zone. Accordingly, due to environmental change and treeline shift, biodiversity has been changing in the subalpine and lower alpine zone in many European mountains. In this paper, we discuss the biodiversity in various trophic levels in treeline ecotones throughout Europe, with particular focus on recent changes in land use and climate in northern and central mountains.

*Corresponding author: f.e.wielgolaski@ibv.uio.no

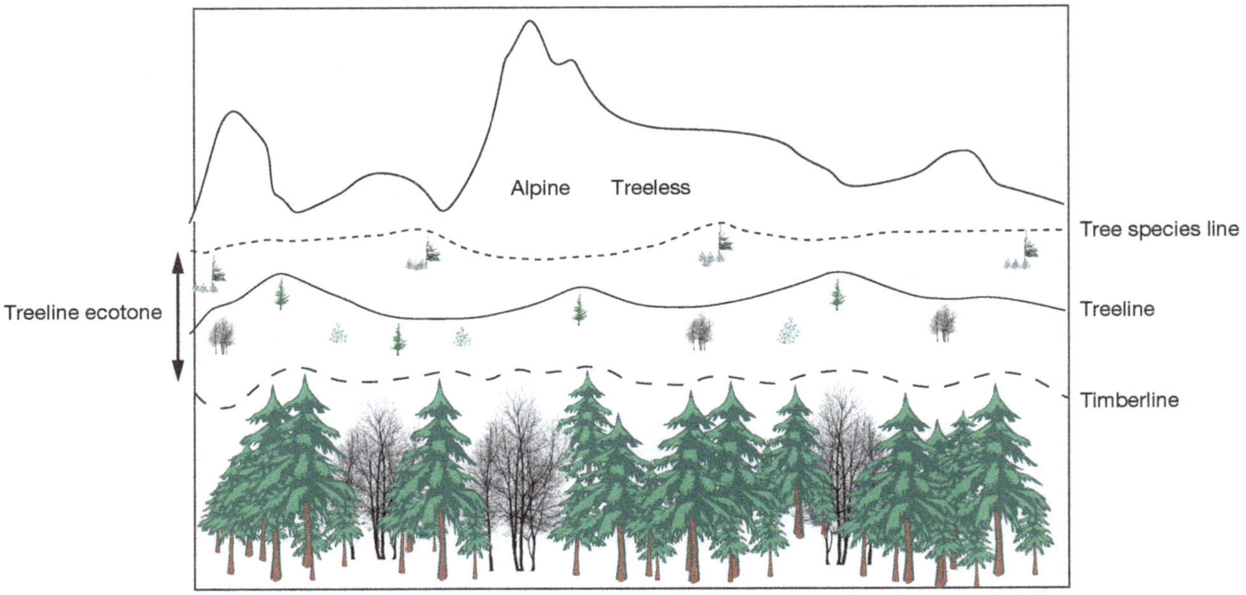

Fig. 1. Characteristics of the treeline ecotone

1.1. Treeline ecotone

The treeline ecotone spans the transition in decreased tree cover and tree height from the upper closed mountain forest to the treeless tundra. This transition boundary includes a number of tree status delineations (e.g. timberline, treeline, and tree species line; Fig. 1) used in studies of treeline ecotone dynamics. The definition of the ecotone and included lines may vary in the published literature (Callaghan et al. 2002a, Holtmeier 2009, Körner 2012, Irl et al. 2016, Holtmeier & Broll 2017, this Special), but are commonly classified regarding causal background, to climatically, topographically, or anthropogenically defined ecotones and lines. The climatic treeline ecotone generally decreases in altitude from southern to northern mountains in Europe and is found from above 2000 m above sea level (a.s.l.) in southern Europe to close to sea level in the northernmost subarctic parts (Cudlín et al. 2017, this Special). The ecotone may be abrupt (e.g. in steep or heavily grazed areas), but is normally a relatively wide boundary, and may thus regionally cover a considerable area of the lower and most productive part of the alpine zone. The vastness of the treeline ecotone, and its conspicuous transition from tree-covered to treeless areas, makes it an important biogeographic component of region-wide ecological, climatic, and socioeconomic relevance (Callaghan et al. 2002b).

In northern Europe, both the alpine and arctic treeline ecotones are normally formed by mountain birch *Betula pubescens* subsp. *tortuosa* (Ledeb.) Nyman (Wielgolaski 2001, 2005), but may regionally also include Norway spruce *Picea abies* L. and Scots pine *Pinus sylvestris* L. In mountains further south, the ecotone is normally formed by conifers such as spruce, pine (e.g. *Pinus sylvestris*, *P. cembra* L., *P. uncinata* Ramond ex DC.) and larch *Larix decidua* L., but regionally also beech *Fagus sylvatica* L. (Wilmanns 1989, Nagy et al. 2003, Holtmeier 2009 for ample references). In the upper part of the ecotone, trees become progressively more stunted and may form extensive areas with scattered individual krummholz or krummholz groups. In some mountain areas of Europe, the 'true krummholz' mountain pine *Pinus mugo* Turra, the usually prostrate and gnarled growth of which is genetically predetermined, dominates above the high-stemmed mountain forests (Holtmeier 1981, 2009).

Historically, extensive changes in the elevation of the treeline ecotone have occurred throughout Europe. For example, in both the Scandes and mountains further south, pollen analyses and radiocarbon dating of tree remains found in mires, alpine sediments, and at retreating glacier fronts have revealed that trees grew at much higher elevation during the early to mid-Holocene than today (Holtmeier 1974, 1993, 2009, Kullman 1995, 2004, Tinner et al. 1996, Allen & Huntley 1999, Kullman & Källgren 2000, Aas & Faarlund 2001, Tinner & Theurillat 2003, Heiri et al. 2006). During the late Holocene and until termination of the Little Ice Age (Grove 1988), treeline ecotone retreat was the dominating trend in many European mountains (Karlén 1976, Shiyatov 1993, Kullman 1995). Later, the trend has reversed in response to both the release from the long-term climatic suppression during the Little Ice Age and the more recent climate warming (Kullman 1986, 2003, Gervais & MacDonald 2000, Shiyatov 2003, Camarero & Gutiérrez 2004, Motta et al. 2006). However, although an advancing trend is dominating, non-advancing trends occur across European mountains (Harsch et al. 2009, Van Bogaert et al. 2011, Dinca et al. 2017, this Special).

In general, thermal deficiency during the growing season is a main factor limiting tree growth and development at high elevation, which has been well known since the beginning of treeline research. However, although southern slopes usually provide favorable thermal conditions, the treeline ecotone may be at a relatively low elevation due to moisture deficiency, as is the case in many summer-dry Mediterranean mountains (e.g. Brandes & Ise 2007, Gonzáles de Andrés et al. 2015). Insufficient moisture supply as a result of summer drought also occurs at the treeline in Central Europe, as for example, in the Sudetes, where it affects tree seedling establishment on southern exposures (Treml & Chuman 2015, Treml et al. 2016). Extreme winds, snow cover, wildfires, etc. may also influence tree growth in European treeline areas (Holtmeier & Broll 2017). In addition, low-elevation ecotones may be due to historical and/or ongoing human activities, e.g. logging and pastoral use, creating anthropogenic elevation of the treeline ecotone. When landscape controlling pressures (e.g. human activities) cease or lessen, the response might be seen as a swift reforestation (Hofgaard 1997a, Bolli et al. 2007, Batllori & Gutiérrez 2008, Bryn 2008). However, summer drought periods or other disturbances may affect tree growth and prevent or delay natural reforestation considerably (Hofgaard 1997a,b, Brandes & Ise 2007, Grunewald & Scheithauer 2008, Gonzáles de Andrés et al. 2015), due to the multitude of abiotic and biotic factors controlling changes in both structure and location of the treeline ecotone (Holtmeier & Broll 2005, Hofgaard et al. 2012, Weisberg et al. 2013, Kulakowski et al. 2016).

The altitude of the treeline ecotone location decreases from central parts of mountain massifs to coastal areas. Central areas have a more continental climate due to protection from cool and moisture-carrying air masses, and thus, normally have higher daytime temperatures during the growing season, compared to heavily dissected and maritime mountain ranges. The 'mass elevation effect' (De Quervain 1904) often overlaps with the influence of the continental climate (Brockmann-Jerosch 1919, Turner 1961, 1970, Holtmeier 2009, Kašpar & Treml 2016). In addition to the north–south and coast–inland gradients, mountain topography strongly influences ecotone elevation and spatial patterns at smaller scales (Holtmeier 2009, Holtmeier & Broll 2010, 2012). This is most evident for steep slopes, where recurrent avalanches often prevent tree establishment and cause topographically defined ecotone location.

1.2. Biodiversity

Transition zones between 2 major biomes generally have high biodiversity. This also applies to the treeline ecotone compared to the forest at lower elevation and the treeless tundra at higher elevation (Hofgaard & Wilmann 2002). High biodiversity in ecotones is mainly caused by an overlapping distribution of species originally belonging to the 2 adjoining biomes. In addition, as mountains are often characterized by a highly varying and rugged topography, the biological richness is high with strong differences occurring at short distances (Huston 1994). Further, a mix of vegetation-covered ground and open patches with high light intensity at the ground is characteristic of the treeline ecotone. Taken together, this provides a wide range of temperature- and moisture-defined microhabitats favorable to high species diversity (Körner 2003, Nagy & Grabherr 2009).

Species richness across European treeline ecotones decreases with latitude, but depends on a large number of abiotic and biotic factors, such as human activities and soil conditions (Callaghan et al. 2004, Vittoz et al. 2010). However, the general trend with decreasing species diversity towards high latitudes or altitudes makes the ecotone an indistinct species boundary. A general decrease in species diversity is accompanied by a strong nutrient and productivity gradient (Callaghan et al. 2004), but there is a lack of evidence for a causal connection between latitudinal decrease in species diversity and productivity (Rohde 1992). This is evidenced by some species groups with a high frequency in the treeline ecotone showing a reversed latitudinal trend, such as willows, wasps, sawflies, aphids, and peatland birds, which has been related to habitat heterogeneity (Kouki 1999).

Scenarios for biodiversity change caused by human activity indicate land use as the most important driver for biodiversity changes in terrestrial ecosystems (Sala et al. 2000), which has strong relevance to the treeline ecotone. Changes in biodiversity affect the functioning of ecosystems, and thereby also society through many ecosystem services (Garcia-Ruiz et al. 1996, Cardinale et al. 2012; this Special: Sarkki et al. 2017a,b, Kyriazopoulos et al. 2017, Fleischer et al. 2017, Nijnik et al. 2017).

2. CHANGES IN PLANT DISTRIBUTION

Although changes in temperature and precipitation (including snow cover) surely alter treeline ecotone locations and species diversity, as apparent from the Holocene period and recent history (Kullman 1995, 2003, Aas & Faarlund 1996, 2001, Körner 2003, Shiyatov 2003, Hofgaard et al. 2013, Mathisen et al. 2014, Schwörer et al. 2014), it has also been shown that land use changes may have stronger impact than climate change (Motta et al. 2006, Gehrig-Fasel et al. 2007, Aune et al. 2011, Callaghan et al. 2013, Grytnes et al. 2014, Strebel & Bühler 2015). This is also indicated in reports from the present SENSFOR study (Sarkki et al. 2016, Cudlín et al. 2017, Kyriazopoulos et al. 2017).

In many parts of Europe, there is a long tradition of grazing cattle, sheep, and goats at and above the treeline. This allows forage around the farms at low elevation to be saved for the winter season (Holtmeier 1974, 1987, 2009, Bryn & Daugstad 2001). In northern Europe, semi-domestic reindeer are similarly moved between alpine summer grazing areas and lower elevation winter grazing areas. Through time, this European-wide practice has created non-climatic ecotone locations and diversity characteristics. Many different human activities has contributed to this throughout history (Emanuelsson 1987, Bryn & Daugstad 2001, Gehrig-Fasel et al. 2007) and in more recent times, tourism and recreation has impacted the treeline ecotone vegetation to an increasing degree (Wielgolaski 1998, Körner 2003, Forbes et al. 2005, Törn et al. 2009, Rixen & Rolando 2013, Sato et al. 2013, Tolvanen & Kangas 2016, Ylisirniö & Allén 2016). Activities with an impact across or directly below the treeline ecotones are, or have been, tree clearing for space, fire wood, fencing, and building purposes, and harvesting of young twigs and leaf material, particularly from deciduous trees, as additional fodder. Litter has been used for bedding in the cattle sheds. The magnitude of an impact is dependent on both direct human activities (e.g. cutting) and on grazer diversity and density (Fig. 2), and in addition, on the duration of summer grazing throughout history (Austrheim et al. 2008, Speed et al. 2010, 2012). In some treeline areas, natural pastures have been cultivated, fertilized and sown with grass of non-alpine origin to increase the amount of the fodder (Fig. 3). These pasture management methods

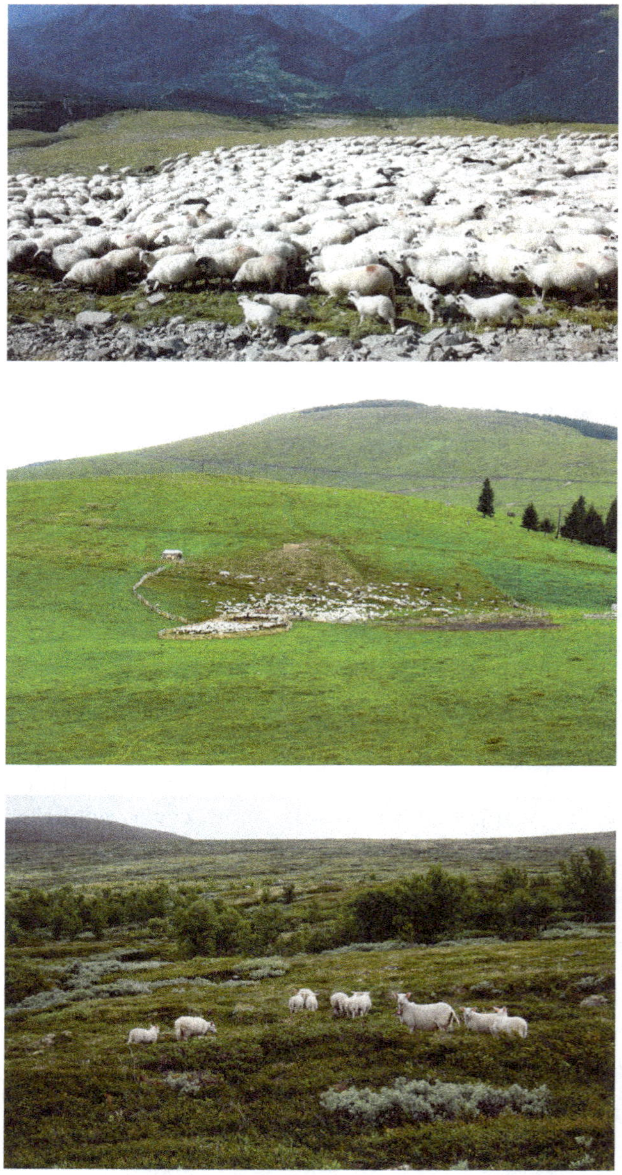

Fig. 2. Pasture husbandry may differ widely between European mountains. In southern and central Europe, sheep are generally kept in dense herds controlled by herders (upper panel), and gathered in fenced areas for the night to avoid carnivore predation (middle panel). In the Scandinavian mountains, a non-herding system is practiced which allows sheep to disperse over the landscape in small groups (lower panel) under occasional monitoring by owners. Upper panel: Pyrenees, Spain. Photography by J. Inkeröinen. Middle and lower panels: Rodnei Mountains, Carpathians, Romania and Dovre mountains, Norway. Photography by A. Hofgaard

Fig. 3. Summer farm with grass production using non-alpine grass species and artificial fertilizers. Flåmsætrin, Central Norway. Photography by A. Hofgaard

have strongly influenced the natural vegetation. Lowland species, including anthropocores, brought to the treeline ecotone areas by these cultivation practices might cause an anomalous increase in species diversity in an area, at least locally and for a short period (Cudlín et al. 2017). However, these non-alpine species could linger in the area and become invasive in future favorable environmental conditions (Crooks & Soulé 1999). As to novel competitive interactions, it may be essential to accurately predict plant species' responses to e.g. climate change (Hofgaard 1999, Alexander et al. 2015). The presence of potentially invasive species make any prediction difficult or impossible (Mooney & Hofgaard 1999, Petitpierre et al. 2016).

During the last 50 to 100 yr, summer farming practice has become strongly reduced both in northern and more southern European mountain areas (Bryn & Daugstad 2001, Tasser & Tappeiner 2002, Camarero & Gutiérrez 2007, Chauchard et al. 2007, Tasser et al. 2007, Batllori & Gutiérrez 2008, Ameztegui et al. 2010, Treml et al. 2016). One reason is the increased importance of imported fodder due to its low cost, while at the same time, the labor-demanding traditional summer farming has become too costly. In some European mountains, however, such as the Alps, farmers are paid by the authorities (Fischer et al. 2008) to continue the traditional pastoral use of elevated mountain areas. This is intended to maintain the character and biodiversity of the cultural landscape that the alpine zone represents. However, in many European mountain areas, grazing by e.g. sheep, is still a normal land use form (Fig. 2), and in others, traditional land use is being replaced by use linked to winter and/or summer tourism (Fig. 4).

Fig. 4. In downhill ski areas, the upper part of the mountain forest and corridors through the forest are managed as open woodlands to favor ski tourism (upper panel: Klosters, Switzerland), and preservation of pastures in the treeline ecotone creates attractive areas for hiking tourism (lower panel: Jaman, Switzerland). Photography by A. Hofgaard

Abandonment or reduction of traditional pastoralism in treeline ecotone areas is normally followed by strong and very fast recolonization and growth of trees, shrubs, and other plant species palatable to domestic and semi-domestic animals. Biodiversity may increase at the early stages of recolonization of former alpine pasture (Strebel & Bühler 2015, Cudlín et al. 2017), but then decrease at later stages due to growing tree and shrub populations outcompeting alpine pasture species (Holtmeier & Broll 2017) (Fig. 5). This basically land use-driven process is evident both in the Scandes, where mountain birch quickly colonizes abandoned alpine pasture fields (Bryn 2008, Bryn & Hemsing 2012, Bryn et al. 2013) (Fig. 6), and in more southern European mountains (Dirnböck et al. 2003, Gehrig-Fasel et al. 2007, Sitko & Troll 2008), where e.g. mountain pine is an efficient colonizer (Fig. 7). However, in northern Fennoscandia, where reindeer husbandry prevails, temperature also appears to be an important factor directly (Karlsen et al. 2017, this Special). In addition, recent increased precipitation in northern regions (Hanssen-Bauer et al. 2015) might also have influenced mountain birch growth (Mathisen et al. 2014), as this species is favored by precipitation (Wielgolaski 2001, 2003, Wielgolaski & Karlsen 2007). The increased growth observed in mountain birch in the Scandes might also indicate increased nutrient supply as a consequence of increased precipitation and temperature, and subsequent increased decomposition (Wielgolaski & Nilsen 2001, Wielgolaski & Karlsen 2007). Similarly, expansion of green alder *Alnus viridis* Chaix in the Alps as a result of reduced land management has an important influence on nitrogen conditions in former nitrogen-poor montane grasslands (Bühlmann et al. 2016). However, as observed in Swiss mountain grasslands, nitrogen deposition might be negatively related to species richness due to increased competition among vascular plants and bryophytes (Roth et al. 2013). Increased nutrient availability in the soil, either through artificial fertilization or increased decomposition, will change the species composition. In particular, the cover and frequency of lichen and bryophyte species are reduced, while graminoids and some deciduous shrubs are generally favored (Klanderud 2008, Olsen & Klanderud 2014). This change in species composition most often results in reduced biodiversity.

Reforestation of husbandry-related grazing lands and other tree colonization in alpine areas requires viable seed production and dispersal from the forest at lower elevation. However, although viable seeds are generally available and easily dispersed by wind

Fig. 6. Re-growth of young birch *Betula pubescens* after abandoned summer grazing practice. The previously open meadows with scattered old birch is now dominated by a dense layer of birch saplings (dark green). Røldal, southern Norway. Photography by N. Eide

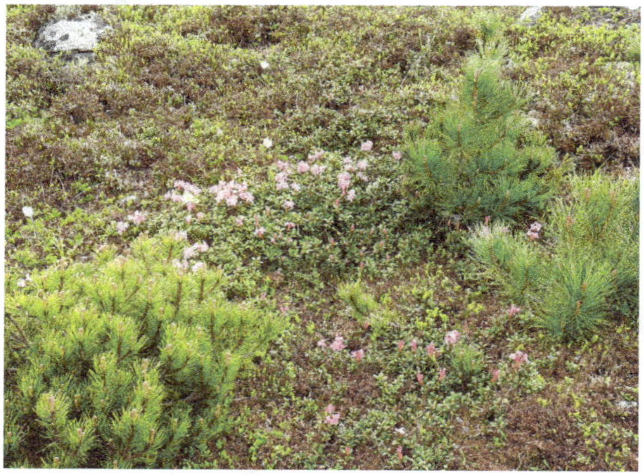

Fig. 5. Recruitment and spread of coniferous shrub (*Pinus mugo*, lower left) and tree (*P. cembra*, right) populations outcompeting alpine pasture species. Rodnei Mountains in the northern Carpathians, Romania. Photography by A. Hofgaard

Fig. 7. Mountain pine *Pinus mugo* quickly colonize abandoned herding areas (right hand side of the ridge) and form a dense shrub cover. In areas with continued herding (left side), the alpine flora prevail. Rodnei Mountains in the northern Carpathians, Romania. Photography by A. Hofgaard

or birds across the treeline ecotone and beyond, this might not necessarily result in recruitment of tree seedlings surviving to sapling and tree size (Aune et al. 2011). Temporal and transitory seedling cohorts are often characteristic of the upper part of the treeline ecotone (Juntunen et al. 2002, Kullman 2002, Aune et al. 2011) making the tree species line (cf. Fig. 1) very dynamic. Survival and growth to sapling and tree size in the open exposed alpine area requires long-term favorable site conditions with regard to a large number of abiotic and biotic factors, such as topography, snow cover and duration, soil, wind, temperature, moisture, plant density, and herbivory, and the interplay between these factors (Cairns & Moen 2004, Holtmeier & Broll 2005, Batllori et al. 2010, Hofgaard et al. 2010). In addition, the importance of these factors is species-specific and varies through time (Holtmeier & Broll 2005, Hofgaard et al. 2012, Wielgolaski & Inouye 2013), and the response is also sensitive to the current ecotone structure (Camarero et al. 2017). Consequently, disentangling the causes and predicting treeline ecotone responses to environmental changes are challenging (Sveinbjörnsson et al. 2002), although the most common determinants for treeline ecotone location are temperature and land use (see Section 1.1. above).

Tree advance is initially associated with a change in height growth of previously established saplings (Kullman 2002, Hofgaard et al. 2009) causing densification of the current scattered tree layer (Batllori & Gutiérrez 2008, Mathisen et al. 2014) and movement of the treeline location (Kullman & Öberg 2009). Further densification and relocation is dependent on new establishment and survival in the ecotone and beyond the current upper sapling cohort (Kullman 2002, Hofgaard et al. 2009). The increased abundance of trees and tree saplings changes the structure of the ecotone, including enhanced snow trapping during winter, and thus further promotes tree growth and establishment through, for example, reducing wind destruction of leading shoots. During the winter season, soil temperatures under a deep snowpack do not drop much below zero. Soil moisture is increased in the early growing season by the meltwater (Sveinbjörnsson et al. 2002, Dalen & Hofgaard 2005, Holtmeier 2009).

As trees and forest advance to higher altitudes and latitudes, increasingly more of the former low-alpine or low-arctic area disappears, leaving less space for tundra species (Dirnböck et al. 2003, Gottfried et al. 2012). Knowledge of the rate of this process is important for predictions of tundra disappearance and associated threats to alpine biodiversity and climate feedbacks (Callaghan et al. 2002b, Pearson et al. 2013). Expansion of forest or shrub cover to areas beyond the current forest at high elevation and high latitude has contrasting climate feedbacks through carbon sequestration (cooling) and reduced surface reflectance (warming) (Bala et al. 2007, Pearson et al. 2013, te Beest et al. 2016). According to estimates for Scandinavian mountain forests, the warming effect is considerably stronger than the cooling, because of the typically low density in mountain forests and the large changes in surface reflectance of snow-covered tundra areas (de Wit et al. 2014). However, the change in reflectance caused by vegetation is a slow process, as the rate of forest migration is low due to the multitude of interacting and counteracting abiotic and biotic environmental factors. The typical advance rate for the warming periods since the late 19th century has been less than 1 m yr^{-1} altitudinally (Kullman & Öberg 2009, Kharuk et al. 2010, Mathisen et al. 2014, Cudlín et al. 2017) and some 10s m yr^{-1} latitudinally (Hofgaard et al. 2013). These empirically based rate estimates represents less than one-tenth of model-based rate estimates (Hofgaard et al. 2013), and it is essential to consider this mismatch when discussing magnitude and time frame of potential threats to alpine biodiversity.

3. ANIMAL IMPACT IN TREELINE AREAS

In addition to domestic animals, wild fauna depend on, interact with, and change the structure and location of the treeline ecotone. This fauna represents a large number of species of mammals, birds, insects, and other vertebrates and invertebrates. Herbivore activities will have a direct impact through both consumption of biomass (browsing, grazing, seed feeding) and other regular or life history related activities such as trampling, digging, gnawing, and girdling. Animal activities also have indirect impact in treeline areas through e.g. carnivore–herbivore interactions and subsequent animal population dynamics (Hambäck et al. 2004), and through decomposition of dead organic matter by soil-dwelling invertebrates. Quantification of animal impact in treeline areas is not straightforward due to the structural heterogeneity of the ecotone and biogeographic differences throughout Europe. The heterogeneity provides diverse macro- and microhabitats supporting a variety of organisms from soil microorganisms that specialize in particular habitats to animals with large ranges that require different habitats for forage and shelter.

Further, animals typically associated with forested areas, such as some corvids, deer, wolverine, and red fox, frequently use treeline areas and the above tundra to search for food and occasionally for breeding. Animals associated with tundra, such as lemmings and reindeer, periodically or annually forage in the ecotone (Fig. 8) (Post et al. 2009, Killengreen et al. 2012). In addition, as discussed in Section 2, the structure and quality of present treeline ecotone habitats have often been, and still are, strongly influenced by human activities. As the type and extent of human influences vary regionally and locally, the possibility of generalization is limited. In the following paragraphs, we outline some examples of animal impact in treeline areas.

In the central European mountains, increased numbers of ungulates such as red deer *Cervus elaphus* have locally suppressed tree regeneration and impaired self-maintenance of tree stands in treeline areas (Loison et al. 2003, Kiffner et al. 2008, Holtmeier 2012). Similarly, but to a lesser extent, chamois *Rupicapra rupicapra* and ibex *Capra ibex* herbivory affect saplings and trees across the ecotone (ten Houte de Lange 1978, Senn 2000). In addition, high ungulate densities may cause severe soil erosion both in the treeline ecotone and in the adjacent alpine tundra (Holtmeier 1967, 2012, 2015). In most cases, habitat fragmentation and inadequate game management are major causes of 'over-sized' ungulate populations.

In the north, reindeer (semi-domesticated and wild) occur regionally in large populations, and throughout history, have profoundly affected the vegetation in their foraging range, including the treeline ecotone vegetation (Oksanen et al. 1995, Kashulina et al. 1997, Mårell et al. 2002, Colpaert et al. 2003, Helle & Kojola 2006, Olofsson et al. 2009). Reindeer grazing may inhibit tree seedling survival and prevent vegetative regeneration from basal shoots of mountain birch (Kaitaniemi et al. 1999, Holtmeier 2002, Cairns & Moen 2004, Neuvonen & Wielgolaski 2005, Solberg et al. 2005), and under high reindeer densities, the grazing areas might become increasingly degraded with regard to species diversity and productivity (Kullman 2005, Holtmeier & Broll 2006, Broll et al. 2007, Käyhkö 2007, Anschlag et al. 2008, Tømmervik et al. 2009). This is particularly evident in winter grazing areas where the main food for reindeer is slow-growing reindeer lichens (Gaare & Skogland 1975). Lichen ground cover has been strongly reduced both by high grazing pressure and by climate change. In addition to reindeer, increasing north Scandinavian populations

Fig. 8. Winter grazing by semi-domestic reindeer in the treeline ecotone of northern Fennoscandia. Photography by K. Laine

of moose *Alces alces* are affecting height growth and survival of young Scots pine in treeline areas (Stöcklin & Körner 1999, Holtmeier & Broll 2011) and in afforestation areas at lower elevation. These impacts by reindeer and moose on tree recruitment may locally or regionally overrule the influence of changing climate (Stöcklin & Körner 1999, Aune et al. 2011, Holtmeier 2012), and in the case of reindeer, the carrying capacity of the landscape has sometimes been questioned (Neuvonen & Wielgolaski 2005, Solberg et al. 2005).

In addition to large mammals, rodents, chiefly microtine rodents, may have a large impact on treeline ecotone vegetation, particularly in north European mountains. In this region, massive population peaks of lemmings *Lemmus lemmus* and voles, e.g. *Microtus agrestis*, are well-known biotic characteristics with a return cycle of approximately 4 yr (Andersson & Jonasson 1986, Henttonen & Wallgren 2001, Ims & Fuglei 2005). The grazing and gnawing during population peak periods reduce moss and dwarf shrub cover locally and over large regions (Olofsson et al. 2012, Kaarlejärvi et al. 2015). This fragmentation of the bottom and field layer might facilitate establishment of new mountain birch seedlings. However, lemmings and voles also damage and feed on birch seedlings, and in general, rodents adversely affect young trees and shrubs (Fig. 9) rather than promoting successful seedling establishment. Thus, field vole feeding may hasten birch decline, in combination with outbreaks of defoliating geometrid moths and subsequent reindeer grazing (see below). Further, but not of large scale

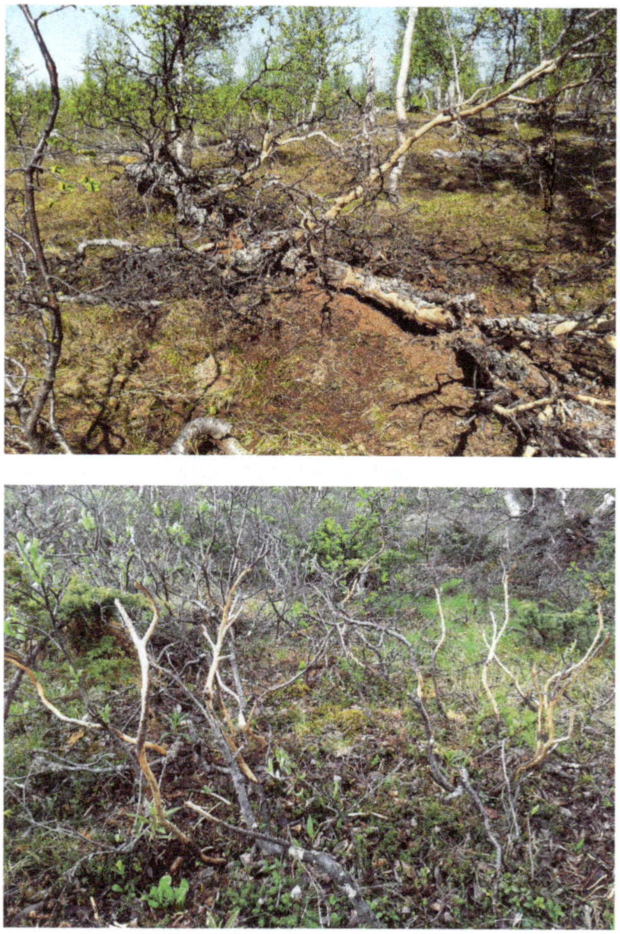

Fig. 9. Lemming and microtine vole forage damage of mountain birch and surrounding woody plants (upper panel) and willow shrubs (*Salix* sp.) (lower panel). The gnawing activity occurred in winter below the snow surface. Upper panel: Tärnafjällen in the Central Scandes Mountains, Sweden. Lower panel: Abisko in the Northern Scandes Mountains, Sweden. Photography by A. Hofgaard

importance, mountain hares *Lepus timidus* reduce growth of mountain birch saplings by heavy browsing and damage trees by gnawing off the bark (Rao et al. 2003, Holtmeier 2012).

In northern mountain birch-dominated ecotones and adjacent forest areas, the natural dynamics are driven by cyclic and abrupt population increases of defoliating insects such as the autumnal moth *Epirrita autumnata* and winter moth *Operoptera brumata* (Tenow 1972, Tenow et al. 2007). The frequency and intensity of these insect outbreaks are linked to the climate at local to regional scales. They can cause large scale stand mortality and former forest might be turned into tundra (Tenow & Nilssen 1990, Neuvonen et al. 2005, Tenow et al. 2007). This tundra produced by defoliators, together with the lowered treeline ecotone, might become a long-term state due to intensified reindeer grazing in the newly deforested areas (Kallio & Lehtonen 1975, Oksanen et al. 1995, Holtmeier et al. 2003, Lempa et al. 2005, Neuvonen & Wielgolaski 2005, Neuvonen et al. 2005, Holtmeier & Broll 2006). In a warming climate, expansion of the outbreak range of defoliating insects is likely (Jepsen et al. 2011), and will include higher altitudes and latitudes (Skre et al. 2017, this Special). This might hinder or counteract climate-driven advance of treeline ecotones (Olofsson et al. 2009, Aune et al. 2011, Hofgaard et al. 2013). A parallel to the autumnal moth outbreaks in northern Europe are the cyclic outbreaks of the larch-bud moth *Zeiraphera diniana* in the European Alps. However, larch stands at treeline ecotone elevations are generally not affected due to the low density of trees. At lower elevations, increased cyclic outbreaks are probably due to human-induced expansion of pure larch forests. In cases of severe defoliation, growth and seed production are reduced (Holtmeier 1974, 2015).

Soil-dwelling invertebrates (e.g. earthworms, enchytraeids, collembola, spiders, tardigrades, woodlice, snails, millipedes, nematodes, dipteral larvae, and ants) play an important role through breaking down dead organic matter, mineral-rich (nitrogen, phosphorous) excretory products, and bioturbation, thus influencing nutrient turnover and plant communities (Broll 1998, Holtmeier 2015). In general, taxa, abundance, biomass, and species richness of soil-dwelling invertebrates decrease with altitude. In the treeline ecotone, however, they are controlled by often sharply contrasting site conditions (geological substrate, soils, microclimates, moisture, vegetation) overlapping with historical human impact (Holtmeier 2009). However, in contrast to mass outbreaks of leaf-eating insects, soil invertebrates do not significantly influence treeline spatial and temporal structures, whereas reforestation of abandoned alpine pastures will probably bring about major changes influencing soil invertebrate fauna and also aboveground insects and ground beetles (*Carabidae*) in the long-term. Predicting possible feedbacks on the treeline, however, is difficult because of the often inscrutable interactions of the numerous environmental factors and their relative implications (Holtmeier 2009, 2015).

Among birds, the Eurasian nutcracker *Nucifraga caryocatactes* is a highly effective agent influencing the tree distribution pattern of stone pine *Pinus cembra* and the dynamics of treeline ecotones in central and eastern European mountain areas due to its seed-caching activities (Zong et al. 2010, Holtmeier 2012, 2015). Surplus stored seeds (i.e. seeds not consumed during the winter) may germinate and result in the establishment of trees in new areas and at higher elevation. This bird-mediated sowing has contributed to the re-establishment of stone pine over large areas formerly cleared by man. Further, without the nutcracker, natural upward advancement of trees in response to climatic warming would be impossible. Birds other than nutcrackers, e.g. grouse and ptarmigan species, and seed-eating birds, usually have a low impact on the treeline ecotone. However, grouse and ptarmigan species might locally limit growth, particularly of young trees, by consuming or destroying buds and terminal shoots (Holtmeier 2012, 2015), and in addition, pastoral abandonment may lead to an overall increase in avian diversity (Laiolo et al. 2004) due to increased shrub and tree cover.

Fig. 10. Wind-eroded convex topography on Koahppeloaivi (northernmost Finnish Lapland). The substrate is rapidly draining sandy-skeletal glacial till. Erosion was initiated by reindeer winter-grazing activities that destroyed the dwarf shrub–lichen vegetation and made the substrate susceptible to deflation. Moisture deficiency is characteristic of such sites and adversely affects tree seedling establishment. The photograph was taken from approximately 250 to 300 m above ground. Photography by F. K. Holtmeier

4. CASCADING EFFECTS OF LAND USE AND CLIMATE CHANGES: A SCENARIO EXAMPLE FROM THE NORTH

The outcome of combined pressure on the natural environment by intense land use and climate changes might be difficult to quantitatively and qualitatively forecast. However, it is necessary to consider the matter for sustainable management reasons. A commonly discussed example is the semi-domestic reindeer herding system in northern Scandinavia. The long history of the herding system has shaped the distribution and abundance of species and thus formed the current cultural landscape. However, the herding system is not static and has to adapt to modern socioeconomic changes and requirements, at the same time as the climate is both highly variable and changing. Here we outline some of the biological complexities involved (see also Sections 2 & 3 above). While reindeer owners may wish to increase their income by allowing more animals within a given area, this of course will result in increased grazing pressure. The new grazing regime might or might not initially affect both summer (herb-dominated) and winter (lichen-dominated) grazing grounds, but winter areas are common bottlenecks in the annual migration practice. Lichens are a major reindeer food source in the winter, and accessibility varies widely between individual winters due to snow quality. Warm winters with rain and icing events force reindeer to use a lot of energy digging for lichens. Due to the very slow growth rate of lichens, considerable time is needed for vegetation recovery following excess removal of lichens by e.g. too high grazing pressure. At the same time, digging for lichens by the reindeer population cause patches of open soil, which facilitate establishment of higher plants, e.g. mountain birch. Birch is further facilitated by the ongoing increased temperature and precipitation in the region, although on well drained substrates exposed by reindeer scraping and trampling, moisture deficiency affects or prevents birch seedling establishment (Fig. 10). Increased temperature and moisture availability may also increase decomposition rate and mineralization of organic material. In these ways, nutrient-demanding plant species may outcompete the slow-growing lichens. Because of diminishing lichen cover, the reindeer population has to adjust its diet to include vascular plant parts such as the young shoots of birch saplings. However, at increased temperatures in winter, the survival rate of the eggs of defoliating insects using birch as host species will increase along with arrival of new defoliating species in the area. Increased frequency of defoliators and increased spatiotemporal outbreak occurrence may transform birch-dominated areas to treeless tundra. The quality of these areas as winter grazing areas is thus further diminished, and the reindeer owners will have to reduce the number of animals using the area or keep the herds in summer grazing areas for prolonged periods. This will, however, cause other cascading effects.

5. CONCLUDING REMARKS AND PERSPECTIVES

While from a continent-wide view, thermal deficiency increasing with elevation and latitude controls treeline ecotone structure and position, many other climatic and biotic factors are involved at regional and smaller scales. Not least, the after effects of historical human impact are often of major importance and may overrule the influence of natural factors. Therefore, assessment of treeline ecotone variety and causation needs a cross-disciplinary complex approach combining natural and socio-economic sciences. Climate change, reduced pastoral use, and increased tourism and other human uses are the main driving factors of current changes across the treeline ecotone. The role of animals (wild and domestic or semi-domestic ungulates, rodents, birds, and insects) in the treeline ecotone needs to be studied in more detail, in particular regarding possible cascading effects such as the presented mountain birch–reindeer scenario. Treeline ecotone change will have far-reaching implications for biodiversity, plant and animal communities, and also for the relative effects of microtopography on site conditions (microclimates, soil ecological conditions) and ecosystem services (e.g. protection from destructive avalanches, prevention/reduction of soil erosion).

Acknowledgements. This article is based on work from COST Action ES 1203 (SENSFOR), supported by COST (European Cooperation in Science and Technology), www.cost.eu. We are in addition thankful for financial support from the CLIMFOR project (grant code EEA-jrp-ro-no-2013-1-0204), the Research Council of Norway (grant nos. 160022/F40 and 244557/E50), the German Research Foundation, and the Lapland Atmosphere-Biosphere Facility (LAPBIAT, EU).

LITERATURE CITED

Aas B, Faarlund T (1996) The present and the Holocene subalpine birch belt in Norway. In: Frenzel B, Birks HH, Alm T, Vorren KD (eds) Holocene treeline oscillations, dendrochronology and palaeoclimate. Paläoklimaforschung — Palaeoclimate Res, Vol 20. Gustav Fischer, Stuttgart, p 19–42

Aas B, Faarlund T (2001) The Holocene history of the Nordic maintain birch belt. In: Wielgolaski FE (ed) Nordic mountain birch ecosystems. Man and the Biosphere Series 27. Parthenon Publishing-UNESCO, Paris, London, New York, p 5–22

Alexander JM, Diez JM, Levine JM (2015) Novel competitors shape species' responses to climate change. Nature 525:515–518

Allen JRM, Huntley B (1999) Estimating past floristic diversity in montane regions from macrofossil assemblages. J Biogeogr 26:55–73

Ameztegui A, Brotons L, Coll L (2010) Land-use changes as major drivers of mountain pine (*Pinus uncinata* Ram.) expansion in the Pyrenees. Glob Ecol Biogeogr 19:632–641

Andersson JM, Jonasson S (1986) Rodent cycles in relation to food resources on an alpine heath. Oikos 46:93–106

Anschlag K, Broll G, Holtmeier FK (2008) Mountain birch seedlings in the treeline ecotone, Subarctic Finland. Variation in above- and below-ground growth in relation to microtopography. Arct Antarct Alp Res 40:609–616

Aune S, Hofgaard A, Söderström L (2011) Contrasting climate and land use driven tree encroachment pattern of sub-arctic tundra in Northern Norway and Kola Peninsula. Can J For Res 41:437–449

Austrheim G, Mysterud A, Pedersen B, Halvorsen R, Hassel K, Evju M (2008) Large scale experimental effects of three levels of sheep densities on an alpine ecosystem. Oikos 117:837–846

Bala G, Caldeira K, Wickett M, Phillips TJ, Lobell DB, Delire C, Mirin A (2007) Combined climate and carbon-cycle effects of large-scale deforestation. Proc Natl Acad Sci USA 104:6550–6555

Batllori E, Gutiérrez E (2008) Regional tree line dynamics in response to global change in the Pyrenees. J Ecol 96: 1275–1288

Batllori E, Camarero JJ, Gutiérrez E (2010) Current regeneration patterns at the tree line in the Pyrenees indicate similar recruitment processes irrespective of the past disturbance regime. J Biogeogr 37:1938–1950

Bolli JC, Rigling A, Bugmann H (2007) The influence of changes in climate and land-use on regeneration dynamics of Norway spruce at the treeline in the Swiss Alps. Silva Fenn 41:55–70

Brandes R, Ise M (2007) Fingerprints of climate change in Mediterranean mountain forests? Observations in Mediterranean fir-species threatened by climate change. Geo-Öko 28:1–26

Brockmann-Jerosch H (1919) Baumgrenze und Klimacharakter. Beiträge zur geobotanischen Landesaufnahme 6, Zürich

Broll G (1998) Diversity of soil organisms in alpine and arctic soils in Europe. Review and research needs. Pireneos 151-152:43–72

Broll G, Holtmeier FK, Anschlag K, Brauckmann HJ, Wald S, Drees B (2007) Landscape mosaic in the treeline ecotone on Mt. Rodjanoaivi, subarctic Finland. Fennia 185:89–105

Bryn A (2008) Recent forest limit changes in south-east Norway: effects of climate change or regrowth after abandoned utilisation? Nor Geogr Tidsskr 62:251–270

Bryn A, Daugstad K (2001) Summer farming in the subalpine birch forest. In: Wielgolaski FE (ed) Nordic mountain birch ecosystems. Man and the Biosphere Series 27. Parthenon Publishing-UNESCO, Paris, London, New York, p 307–315

Bryn A, Hemsing LO (2012) Impacts of land use on the vegetation in three rural landscapes of Norway. Int J Biodivers Sci Ecosyst Serv Manag 8:360–371

Bryn A, Dourojeanni P, Hemsing LO, O'Donnell S (2013) A high-resolution GIS null model of potential forest expansion following land use changes in Norway. Scand J For Res 28:81–98

Bühlmann T, Körner C, Hiltbrunner E (2016) Shrub expansion of *Alnus viridis* drives former montane grassland into nitrogen saturation. Ecosystems 19:968–985

Cairns D, Moen J (2004) Herbivory influences tree lines. J Ecol 92:1019–1024
Callaghan TV, Werkman BR, Crawford RMM (2002a) The tundra-taiga interface and its dynamics: concepts and applications. Ambio Spec Rep 12:6–14
Callaghan TV, Crawford RMM, Eronen M, Hofgaard A and others (2002b) The dynamics of the tundra taiga boundary: an overview and a co-ordinated and integrated approach to research. Ambio Spec Rep 12:3–5
Callaghan TV, Björn LO, Chernov Y, Chapin T and others (2004) Biodiversity, distributions and adaptations of Arctic species in the context of environmental change. Ambio 33:404–417
Callaghan TV, Jonasson C, Thierfelder T, Yang Z and others (2013) Ecosystem change and stability over multiple decades in the Swedish subarctic: complex processes and multiple drivers. Phil Trans R Soc B 368:20120488
Camarero JJ, Gutiérrez E (2004) Pace and pattern of recent treeline dynamics: response of ecotones to climatic variability in the Spanish Pyrenees. Clim Change 63: 181–200
Camarero JJ, Gutiérrez E (2007) Response of *Pinus uncinata* recruitment to climate warming and changes in grazing pressure in an isolated population of the Iberian system (NE Spain). Arct Antarct Alp Res 39:210–217
Camarero JJ, Linares JC, García-Cervigón AI, Batllori E, Martínez I, Gutiérrez E (2017) Back to the future: the responses of alpine treelines to climate warming are constrained by the current ecotone structure. Ecosystems 20: 683–700
Cardinale BJ, Duffy JE, Gonzales A, Hooper DU and others (2012) Biodiversity loss and its impact on humanity. Nature 486:59–67
Chauchard S, Carcaillet C, Guibal F (2007) Patterns of land-use abandonment control tree-recruitment and forest dynamics in Mediterranean mountains. Ecosystems 10: 936–948
Colpaert A, Kumpula J, Nieminen M (2003) Reindeer pasture biomass assessment using satellite remote sensing. Arctic 56:147–158
Crooks JA, Soulé ME (1999) Lag times in population explosions of invasive species: causes and implications. In: Sandlund OT, Schei PJ, Viken Å (eds) Invasive species and biodiversity management. Kluwer, Dordrecht, p 103–125
Cudlín P, Klopčič M, Tognetti R, Malis F and others (2017) Drivers of treeline shift in different European mountains. Clim Res 73:135–150
Dalen L, Hofgaard A (2005) Differential regional treeline dynamics in the Scandes Mountains. Arct Antarct Alp Res 37:284–296
De Quervain A (1904) Die Hebung der atmosphärischen Isothermen in den Schweizer Alpen und ihre Beziehung zu den Höhengrenzen. Gerlands Beitr Geophys 6:481–533
de Wit HA, Bryn A, Hofgaard A, Karstensen J, Kvalevåg M, Peters G (2014) Climate warming feedback from mountain birch forest expansion: reduced albedo dominates carbon uptake. Glob Chang Biol 20:2344–2355
Dinca L, Nita MD, Hofgaard A, Alados CL and others (2017) Forests dynamics in the montane–alpine boundary: a comparative study using satellite imagery and climate data. Clim Res 73:97–110
Dirnböck T, Dullinger S, Grabherr G (2003) A regional impact assessment of climate and land-use change on alpine vegetation. J Biogeogr 30:401–417
Emanuelsson U (1987) Human influence on vegetation in the Torneträsk area during the last three centuries. Ecol Bull 38:95–111
Fischer M, Rudmann-Maurer K, Weyand A, Stöcklin J (2008) Agricultural land use and biodiversity in the Alps. Mt Res Dev 28:148–155
Fleischer P, Pichler V, Fleischer P Jr, Holko L and others (2017) Forest ecosystem services affected by natural disturbances, climate and land-use changes in the Tatra Mountains. Clim Res 73:57–71
Forbes BC, Tolvanen A, Wielgolaski FE, Laine K (2005) Rates and processes of natural regeneration in disturbed habitats. In: Wielgolaski FE (ed) Plant ecology, herbivory, and human impact in Nordic mountain birch forests. Ecological Studies 180. Springer-Verlag, Berlin, p 193–202
Gaare E, Skogland T (1975) Wild reindeer food habits and range use at Hardangervidda. In: Wielgolaski FE (ed) Fennoscandian tundra ecosystems, Part 2, Animals and systems analysis. Ecological Studies 17. Springer-Verlag, Berlin, p 195–205
Garcia-Ruiz J, Lasanta T, Ruiz-Flano P, Ortigosa L, White S, González C, Martí C (1996) Land-use changes and sustainable development in mountain areas: a case study in the Spanish Pyrenees. Landsc Ecol 11:267–277
Gehrig-Fasel J, Gusian A, Zimmermann N (2007) Tree line shifts in the Swiss Alps: Climate change or land abandonment? J Veg Sci 18:571–582
Gervais BR, MacDonald GM (2000) A 403-year record of July temperatures and treeline dynamics of *Pinus sylvestris* from the Kola Peninsula, Northwest Russia. Arct Antarct Alp Res 32:295–302
Gonzáles de Andrés E, Camarero JJ, Büntgen U (2015) Complex climate constraints of upper treeline formation in the Pyrenees. Trees 29:941–952
Gottfried M, Pauli H, Futschik A, Akhalkatsi M and others (2012) Continent-wide response of mountain vegetation to climate change. Nat Clim Chang 2:111–115
Grove JM (1988) The Little Ice Age. Methuen, London
Grunewald K, Scheithauer J (2008) Untersuchungen an der alpinen Waldgrenze im Piringebirge (Bulgarien). Geo-Öko 29:1–32
Grytnes JA, Kapfer J, Jurasinski G, Birks HH and others (2014) Identifying the driving factors behind observed elevational range shifts on European mountains. Glob Ecol Biogeogr 23:876–884
Hambäck PA, Oksanen L, Ekerholm P, Lindgren Å, Oksanen T, Schneideret M (2004) Predators indirectly protect tundra plants by reducing herbivore abundance. Oikos 106:85–92
Hanssen-Bauer I, Førland EJ, Haddeland I, Hisdal H and others (2015) Klima i Norge 2100. Kunnskapsgrunnlag for klimatilpasning oppdatert 2015. Norsk klimaservicesenter rapport, no. 2/2015
Harsch MA, Hulme PE, McGlone M, Duncan RP (2009) Are treelines advancing? A global meta-analysis of treeline response to climate warming. Ecol Lett 12:1040–1049
Heiri C, Bugmann H, Tinner W, Heiri O, Lischke H (2006) A model-based reconstruction of Holocene treeline dynamics in the Central Swis Alps. J Ecol 94:206–216
Helle T, Kojola I (2006) Population trends of semi-domesticated reindeer in Fennoscandia — evaluation of explanations. In: Forbes BC, Bölter M, Müller-Wille L, Hukkinen J, Müller F, Gunslay N, Konstantino Y (eds) Reindeer management in northernmost Europe: Linking

practical and scientific knowledge in socio-ecological systems. Ecological Studies 148. Springer-Verlag, Berlin, p 319–339

Henttonen H, Wallgren H (2001) Rodent dynamics and communities in the birch forest zone of northern Fennoscandia. In: Wielgolaski FE (ed) Nordic mountain birch ecosystems. Man and the Biosphere Series 27, Parthenon Publishing-UNESCO, Paris, p 261–278

Hofgaard A (1997a) Structural changes in the forest–tundra ecotone: a dynamic process. In: Huntley B, Cramer W, Morgan AV, Prentice HC, Allen JRM (eds) Past and future rapid environmental changes: the spatial and evolutionary responses of terrestrial biota. NATO ASI Series, Vol I 47, Springer-Verlag, Berlin, p 255–263

Hofgaard A (1997b) Inter-relationships between treeline position, species diversity, land-use and climate change, in the central Scandes Mountains of Norway. Glob Ecol Biogeogr Lett 6:419–429

Hofgaard A (1999) The role of 'natural' landscapes influenced by man in predicting responses to climate change. Ecol Bull 47:160–167

Hofgaard A, Wilmann B (2002) Plant distribution patterns across the forest–tundra ecotone. The importance of treeline position. Ecoscience 9:375–385

Hofgaard A, Dalen L, Hytteborn H (2009) Tree recruitment above the treeline and potential for climate driven treeline change. J Veg Sci 20:1133–1144

Hofgaard A, Løkken JO, Dalen L, Hytteborn H (2010) Comparing warming and grazing effects on birch growth in an alpine environment — a 10 year experiment. Plant Ecol Divers 3:19–27

Hofgaard A, Harper KA, Golubeva E (2012) The role of the circumarctic forest–tundra ecotone for arctic biodiversity. Biodiversity (Nepean) 13:174–181

Hofgaard A, Tømmervik H, Rees G, Hanssen F (2013) Latitudinal forest advance in northernmost Norway since the early 20th century. J Biogeogr 40:938–949

Holtmeier FK (1967) Das Steinwild in der Landschaft von Pontresina. Natur Mus 99:15–24

Holtmeier FK (1974) Geoökologische Beobachtungen und Studien an der subarktischen und alpinen Waldgrenze in vergleichender Sicht (nördliches Fennoskandie/Zentralalpen). Erdwissenschaftliche Forschung 8. Franz Steiner Verlag, Wiesbaden

Holtmeier FK (1981) What does the term 'krummholz' really mean? Observations with special reference to the Alps and the Colorado Front Range. Mt Res Dev 1:253–260

Holtmeier FK (1987) Human impacts in high altitude forests and upper timberline with special reference to middle latitudes. In: Fujimori T, Kimura M (eds) Human impacts and management of mountain forests. Forest Products Research Institute, Ibaraki, p 9–20

Holtmeier FK (1993) Timber lines as indicators of climatic changes: problems and research needs. In: Frenzel B (ed) Oscillations of the alpine and polar tree limits in the Holocene. Paläoklimaforschung – Palaeoclimate Res, Vol 9. Gustav Fischer, Stuttgart, p 211–222

Holtmeier FK (2002) Tiere in der Landschaft. Einfluß und ökologische Bedeutung. Ulmer, Stuttgart

Holtmeier FK (2009) Mountain timberlines. Ecology, patchiness, and dynamics. Advances in Global Change Research 36. Springer, Dordrecht

Holtmeier FK (2012) Impact of wild herbivorous mammals and birds on the altitudinal and northern treeline ecotones. Landsc Online 30:1–28

Holtmeier FK (2015) Animals' influence on the landscape and ecological importance. Natives, newcomers, homecomers. Springer, Dordrecht

Holtmeier FK, Broll G (2005) Sensitivity and response of northern hemisphere altitudinal and polar treelines to environmental change at landscape and local scales. Glob Ecol Biogeogr 14:395–410

Holtmeier FK, Broll G (2006) Radiocarbon-dated peat and wood remains from the Finnish Subarctic: evidence of treeline and landscape history. Holocene 16:743–751

Holtmeier FK, Broll G (2010) Altitudinal and polar treelines in the northern hemisphere — causes and response to climate change. Polarforschung 79:139–153

Holtmeier FK, Broll G (2011) Response of Scots pine (*Pinus sylvestris*) to warming climate at its altitudinal limit in northernmost subarctic Finland. Arctic 64:269–280

Holtmeier FK, Broll G (2012) Landform influences on treeline patchiness and dynamics in a changing climate. Phys Geogr 35:430–437

Holtmeier FK, Broll G (2017) Feedback effects of clonal groups and tree clusters on site conditions at the treeline: implications for treeline dynamics. Clim Res 73:85–96

Holtmeier FK, Broll G, Müterthies A, Anschlag K (2003) Regenertaion of trees in the treeline ecotone: northern Finnish Lapland. Fennia 181:103–128

Huston MA (1994) Biological diversity: the coexistence of species on a changing landscape. Cambridge University Press, Cambridge

Ims RA, Fuglei E (2005) Trophic interaction cycles in tundra ecosystems and the impact of climate change. Bioscience 55:311–322

Irl SDH, Anthelme F, Harter DEV, Jentsch A, Lotter E, Steinbauer MJ, Beirkuhnlein C (2016) Patterns of island treeline elevation — a global perspective. Ecography 39: 427–436

Jepsen JU, Kapari L, Hagen SB, Schott T, Vindstad OPL, Nilssen AC, Ims RA (2011) Rapid northwards expansion of a forest insect pest attributed to spring phenology matching with sub-Arctic birch. Glob Change Biol 17: 2071–2083

Juntunen V, Neuvonen S, Norokorpi Y, Tasanen T (2002) Potential for timberline advance in northern Finland, as revealed by monitoring during 1983–99. Arctic 55: 348–361

Kaarlejärvi E, Hoset KS, Olofsson J (2015) Mammalian herbivores confer resilience of Arctic shrub-dominated ecosystems to changing climate. Glob Chang Biol 21: 3379–3388

Kaitaniemi P, Neuvonen S, Nyyssönen T (1999) Effects of cumulative defoliation on growth, reproduction, and insect resistance in the mountain birch. Ecology 80: 524–532

Kallio P, Lehtonen J (1975) On the ecocatastophe of birch forests cause by *Oporinia autumnata* (BKH) and the problem of reforestation. In: Wielgolaski FE (ed) Fennoscandian tundra ecosystems. 2. Ecological Studies 17. Springer-Verlag, Berlin, p 174–180

Karlén W (1976) Lacustrine sediments and tree-limit variations as indicators of Holocene climatic fluctuations in Lappland: northern Sweden. Geogr Ann 58:1–34

Karlsen SR, Tømmervik H, Johansen B, Riseth JÅ (2017) Future forest distribution on Finnmarksvidda, North Norway. Clim Res 73:125–133

Kashulina G, Reimann C, Finne TE, Halleraker JH, Äyräs M, Chekushin VA (1997) The ecosystem of the central Barents Region: scale, factors and mechanisms of disturbance. Sci Total Environ 206:203–225

Kašpar J, Treml V (2016) Thermal characteristics of alpine treelines in Central Europe north of the Alps. Clim Res 68:1–12

Käyhkö J (2007) Aeolian blowout dynamics in subarctic Lapland based on decadal levelling investigations. Geogr Ann 89:65–81

Kharuk VI, Ranson KJ, Im ST, Vdovin AS (2010) Spatial distribution and temporal dynamics of high-elevation forest stands in southern Siberia. Glob Ecol Biogeogr 19: 822–830

Kiffner C, Rössiger E, Trisl O, Schulz R, Rühe F (2008) Probability of recent bark stripping damage by red deer (*Cervus elaphus*) on Norway spruce (*Picea abies*) in a low mountain range in Germany — a preliminary analysis. Silva Fenn 42:125–134

Killengreen ST, Strømseng E, Yoccoz NG, Ims RA (2012) How ecological neighbourhoods influence the structure of the scavanger guild in low arctic tundra. Divers Distrib 18:563–574

Klanderud K (2008) Species-specific responses of an alpine plant community under simulated environmental change. J Veg Sci 19:363–372

Körner C (2003) Alpine plant life. Springer-Verlag, Berlin

Körner C (2012) Alpine treelines. Springer, Basel

Kouki J (1999) Latitudinal gradients in species richness in northern areas: some exceptional patterns. Ecol Bull 47: 30–37

Kulakowski D, Barbeito I, Casteller A, Kaczka R, Bebi P (2016) Not only climate: Interacting drivers of treeline change in Europe. Geogr Pol 89:7–15

Kullman L (1986) Recent tree-limit history of *Picea abies* in the southern Swedish Scandes. Can J Res 16:761–771

Kullman L (1995) Holocene tree-limit and climate history from the Scandes Mountains, Sweden. Ecology 76: 2490–2502

Kullman L (2002) Rapid recent range-margin rise of tree and shrub species in the Swedish Scandes. J Ecol 90:68–77

Kullman L (2003) Recent reversal of Neoglacial climate cooling trend in the Swedish Scandes as evidenced by mountain birch tree-limit rise. Global Planet Change 36:77–88

Kullman L (2004) Early Holocene appearance of mountain birch (*Betula pubescens* ssp. *tortuosa*) at unprecedented high elevations in the Swedish Scandes: megafossil evidence exposed by recent snow and ice recession. Arct Antarct Alp Res 36:172–180

Kullman L (2005) Wind-conditioned 20th century decline of birch treeline vegetation in the Swedish Scandes. Arctic 58:286–294

Kullman L, Källgren L (2000) A coherent postglacial tree-limit chronology (*Pinus sylvestris* L.) for the Swedish Scandes: aspects of paleoclimate and 'recent warming' based on megafossil evidence. Arct Antarct Alp Res 32: 419–428

Kullman L, Öberg L (2009) Post-Little Ice Age tree line rise and climate warming in the Swedish Scandes: a landscape ecological perspective. J Ecol 97:415–429

Kyriazopoulos AP, Skre O, Sarkki S, Wielgolaski FE, Abraham EM, Ficko A (2017) Human–environment dynamics in European treeline ecosystems: a synthesis based on the DPSIR framework. Clim Res 73:17–29

Laiolo P, Dondero F, Ciliento E, Rolando A (2004) Consequences of pastoral abandonment for the structure and diversity of the alpine avifauna. J Appl Ecol 41:294–304

Lempa K, Neuvonen S, Tømmervik H (2005) Effect of reindeer grazing on pastures. A necessary basis for sustainable reindeer herding. In: Wielgolaski FE (ed) Plant ecology, herbivory, and human impact in Nordic mountain birch forests. Ecological Studies 180. Springer-Verlag, Berlin, p 159–164

Loison A, Toïgo C, Gaillard JM (2003) Large herbivores in European alpine ecosystems: current status and challenges for the future. In: Nagy L, Grabherr G, Körner C, Thompson DBA (eds) Alpine biodiversity in Europe. Ecological Studies 167. Springer-Verlag, Berlin, p 351–366

Mårell A, Ball JP, Hofgaard A (2002) Foraging and movements paths of female reindeer: insights from fractal analysis, correlated random walks, and Lévy flights. Can J Zool 80:854–865

Mathisen IE, Mikheeva A, Tutubalina OV, Aune S, Hofgaard A (2014) Fifty years of tree line change in Khibiny Mountains, Russia: advantages of combined remote sensing and dendroecological approaches. Appl Veg Sci 17:6–16

Mooney HA, Hofgaard A (1999) Biological invasions and global change. In: Sandlund OT, Schei PJ, Viken Å (eds) Invasive species and biodiversity management. Kluwer, Dordrecht, p 139-148

Motta R, Morales M, Nola P (2006) Human land-use, forest dynamics and tree growth at the treeline in the Western Italian Alps. Ann Sci 63:739–747

Nagy L, Grabherr G (2009) The biology of alpine habitats. Oxford University Press, New York, NY

Nagy L, Grabherr G, Körner C, Thompson DBA (2003) Alpine biodiversity in space and time: a synthesis. In: Nagy L, Grabherr G, Körner C, Thompson DBA (eds) Alpine biodiversity in Europe. Ecological Studies 167. Springer-Verlag, Berlin, p 453–464

Neuvonen S, Wielgolaski FE (2005) Herbivory in northern birch forests. In: Wielgolaski FE (ed) Plant ecology, herbivory, and human impact in Nordic mountain birch forests. Ecological Studies 180. Springer-Verlag, Berlin, p 183–189

Neuvonen S, Bylund H, Tømmervik H (2005) Forest defoliation risks in birch forest by insects under different climate and land use scenarios in northern Europe. In: Wielgolaski FE (ed) Plant ecology, herbivory, and human impact in Nordic mountain birch forests. Ecological Studies 180. Springer-Verlag, Berlin, p 125–138

Nijnik A, Nijnik M, Kopiy S, Zahvoyska L, Sarkki S, Kopiy L, Miller D (2017) Identifying and understanding attitudinal diversity on multi-functional changes in woodlands of the Ukrainian Carpathians. Clim Res 73:45–56

Oksanen L, Moen J, Helle T (1995) Timberline patterns in northernmost Fennoscandia. Relative importance of climate and grazing. Acta Bot Fenn 153:93–105

Olofsson J, Oksanen L, Callaghan T, Hulme PE, Oksanen T, Suominen O (2009) Herbivores inhibit climate-driven shrub expansion on the tundra. Glob Chang Biol 15: 2681–2693

Olofsson J, Tømmervik H, Callaghan TV (2012) Vole and lemming activity observed from space. Nat Clim Chang 2:880–883

Olsen SL, Klanderud K (2014) Exclusion of herbivores slows down recovery after experimental warming and nutrient addition in an alpine plant community. J Ecol 102: 1129–1137

Pearson RG, Phillips SJ, Loranty MM, Beck PSA, Damoulas T, Knight SA, Goetz SJ (2013) Shifts in Arctic vegetation and associated feedbacks under climate change. Nat Clim Chang 3:673–677

Petitpierre B, McDougall K, Seipel T, Broennimann O, Guisan A, Kueffer C (2016) Will climate change increase risk of plant invasions into mountains? Ecol Appl 26:530–544

Post E, Forchhammer MC, Bret-Harte MS, Callaghan TV and others (2009) Ecological dynamics across the Arctic associated with recent climate change. Science 325: 1355–1358

Rao SK, Iason GR, Hulbert IAR, Elston DA, Pracey PA (2003) The effect of sapling density, heather height and season of browsing by mountain hares on birch. J Ecol 40: 626–638

Rixen C, Rolando A (eds) (2013) The impacts of skiing and related winter recreational activities on mountain environments. Bentham e Books

Rohde K (1992) Latitudinal gradients in species diversity: the search for the primary cause. Oikos 65:514–527

Roth T, Kohli L, Rihm B, Achermann B (2013) Nitrogen deposition is negatively related to species richness and species composition of vascular plants and bryophytes in Swiss mountain grassland. Agric Ecosyst Environ 178: 121–126

Sala OE, Chapin FS III, Armesto JJ, Berlow E and others (2000) Global biodiversity scenarios for the year 2100. Science 287:1770–1774

Sarkki S, Ficko A, Grunewald K, Nijnik M (2016) Benefits from and threats to European treeline ecosystem services: an exploratory study of stakeholders and governance. Reg Environ Change 16:2019–2032

Sarkki S, Ficko A, Wielgolaski FE, Abraham EM and others (2017a) Assessing the resilient provision of ecosystem services by social-ecological systems: introduction and theory. Clim Res 73:7–16

Sarkki S, Jokinen M, Nijnik M, Zahvoyska L and others (2017b) Social equity in governance of ecosystem services: synthesis from European treeline areas. Clim Res 73:31–44

Sato CF, Wood JT, Lindenmayer DB (2013) The effects of winter recreation on alpine and subalpine fauna: a systematic review and meta-analysis. PLOS ONE 8:e64282

Schwörer C, Henne PD, Tinner W (2014) A model-data comparison of Holocene timberline changes in the Swiss Alps reveals past and future drivers of mountain forest dynamics. Glob Chang Biol 20:1512–1526

Senn J (2000) Huftiere und Verjüngung im Gebirgswald: eine Geschichte mit vielen Variablen und noch mehr Interaktionen. Schweiz Z Forstwes 151:99–106

Shiyatov SG (1993) The upper timberline dynamics during the last 1100 years in the Polar Ural Mountains. In: Frenzel B (ed) Oscillations of the alpine and polar tree limits in the Holocene. Paläoklimaforschung – Palaeoclimate Res, Vol 9. Gustav Fischer, Stuttgart, p 195–203

Shiyatov SG (2003) Rates of change in the upper treeline ecotone in the Polar Ural mountains. PAGES News 11:8–10

Sitko I, Troll M (2008) Timberline changes in relation to summer farming in the western Chornohora (Ukrainian Carpathians). Mt Res Dev 28:263–271

Skre O, Wertz B, Wielgolaski FE, Szydlowska P, Karlsen SR (2017) Bioclimatic effects on different mountain birch populations in Fennoscandia. Clim Res 73:111–124

Solberg B, Tømmervik H, Thannheiser D, Neuvonen S (2005) Economoc limits and possibilities for sustainable utilization of northern birch forests. In: Wielgolaski FE (ed) Plant ecology, herbivory and human impact in Nordic mountain birch forests. Ecological Studies 180. Springer-Verlag, Berlin, p 219–233

Speed JDM, Austrheim G, Hester AJ, Mysterud A (2010) Experimental evidence for herbivore limitation of the treeline. Ecology 91:3414–3420

Speed JDM, Austrheim G, Hester AJ, Mysterud A (2012) Elevational advance of alpine plant communities is buffered by herbivory. J Veg Sci 23:617–625

Stöcklin J, Körner C (1999) Recruitment and mortality of *Pinus sylvestris* near the Nordic treeline: the role of climate change and herbivory. Ecol Bull 47:168–177

Strebel N, Bühler C (2015) Recent shifts in plant species suggest opposing land-use changes in alpine pastures. Alp Bot 125:1–9

Sveinbjörnsson B, Hofgaard A, Lloyd A (2002) The natural causes of the taiga–tundra boundary. Ambio Spec Rep 12:23–29

Tasser E, Tappeiner U (2002) Impact of land use changes on mountain vegetation. Appl Veg Sci 5:173–184

Tasser E, Walde J, Tappeiner U, Teutsch A, Noggler W (2007) Land-use changes and natural reforestation in the Eastern Central Alps. Agric Ecosyst Environ 118:115–129

te Beest M, Sitters J, Ménard CB, Olofsson J (2016) Reindeer grazing increases summer albedo by reducing shrub abundance in Arctic tundra. Environ Res Lett 11:125013

ten Houte de Lange SM (1978) Zur Futterwahl des Alpensteinbocks (*Capra ibex* L.). Eine Untersuchung an der Steinwildkolonie am Piz Albris bei Pontresina. Z Jagdwiss 24:113–138

Tenow O (1972) The outbreaks of *Oporinia autumnata* Bkh. and *Operophthera* spp. (Lep., Geometridae) in the Scandinavian mountain chain and northern Finland 1862–1968. Zoologiska Bidrag från Uppsala Suppl 2

Tenow O, Nilssen AC (1990) Egg cold hardiness and topoclimatic limitations to outbreaks of *Epirrita autumnata* in northern Fennoscandia. J Appl Ecol 27:723–734

Tenow O, Nilssen AC, Bylund H, Hogstad O (2007) Waves and synchrony in *Epirrita autumnata/Operophtera brumata* outbreaks. I. Lagged synchrony: regionally, locally and among species. J Anim Ecol 76:258–268

Tinner W, Theurillat JP (2003) Uppermost limit, extent, and fluctuations of the timberlines and treeline ecocline in the Swiss Central Alps during the past 11500 years. Arct Antarct Alp Res 35:158–169

Tinner W, Ammann B, German P (1996) Treeline fluctuations recorded for 12500 years by soil profiles, pollen, and plant macrofossils in the Central Swiss Alps. Arct Alp Res 28:131–147

Tolvanen A, Kangas K (2016) Tourism, biodiversity and protected areas — Review from northern Fennoscandia. J Environ Manage 169:58–66

Tømmervik H, Johansen B, Riset JÅ, Karlsen SR, Solberg B, Høgda KA (2009) Above ground biomass changes in the mountain birch forests and mountain heaths of Finnmarksvidda, northern Norway, in the period 1957–2006. For Ecol Manage 257:244–257

Törn A, Tolvanen A, Norokorpi Y, Tervo R, Siikamäki P (2009) Comparing the impacts of hiking, skiing and horse riding on trail and vegetation in different types of forest. J Environ Manage 90:1427–1434

Treml V, Chuman T (2015) Ecotonal dynamics of the altitudinal forest limit are affected by terrain and vegetation structure variables: an example from the Sudetes Mountains in Central Europe. Arct Antarct Alp Res 47:133–146

Treml V, Šenfeldr M, Chuman T, Ponocna T, Decová K

(2016) Twentieth century treeline ecotone advance in the Sudetes Mountains (Central Europe) was induced by agricultural land abandonment rather than climate change. J Veg Sci 27:1209–1221

Turner H (1961) Die Niederschlags- und Schneeverhältnisse. Mitteilungen der forstlichen Bundes-Versuchsanstalt Mariabrunn 59:265–315

Turner H (1970) Grundzüge der Hochgebirgsklimatologie. In: Ladurner J, Purtscheller F, Reisigl H, Tratz E (eds) Die Welt der Alpen. Pinguin Verlag, Innsbruck, Umschau Verlag, Frankfurt, p 170–182

Van Bogaert R, Haneca K, Hoogesteger J, Jonasson C, De Papper M, Callaghan TV (2011) A century of tree line changes in sub-arctic Sweden shows local and regional variability and only minor influence of 20th century climate warming. J Biogeogr 38:907–921

Vittoz P, Camenisch M, Mayor R, Miserere L, Vust M, Theurillat JP (2010) Subalpine-nival gradient of species richness for vascular plants, bryophytes and lichens in the Swiss Inner Alps. Bot Helv 120:139–149

Weisberg PJ, Shandra O, Becker ME (2013) Landscape influences on recent timberline shifts in the Carpathian Mountains: abiotic influences modulate effects of land-use change. Arct Antarct Alp Res 45:404–414

Wielgolaski FE (1998) Twenty-two years of plant recovery after severe trampling by man through five years in three vegetation types at Hardangervidda. NTNU Vitenskapsmuseet Rapport Botanisk Serie 4:26–29

Wielgolaski FE (ed) (2001) Nordic mountain birch ecosystems. Man and the Biosphere Series 27. Parthenon Publishing-UNESCO, Paris

Wielgolaski FE (2003) Climatic factors governing plant phenological phases along a Norwegian fjord. Int J Biometeorol 47:213–220

Wielgolaski FE (ed) (2005) Plant ecology, herbivory, and human impact in Nordic mountain birch forests. Ecological Studies 180. Springer-Verlag, Berlin

Wielgolaski FE, Inouye DW (2013) Phenology at high latitudes. In: Schwartz MD (ed) Phenology: an integrative environmental science. Springer-Verlag, New York, NY, p 225–247

Wielgolaski FE, Karlsen SR (2007) Some views on plants in polar and alpine regions. Rev Environ Sci Biotechnol 6: 33–45

Wielgolaski FE, Nilsen J (2001) Coppicing and growth of various provenances of mountain birch in relation to nutrients and water. In: Wielgolaski FE (ed) Nordic mountain birch ecosystems. Man and the Biosphere Series 27. Parthenon Publishing-UNESCO, Paris, p 77–92

Wilmanns O (1989) Die Buchen und ihre Lebensräume. Ber Reinh-Tüxen-Ges 1:47–72

Ylisirniö AL, Allén A (2016) Plant communities of Fennoscandian subarctic mountain ecosystems 60 years after human disturbance. Arct Antarct Alp Res 48:469–483

Zong C, Wauters LA, Dongen SV, Mari V and others (2010) Annual variation in predation and dispersal of Arolla pine (*Pinus cembra* L.) seeds by Eurasian red squirrels and other seed-eaters. For Ecol Manage 260:587–594

Permissions

All chapters in this book were first published in CR, by Inter-Research; hereby published with permission under the Creative Commons Attribution License or equivalent. Every chapter published in this book has been scrutinized by our experts. Their significance has been extensively debated. The topics covered herein carry significant findings which will fuel the growth of the discipline. They may even be implemented as practical applications or may be referred to as a beginning point for another development.

The contributors of this book come from diverse backgrounds, making this book a truly international effort. This book will bring forth new frontiers with its revolutionizing research information and detailed analysis of the nascent developments around the world.

We would like to thank all the contributing authors for lending their expertise to make the book truly unique. They have played a crucial role in the development of this book. Without their invaluable contributions this book wouldn't have been possible. They have made vital efforts to compile up to date information on the varied aspects of this subject to make this book a valuable addition to the collection of many professionals and students.

This book was conceptualized with the vision of imparting up-to-date information and advanced data in this field. To ensure the same, a matchless editorial board was set up. Every individual on the board went through rigorous rounds of assessment to prove their worth. After which they invested a large part of their time researching and compiling the most relevant data for our readers.

The editorial board has been involved in producing this book since its inception. They have spent rigorous hours researching and exploring the diverse topics which have resulted in the successful publishing of this book. They have passed on their knowledge of decades through this book. To expedite this challenging task, the publisher supported the team at every step. A small team of assistant editors was also appointed to further simplify the editing procedure and attain best results for the readers.

Apart from the editorial board, the designing team has also invested a significant amount of their time in understanding the subject and creating the most relevant covers. They scrutinized every image to scout for the most suitable representation of the subject and create an appropriate cover for the book.

The publishing team has been an ardent support to the editorial, designing and production team. Their endless efforts to recruit the best for this project, has resulted in the accomplishment of this book. They are a veteran in the field of academics and their pool of knowledge is as vast as their experience in printing. Their expertise and guidance has proved useful at every step. Their uncompromising quality standards have made this book an exceptional effort. Their encouragement from time to time has been an inspiration for everyone.

The publisher and the editorial board hope that this book will prove to be a valuable piece of knowledge for researchers, students, practitioners and scholars across the globe.

List of Contributors

Jon M. Honea
Conservation Biology Division, Northwest Fisheries Science Center, National Marine Fisheries Service, Emerson College, 120 Boylston Street, Boston, MA 02116, USA

Michelle M. McClure
Fishery Resource Analysis and Monitoring Division, Northwest Fisheries Science Center,National Marine Fisheries Service, National Oceanic and Atmospheric Administration, 2725 Montlake Blvd E., Seattle, WA 98112, USA

Jeffrey C. Jorgensen
Conservation Biology Division, Northwest Fisheries Science Center, National Marine Fisheries Service, National Oceanic and Atmospheric Administration, 2725 Montlake Blvd E., Seattle, WA 98112, USA
Ocean Associates, under contract to Northwest Fisheries Science Center, National Oceanic and Atmospheric Administration, 2725 Montlake Blvd E., Seattle, WA 98112, USA

Mark D. Scheuerell
Fish Ecology Division, Northwest Fisheries Science Center, National Marine Fisheries Service,National Oceanic and Atmospheric Administration, 2725 Montlake Blvd E., Seattle, WA 98112, USA

Satyaban B. Ratna, J. V. Ratnam S. K. Behera and T. Yamagata
Application Laboratory, JAMSTEC, 3173-25 Showa-machi, Kanazawa-ku, Yokohama, Kanagawa, 236-0001, Japan

Fredolin T. Tangang
School of Environmental and Natural Resource Sciences, Faculty of Science and Technology,the National University of Malaysia, Malaysia

Changsub Shim, Jihyun Seo, Jihyun Han, Jongsik Ha and Tae Ho Ro
Korea Environment Institute, 30147 Sejong, ROK

Yun Seop Hwang
Department of International Business and Trade, Kyung Hee University, 02453 Seoul, ROK

Jung Jin Oh
Department of Chemistry, Sookmyung Women's University, 04312 Seoul, ROK

Ewa B. Łupikasza
Department of Climatology, Faculty of Earth Sciences, University of Silesia in Katowice, 41-200 Sosnowiec, Poland

Aristita Busuioc, Madalina Baciu, Traian Breza, Alexandru Dumitrescu, Cerasela Stoica and Nina Baghina
National Meteorological Administration, Sos. Bucuresti-Ploiesti 97, Sect.1, Bucharest 013686, Romania

F. E. Wielgolaski
Department of Bioscience, University of Oslo, Blindern, 0316 Oslo, Norway

K. Laine and J. Inkeröinen
Thule Institute, University of Oulu, 90014 University of Oulu, Finland

O. Skre
Skre Nature and Environment, Fanaflaten 4, 5244 Fana, Norway

A. P. Kyriazopoulos
Department of Forestry and Management of the Environment and Natural Resources, Democritus University of Thrace, 193 Pantazidou str., 68200 Orestiada, Greece

S. Sarkki
Cultural Anthropology, Faculty of Humanities, PO Box 1000, University of Oulu, 90014 Oulu, Finland

E. M. Abraham
Laboratory of Range Science, Department of Forestry and Natural Environment, Aristotle University of Thessaloniki, 54124 Thessaloniki, Greece

A. Ficko
Biotechnical Faculty, Department of Forestry and Renewable Forest Resources, University of Ljubljana, Vecna pot 83,1000 Ljubljana, Slovenia

Albert Nijnik
Environmental Network Limited, The Hillocks, Tarland, Aboyne AB34 4TJ, Scotland, UK

Maria Nijnik and David Miller
The James Hutton Institute, Craigiebuckler, Aberdeen AB15 8QH, Scotland, UK

Serhiy Kopiy, Lyudmyla Zahvoyska and Leonid Kopiy
Ukrainian National Forestry University, Gen. Chuprynky 103, Lviv 79057, Ukraine

Simo Sarkki
Thule Institute, 90014 University of Oulu, Finland

Peter Fleischer
Faculty of Forestry, Technical University in Zvolen, 96053 Zvolen, Slovakia
Research Station of TANAP, State Forest of TANAP, 059 60 Tatranská Lomnica, Slovakia

Viliam Pichler, Peter Fleischer Jr., Erika Gömöryová, Jaroslav Škvarenina, Katarína Střelcová and Pavol Hlaváč
Faculty of Forestry, Technical University in Zvolen, 96053 Zvolen, Slovakia

Ladislav Holko
Institute of Hydrology, Slovak Academy of Sciences, 031 05 Liptovský Mikuláš, Slovakia

František Máliš
Faculty of Forestry, Technical University in Zvolen, 96053 Zvolen, Slovakia
Forest Research Institute Zvolen, National Forest Centre, 960 92 Zvolen, Slovakia

Pavel Cudlín
Institute of Systems Biology and Ecology, Academy of Sciences, 370 05 České Budejovice, Czech Republic

Jan Holeksa
Faculty of Biology, Adam Mickiewicz University, 61 614 Poznań, Poland

Zuzana Michalová
Faculty of Forestry and Wood Technology, Czech University of Life Sciences, 16 521 Prague, Czech Republic

Zuzana Homolová
Research Station of TANAP, State Forest of TANAP, 059 60 Tatranská Lomnica, Slovakia

M. Cristina Moscatelli, Tommaso Chiti and Guido Pellis
Department of Innovation in Biological, Agrofood and Forest systems (DIBAF), University of Tuscia, Viterbo, Italy

Eleonora Bonifacio
Department of Agricultural, Forest and Food Sciences (DISAFA), University of Torino, largo P. Braccini 2, 10095 Grugliasco, Italy

Pavel Cudlín
Global Change Research Centre, Academy of Sciences of the Czech Republic, Lipová 1789/9, České Budjovice 370 05,Czech Republic

Lucian Dinca
National Forest Research-Development Institute I.N.C.D.S. Brasov, 13 Closca St., 500030 Brasov, Romania

Erika Gömöryova
Technical University in Zvolen, Faculty of Forestry, T. G. Masaryka 24, 960 53 Zvolen, Slovakia

Stefano Grego
Department of Science and Technology for Agriculture, Forestry, Nature and Energy (DAFNE), Tuscia University, Viterbo, Italy

Nicola La Porta
Research and Innovation Centre, Fondazione Edmund Mach (FEM) and MOUNTFOR Project Centre,European Forest Institute, Via E. Mach 1, 38010 San Michele all Adige (Trento), Italy

Leszek Karlinski and Maria Rudawska
Institute of Dendrology, ul. Parkowa 5, 62-035 Kórnik, Poland

Andrea Squartin
Department of Agricultural Biotechnology, University of Padua, Agripolis, Viale dell 'Università 16, 35020 Legnaro, Padua, Italy

Miglena Zhiyanski
Forest Research Institute – Bulgarian Academy of Sciences, 132 'Kl. Ohridski' Blvd., 1756 Sofia, Bulgaria

Gabriele Broll
Institute of Geography, University of Osnabrueck, Seminarstr. 19, 49074 Osnabrueck, Germany

Friedrich-Karl Holtmeier
Institute of Landscape Ecology, Heisenbergstraße 2, Westfälische Wilhelms-Universität, 48149 Münster, Germany

Gabriele Broll
Institute of Geography, University of Osnabrück, Seminarstr. 19 a/b, 49074 Osnabrück, Germany

Lucian Dinca
National Forest Research-Development Institute, Cloşca 13, Braşov, 500040, Romania

Mihai Daniel Nita and Stelian Alexandru Borz
Faculty of Silviculture and Forest Engineering, Transilvania University of Brasov, Ludwig van Beethoven 1, Brașov, 500123,Romania

Annika Hofgaard
Norwegian Institute for Nature Research, Høgskoleringen 9, 7034 Trondheim, Norway

Concepcion L. Alados
Pyrenean Institute of Ecology (CSIC), Av. Ntra. Sra. de la Victoria, Huesca, 22700, Spain

Gabriele Broll
University of Osnabrück, Neuer Graben, 49074 Osnabrück, Germany

Bogdan Wertz
University of Agriculture, Faculty of Forestry, Adama Mickiewicza 21, Kraków, 30-001, Poland

Antonio T. Monteiro
Predictive Ecology Group, Research Center on Biodiversity and Genetic Resources, CIBIO-InBIO, Rua Padre Armando Quintas 7, Vairão, 4485-661, Portugal

Oddvar Skre
Skre Nature and Environment, Fanaflaten 4, 5244 Fana, Norway

Bogdan Wertz and Paulina Szydlowska
University of Agriculture in Krakow, Dept. of Biometry and Forest Productivity, Al. 29 Listopada 46, 31-425 Krakow, Poland

Frans E. Wielgolaski
Department of Bioscience, University of Oslo, Blindern, 0316 Oslo, Norway

Stein-Rune Karlsen
NORUT Northern Research Institute, 9294 Tromsø, Norway

Stein Rune Karlsen, Bernt Johansen and Jan Åge Riseth
Norut Northern Research Institute, 9294 Tromsø, Norway

Hans Tømmervik
Norwegian Institute for Nature Research, FRAM – High North Centre for Climate and the Environment, Langnes, 9296 Tromsø, Norway

Pavel Cudlín and Magda Edwards-Jonášová
Global Change Research Institute CAS, Academy of Sciences of the Czech Republic, Ceské Budějovice 370 05, Czech Republic

Matija Klopčič
University of Ljubljana, Biotechnical Faculty, Department of Forestry and Renewable Forest Resources, Slovenia

Roberto Tognetti
Dipartimento di Bioscienze e Territorio, Iniversità degli Studio del Molise, Contrada Fonte Lappone, 86090 Pesche, Italy
MOUNTFOR Project Centre, European Forest Institute, 38010 San Michele all Adige (Trento), Italy

Frantisek Máliš
Technical University Zvolen, Faculty of Forestry, 960 53 Zvolen, Slovakia
National Forest Centre, Forest Research Institute Zvolen, 960 92 Zvolen, Slovakia

Concepción L. Alados
Pyrenean Institute of Ecology (CSIC), Apdo. 13034, 50080 Zaragoza, Spain

Peter Bebi
WSL Institute for Snow and Avalanche Research SLF, 7260 Davos Dorf, Switzerland

Karsten Grunewald
Leibniz Institute of Ecological Urban and Regional Development, 01217 Dresden, Germany

Miglena Zhiyanski
Forest Research Institute, BAS 132, Kl. Ohridski Blvd. 1756 Sofia, Bulgaria

Vlatko Andonowski
Faculty of Forestry, University Ss. Cyril and Methodius, Skopje, Macedonia

Nicola La Porta
Research and Innovation Centre, Fondazione Edmund Mach (FEM) and MOUNTFOR Project Centre,European Forest Institute, 38010 San Michele all Adige (Trento) Italy

Svetla Bratanova-Doncheva and Eli Kachaunova
Division of Ecosystem Research, IBER-Bulgarian Academy of Sciences, 1113 Sofia, Bulgaria

Josep Maria Ninot
University of Barcelona, Department of Plant Biology, 08028 Barcelona, Spain

Andreas Rigling
WSL Swiss Federal Institute for Forest, Snow and Landscape Research, Zürcherstrasse 111, 8903 Birmensdorf, Switzerland

Annika Hofgaard
Norwegian Institute of Nature Research, 7485 Trondheim, Norway

Tomáš Hlásny
Faculty of Forestry and Wood Sciences, Czech University of Life Sciences, 165000 Prague, Czech Republic

Petr Skalák
Global Change Research Institute CAS, Academy of Sciences of the Czech Republic, České Bude˘jovice 370 05, Czech Republic
Czech Hydrometeorological Institute, 143 06 Prague, Czech Republic

Frans Emil Wielgolaski
University of Oslo, 0316 Oslo, Norway

A. Hofgaard
Norwegian Institute for Nature Research, Sluppen, 7485 Trondheim, Norway

F. K. Holtmeier
Institute of Landscape Ecology, Heisenbergstrasse 2, Westfälische Wilhelms-Universität, 48149 Münster, Germany

Index